Plumbing and Mechanical Services:

Book 2

A. H. Masterman and R. M. Boyce

Stanley Thornes (Publishers) Ltd

First published in 1990 by:
Stanley Thornes (Publishers) Ltd
Ellenborough House
Wellington Street
CHELTENHAM GL50 1YW
United Kingdom

98 99 00 / 10 9 8 7 6 5

British Library Cataloguing in Publication Data

Masterman, A. H. (Arnold H.)
Plumbing and mechanical services 2.
1. Buildings. Plumbing. Manuals
I. Title II. Boyce, R. M. (Robert M.)
696'.1

ISBN 0 7487 0232 6

Typeset in Times by Acorn Bookwork Ltd, Salisbury, Wilts
Printed and bound in Great Britain by Redwood Books, Trowbridge, Wiltshire

Contents

Preface

Plumbing and Mechanical Services Book 2 provides a learning resource for students of plumbing, heating, gas and allied industries studying for qualification at City & Guilds Craft Certificate Level.

The text closely follows the requirements of the syllabus in craft technology, working processes and associated subjects, the primary aim being to assist the student towards that qualification.

Although mainly written for those commencing a career in the mechanical services industry this book will contain much of interest for the mature craftsman and those who wish to keep abreast of the changing techniques and advancing technology in this section of the construction industry. In addition to those mentioned above, this book should prove invaluable for technician students and for students attending Links, Foundation, Youth Training Scheme and Employment Training programmes.

Acknowledgements

The authors and publishers are grateful to the following for permission to reproduce textual material and illustrations:

Burn Bros (London) Ltd; Glynnwed Ltd; Esse Ltd; Landis and Gyr; Honeywell; Aqualisa; Trianco Ltd; Polystel; Range Boilers; Polystel Ltd; Worcester Engineering Co Ltd; Grundfoss; Armitage Shanks Ltd; Hillmoor; Rothenberger; Caradon Mira Ltd; Potterton International Ltd.

Ideal-Standard Limited are gratefully acknowledged for supplying the cover photograph of a Trevi overhead shower.

Every effort has been made to reach copyright holders, but the publishers would be grateful to hear from any source whose copyright they may unwittingly have infringed.

1 Domestic hot water and heating

After reading this chapter you should be able to:

1 Identify and name types of domestic hot water and heating system.

2 Recognise different types of hot water storage vessel.

3 Demonstrate knowledge of types of boiler and their operation.

4 Identify and name the component parts of hot water and heating systems.

5 State the function of the control components of hot water and heating systems.

6 Understand the methods of heat transfer related to different forms of heat emitter and radiator.

7 Identify and name different types of thermal insulating material.

8 Describe the fitting and application of thermal insulating materials.

9 Understand and describe the basic methods of water heating by electricity.

10 Demonstrate knowledge of installation procedures related to hot water and heating systems.

Introduction

Before we discuss domestic central heating systems, it may be advantageous to clarify what is meant by:

1 Direct hot water supply systems,
2 Indirect hot water supply systems (conventional),
3 Indirect hot water supply systems (single-feed cylinder).

Some of these systems are shown in detail in Book I and so consequently will be only briefly described in Book II.

Direct hot water supply system

This system comprises three main components: (1) the feed cistern from which the cold water is supplied; (2) the hot storage vessel, usually a copper cylinder; (3) the boiler where the water is heated. The three components are connected together by the appropriate pipework to form a simple and popular common system.

The direct system is defined as one in which 'The water that is heated in the boiler rises by convectional currents into the hot storage vessel to be stored until required. As that heated water is drawn off at the taps, the system is refilled by fresh raw water from the feed cistern.'

Advantages:

1 Simple system,
2 Inexpensive to install,
3 Fairly quick recovery period.

Disadvantages:

1 Not suitable where domestic space heating is required from same boiler,
2 Not suitable where the water is temporarily hard,
3 Precautions to be taken where the water is very soft.

Domestic hot water and heating combined

Where a hot water system is combined with a heating system, several different metals are used to construct that system, with the result that corrosion takes place.

The different materials used may include:

1 Piping and fittings: copper, brass, gunmetal;
2 Cylinder: copper, steel;
3 Cisterns: galvanised steel, plastics;
4 Boiler: cast iron, steel, copper;
5 Radiators: steel, cast iron;
6 Pump: cast iron, steel.

The use of dissimilar metals may cause corrosion by 'electrolytic action', and this action is quite common in most domestic systems. There may also be a problem of de-zincification.

The greatest problem is the formation of 'iron oxide'. This is the result of an attack on iron and steel components. A black oxide sludge forms, which can lead to:

1 Pump malfunction (seizure),
2 Partial blockage in radiators causing poor circulation and possible cold areas,
3 Perforation of the mild steel sheet used to construct radiators or boilers,
4 Excessive gas formation affecting the efficiency of radiators.

Each of the above-mentioned terms will be dealt with later. To keep the discoloured and contaminated water in the radiators separate from the clean water required for domestic purposes, some form of indirect hot water system must be used.

Indirect hot water supply (conventional)

In this system there are two cold water cisterns, a calorifier and a boiler, plus all the appropriate pipework. Essentially there are two separate systems in one (see Figure 1.1).

One cistern supplies cold water to the boiler and the calorifier. A calorifier is a hot storage vessel (cylinder) with another vessel (either cylinder or coil) inside. This part of the system is also connected to the space heaters (i.e. radiators).

The second cistern supplies water to the outer cylinder which in turn supplies the water to the hot discharge points (i.e. appliances). The water in the outer cylinder is heated *indirectly* by means of hot water circulating between the boiler and inner cylinder or coil. Therefore the water that is heated in the boiler is used only to heat *indirectly* the water that is drawn off at the appliances plus any space heating. Only a small amount of fresh water is added to the heating circuit to replenish that which may be boiled away, or lost by evaporation from the feed and expansion cistern. The indirect hot water system is used:

1 Where the water is temporarily of a hard nature,
2 Where the heating medium is steam (not covered in this book),
3 Where both domestic hot water and heating systems are fed from the same boiler.

Indirect hot water supply (single-feed cylinder)

Mention has already been made of the fact that where domestic hot water and domestic space heating are required from the same boiler, some form of indirect system must be installed. Where a dwelling with an ordinary hot water supply system is to have a space heating system added, a considerable saving in cost can be achieved by the adaptation of the existing system to incorporate a single-feed cylinder (see Figure 1.2a and b). The

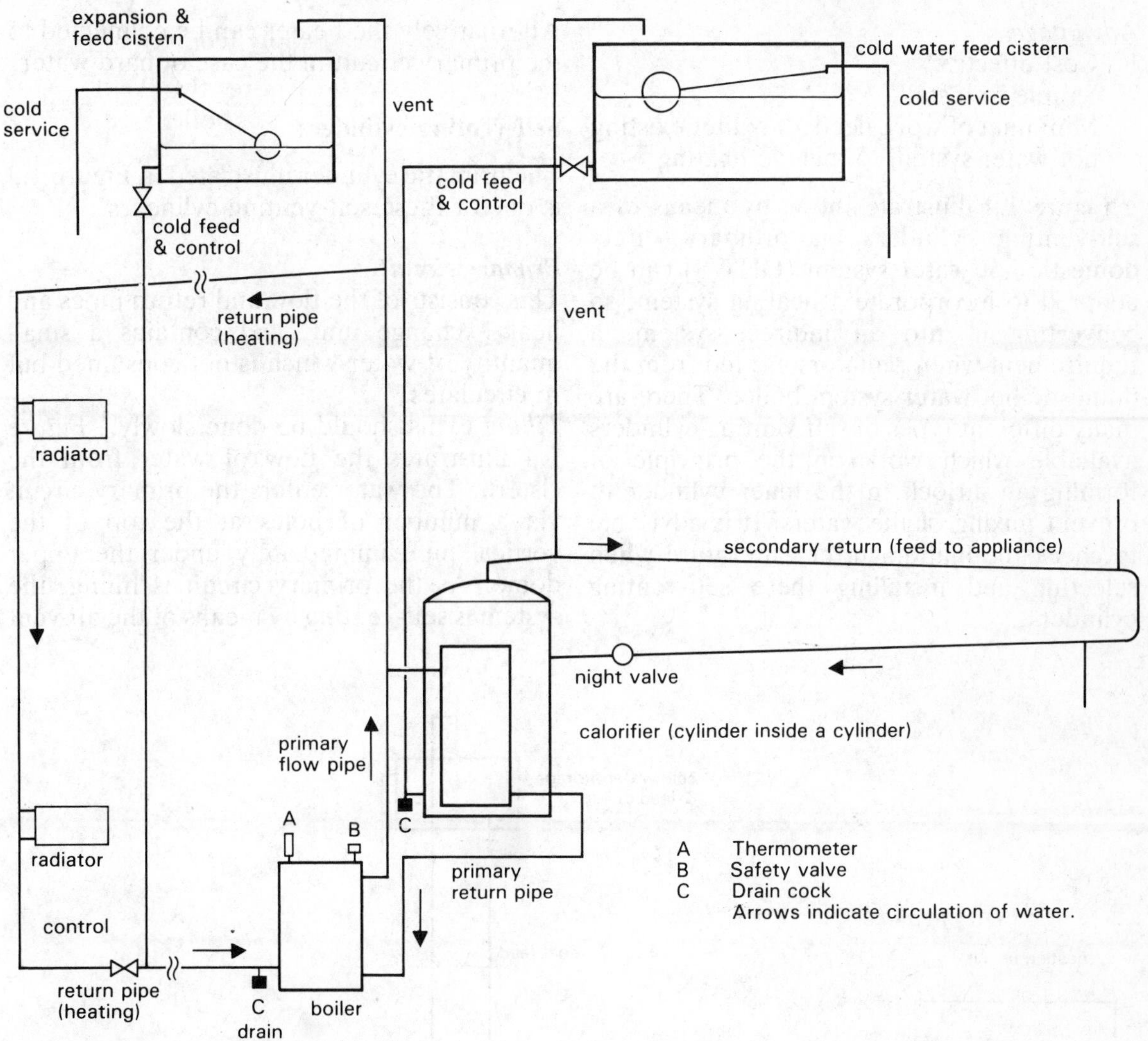

Figure 1.1 *Indirect hot water supply (conventional and secondary circuit)*

single-feed system is exactly the same as the ordinary direct hot water system except that, in the place of the direct cylinder, a special (patented) self-venting cylinder is fitted.

There are several types of self-venting cylinders on the market: although they may differ slightly in appearance, basically they operate on the same principle. You may recall that in an indirect system there are two separate systems, and that the waters in the two systems must be kept separate. In the single-feed system, the waters are prevented from mixing by an airlock in the heat exchange unit, between the heating water in the primary circuit and the domestic supply to the appliances. Although the single-feed system operates as an indirect system it is cost effective because it does not require the additional feed and expansion cistern and appropriate pipework, i.e. the additional cold feed, vent pipes, cold supply and overflow.

Advantages:

1 Cost effective,
2 Simple to install,
3 Minimum of work needed to alter existing hot water system to include heating.

Figure 1.2 illustrates how, by means of a self-venting cylinder, an ordinary direct domestic hot water system (DHWS) can be adapted to incorporate a heating system, so converting it into an indirect system, a requirement when radiators are fed from the domestic hot water system boiler. There are many different types of self-venting cylinders available which work on the principle of forming an airlock in the inner cylinder to prevent mixing of the waters. It is advisable to check the manufacturer's literature when selecting and installing these self-venting cylinders.

Alternatively the heater can be connected to the primary circuit in the case of hard water.

Self-venting cylinders

The primatic cylinder illustrated in Figure 1.3 is one of these self-venting cylinders.

Primary circuit

This consists of the flow and return pipes and heat exchange unit, and contains a small quantity of water which is not consumed but re-circulates.

Filling (This should be done slowly.) Figure 1.4 illustrates the flow of water from the cistern. The water enters the primary circuit via a number of holes at the top of the vertical pipe immediately under the upper dome. As the primary circuit is filling, the system is self-venting by means of the air vent

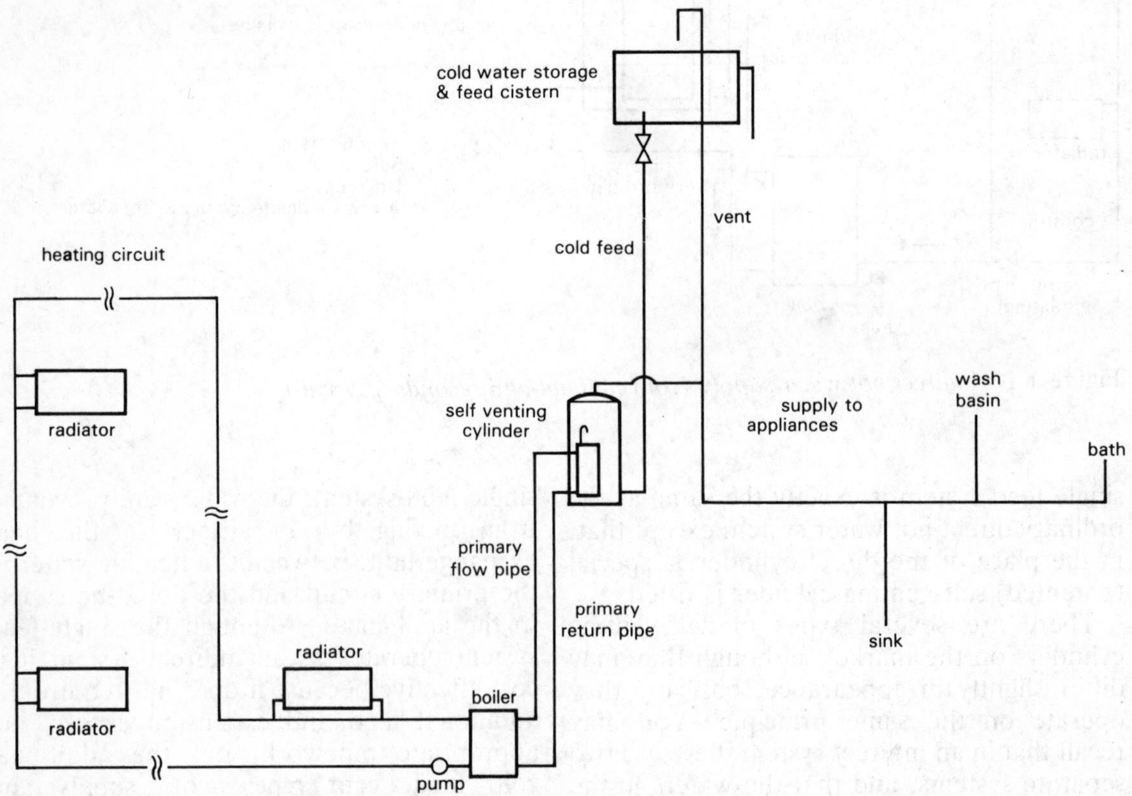

Figure 1.2a *Domestic hot water supply and heating system (indirect single feed cylinder)*

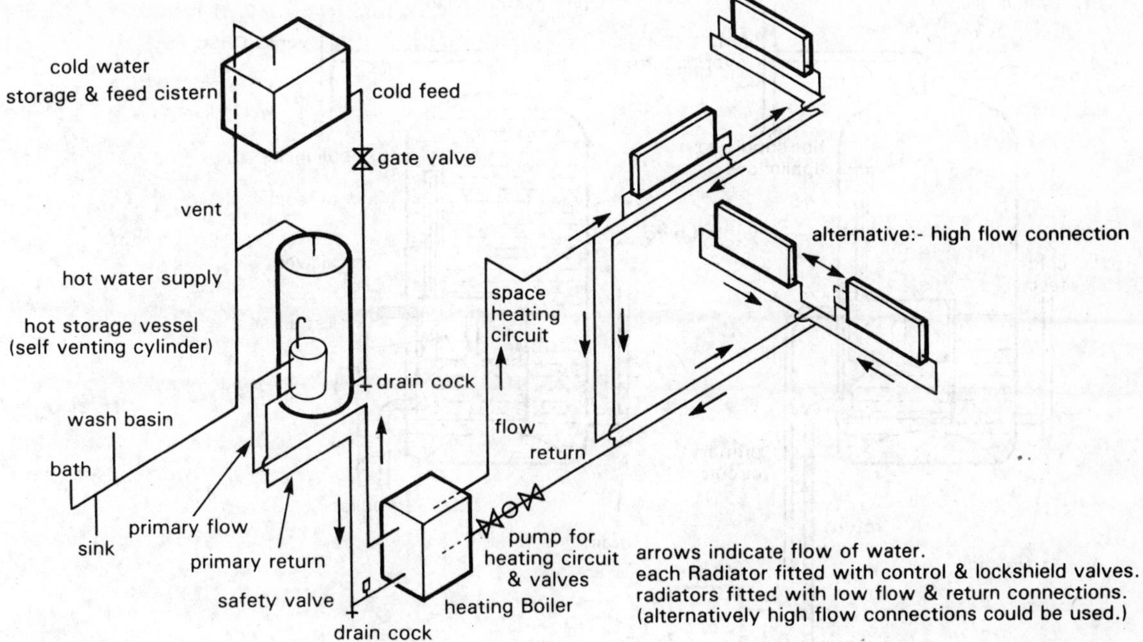

Figure 1.2b *Domestic hot water supply and heating system (indirect single feed cylinder)*

pipe which is incorporated. When the primary circuit is filled, the next stage is to fill the secondary supply, that which is drawn off at the appliances. When this is complete two air seals are automatically formed and permanently maintained, being self-charging during operation by air liberated from the

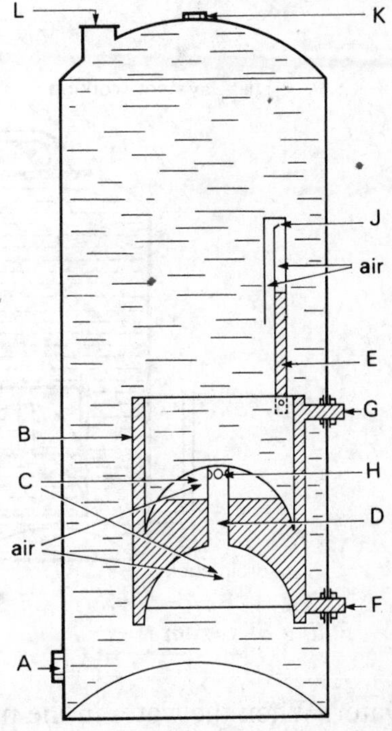

Key

A Cold feed connection
B Heat exchange
C Air locks
D Tube connecting air lock spaces
E Vent pipe
F Primary return pipe (to boiler)
G Primary flow pipe (from boiler)
H Holes to allow initial filling and later venting
J Inverted air trap
K Hot water to appliances
L Immersion heater connection

Figure 1.3 *The primatic self-venting indirect hot water cylinder*

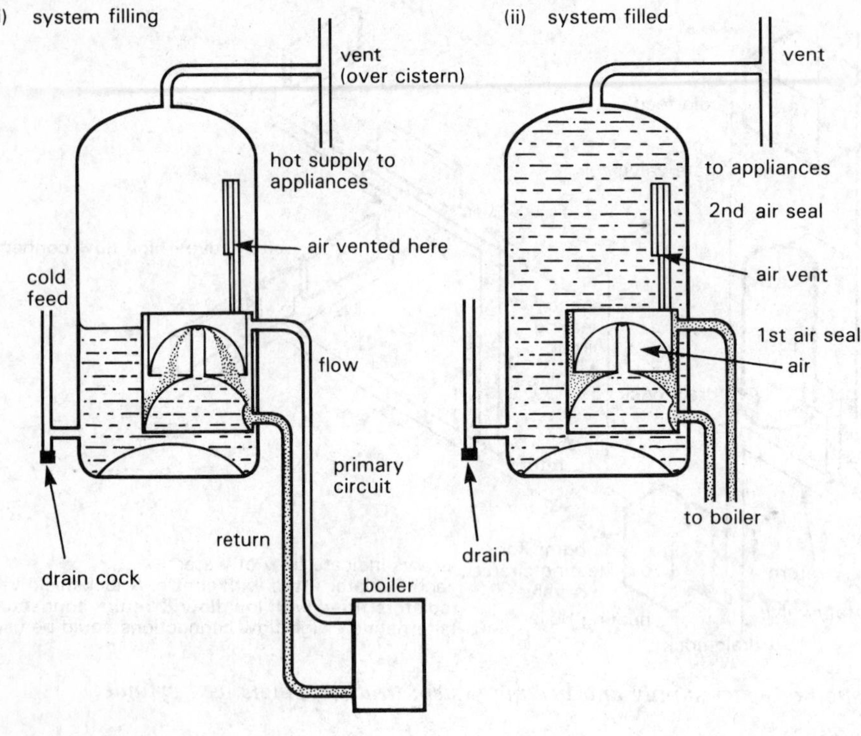

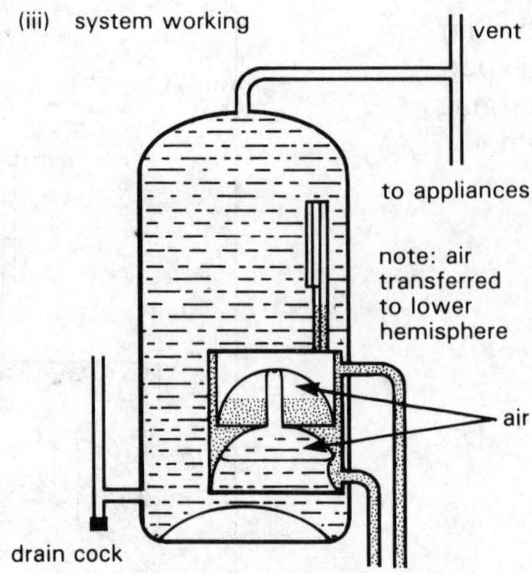

Figure 1.4 *Filling of system*

heated water. When the water in the primary circuit is heated through the normal temperature range, the increase in volume of about 4% is accommodated in the unit by displacing air from the upper dome into the lower one.

Draining The normal provision of fitting a

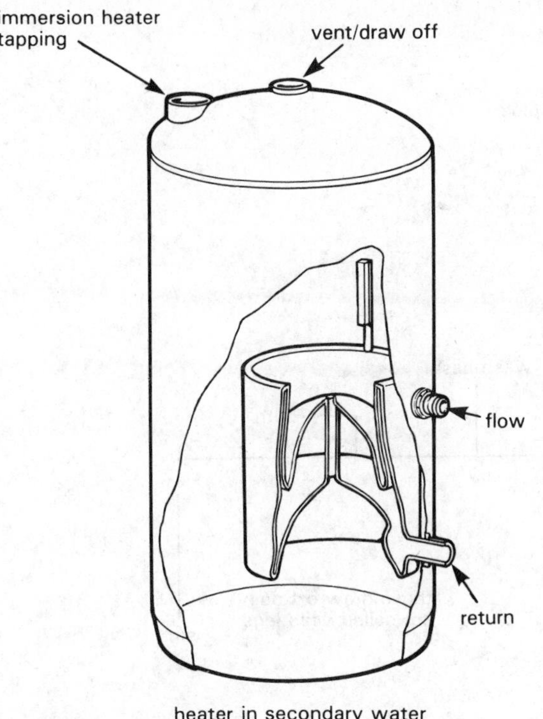

Figure 1.5 *Standard tapping*

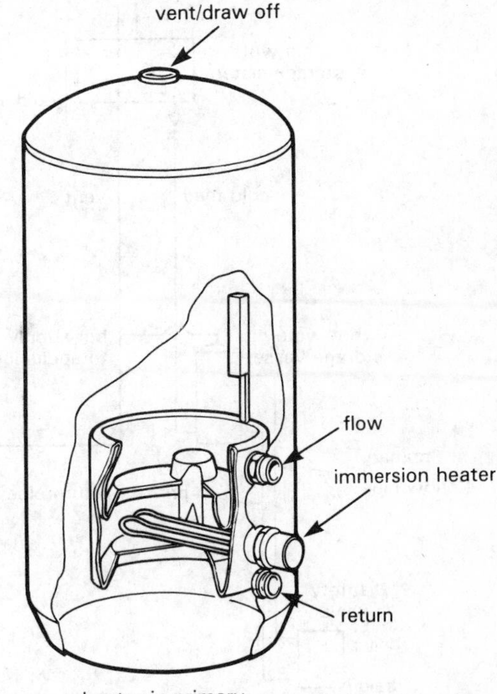

Figure 1.6 *Primary circuit tapping*

drain cock to the boiler should be carried out. In addition to this, a further drain cock must be provided on the cylinder or cold feed to enable the whole system to drain.

Immersion heaters The standard tapping for installing an immersion heater is usually at the top of the cylinder, Figure 1.5. In this case the heater is located in the secondary water.

In the case of temporary hard water, with the danger of lime encrustation taking place, the heater can be located in the primary water inside the unit, Figure 1.6. Because the primary water is not consumed the initial lime deposit, which is small, is insignificant.

Dead legs Many dwellings have hot water systems in which some appliances are fitted a long way from the hot storage vessel. This results in long lengths of pipe, known as 'dead legs', being necessary to convey the water to the appliances, Figure 1.7. Regulations state that any pipe conveying hot water to an appliance must not exceed a set limit, see Table 1.1.

The reason for this regulation is to prevent the unnecessary waste of water which occurs when hot water is required at a distant point and a considerable amount of cooled water must be drawn and run to waste before hot water is delivered at the appliance. When the required quantity of hot water is obtained and the tap closed, the pipe remains full of hot water which, in time, loses its heat. The cycle repeats itself when hot water is again required.

From the householder's point of view this waste of water means a waste of money. Therefore the grouping of appliances, and short pipe runs, are factors of good design. Where such considerations are not possible, the problem can be overcome by the introduction of secondary circulation, see Figure 1.8.

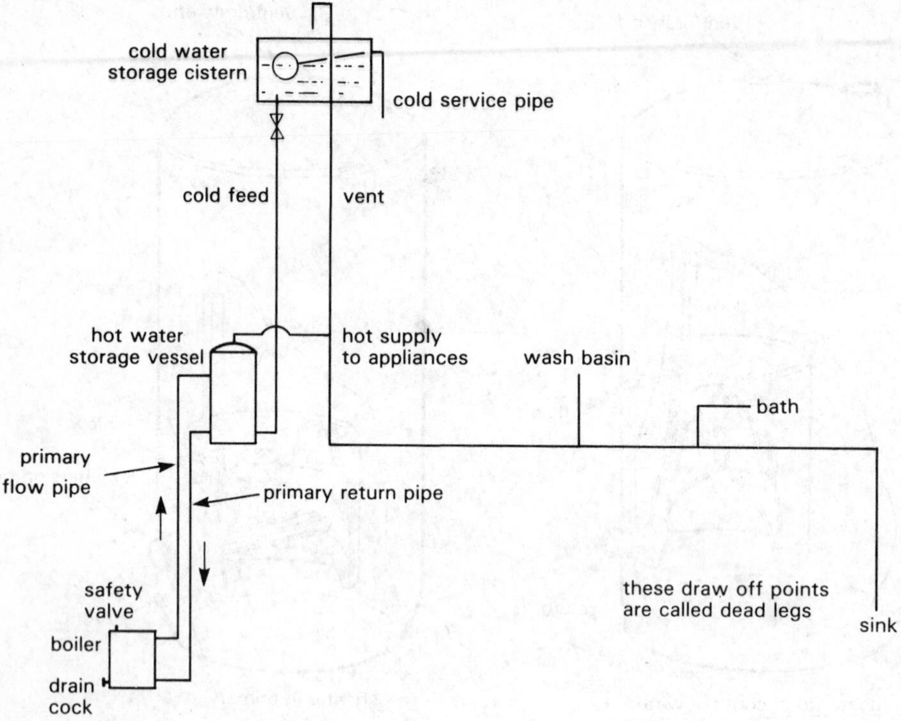

Figure 1.7 *Direct hot water supply with dead legs*

Secondary circulation

We have already discussed the primary circulation of hot water between the boiler and the hot storage vessel in Book 1. Secondary circulation is the circulation of hot water from the top of the hot storage vessel or vent pipe to or near to the appliances, reconnecting into the hot storage vessel. This return connection should be in the top third of the vessel.

It must be appreciated that water will constantly be circulating around this circuit so long as the water is hot. There will, therefore, be a cooling of the hot water and a subsequent and expensive loss of heat. This loss of heat can be curtailed at night time, when the water would not normally be required, by the fixing of a night valve. The night valve is simply a valve fitted in the circuit which can be manually or automatically operated to stop the circulation when demand is low or nil, as shown in Figures 1.8 and 1.9.

Although it is desirable that the secondary return should be re-connected to the hot water system in the top third of the hot storage vessel (cylinder), there are some occasions in practice where this is not possible. Figure 1.9 clearly shows how this problem is solved by the fitting of a non-return valve in the secondary circuit at a point between the *last* draw off and the point where the secondary circuit rejoins the primary circuit.

Table 1.1 *Permissible lengths of pipe*

Diameter of pipe	*Length*
up to 22 mm	12.0 m
22 mm to 28 mm	7.5 m
over 28 mm	3.0 m

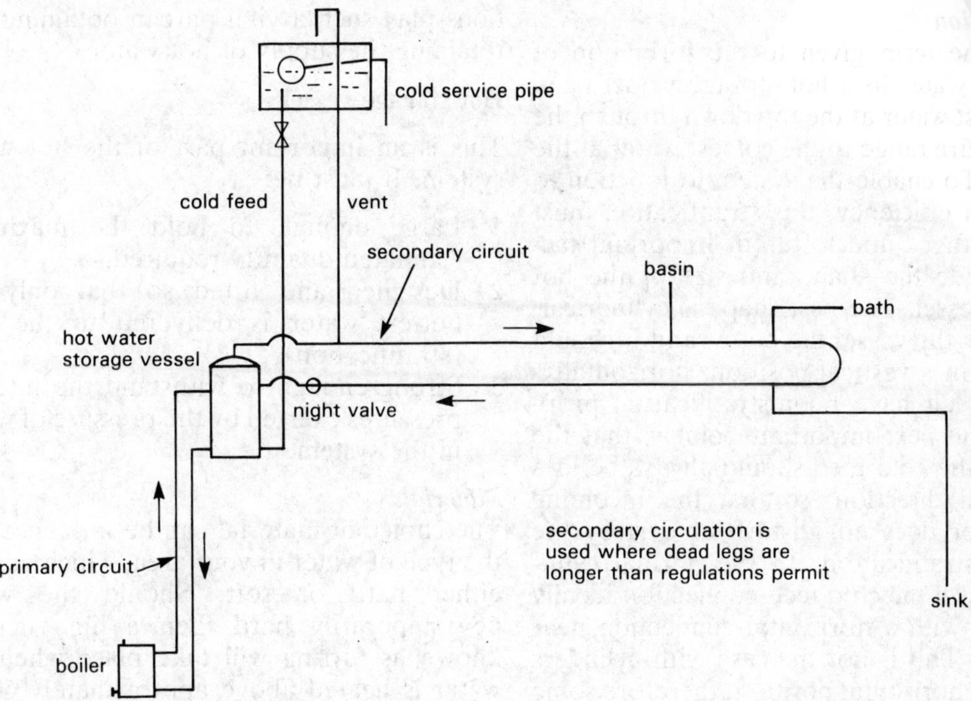

Figure 1.8 *Direct hot water with secondary circuit*

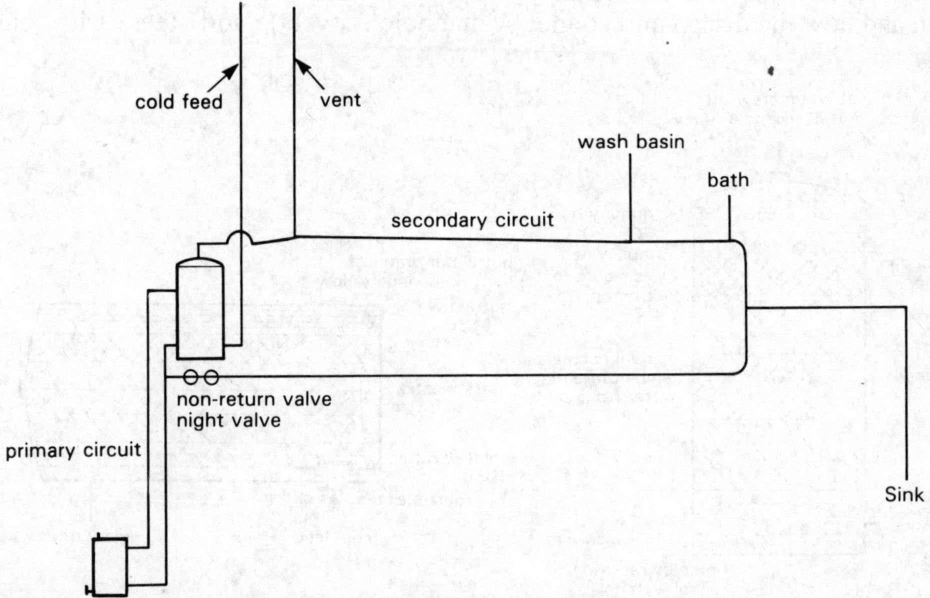

Figure 1.9 *Method of connecting secondary return below hot storage vessel*

the formation of
torage vessel from
top down through the
to the coldest water at the
able the system to function to
efficiency, this stratification must
ouraged and fostered. Important fac-
include the shape and size of the hot storage vessel. The best shape is cylindrical. The taller the vessel the better and it should be fitted in a vertical position: horizontally-fitted vessels have poor stratification properties. The next important point is that the entry of the cold feed should always be in a horizontal direction, so that the incoming cold water does not disturb or destroy the existing stratification. In the normal cylindrical vessel the cold feed connection ideally is placed with a horizontal connection near the base. This is not the case with cylinders fixed in a horizontal position, therefore some modification is necessary. The modification takes the form of a spreader tee fitted inside the vessel, which diverts the flow of water from a vertical to a horizontal direction. Figure 1.10 clearly indicates what is meant by stratification and how the design and connections play such a vital part in obtaining and retaining the supply of hot water.

Hot storage vessel

This is an important part of the hot water system. It must be:

1 Large enough to hold the maximum estimated quantity required,
2 Designed and fitted so that only the hottest water is delivered to the taps (stratification),
3 Strong enough to withstand the internal pressures exerted by the pressure of water in the system.

Materials

The choice of material can be influenced by the type of water in your area, which may be either hard or soft. Should the water be temporarily hard then a lime deposit known as furring will take place when the water is heated above approximately 60 °C. For this reason, copper and steel storage vessels and pipes are generally fitted. Adequate provisions should be made to facilitate the cleaning out of the component parts of the system, such as access covers (hand or manhole covers) and tees with cleaning

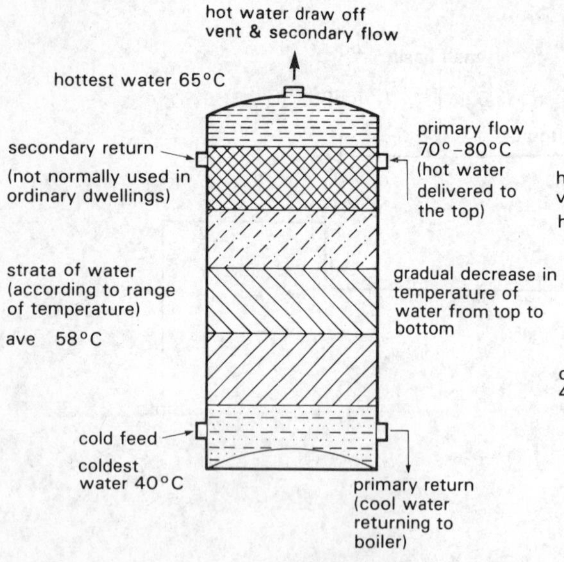

Figure 1.10a *Stratification (vertical)*

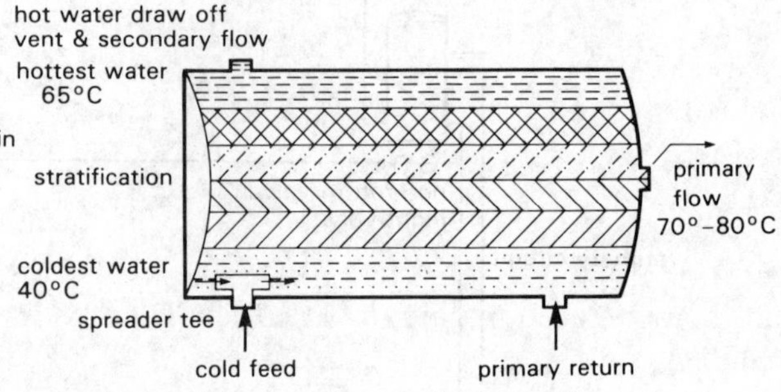

Figure 1.10b *Stratification (horizontal)*

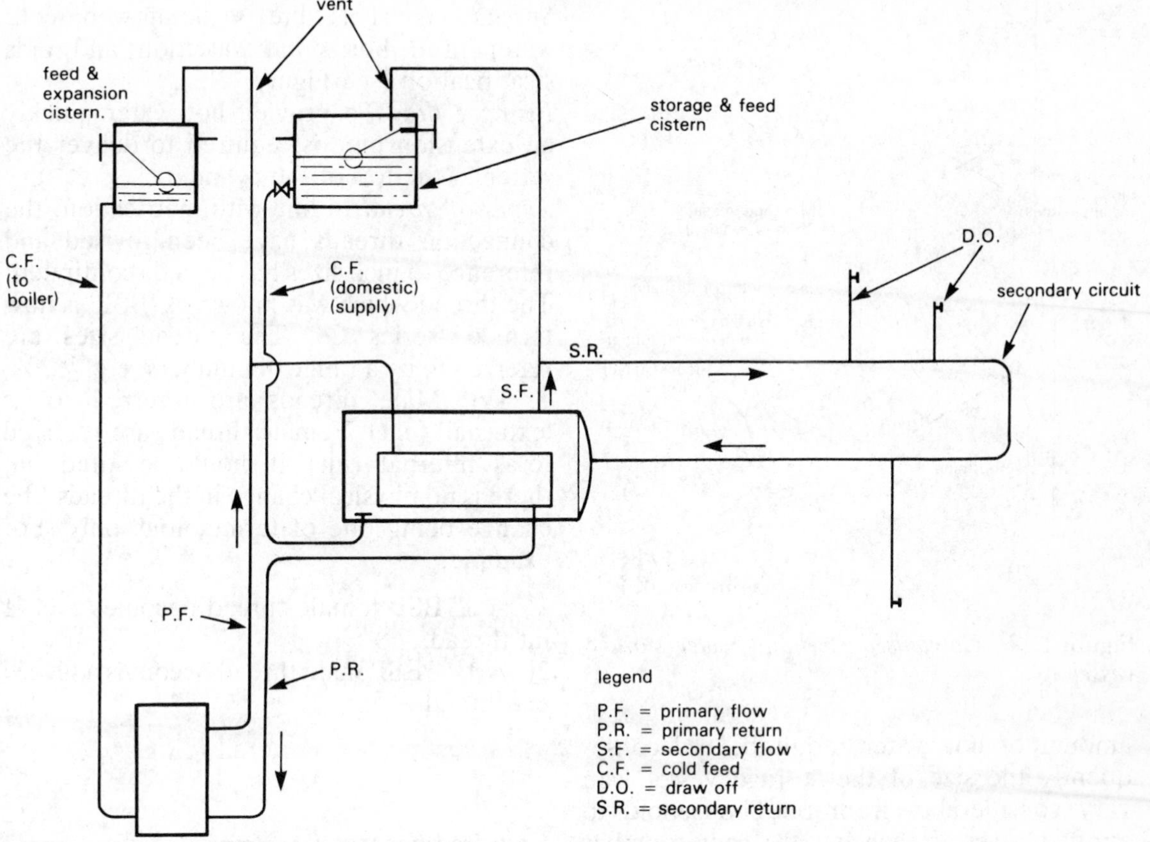

Figure 1.11 *Indirect hot water system with horizontal cylinder and secondary circuit*

access at change of direction. There would be no adverse affect if copper was used although it may suffer some damage from frequent cleaning due to not being as strong as low carbon steel. In districts where the water is of a soft nature it would be advisable to fit copper pipes and hot storage vessels.

Design

The size and shape of the vessel is extremely important. The size must be large enough to hold the estimated maximum requirement of water, plus an additional amount for extra demand. For ordinary domestic work the size of vessel is not calculated but is selected using the knowledge and experience of the installer.

When calculating the size of hot storage vessels, the following facts must be known:

1 Number of occupants,
2 Number and type of appliances,
3 Temperature of water required,
4 Frequency of use.

It will be readily appreciated that the

Table 1.2 *Hot water requirements for domestic appliances*

Bath	70 litres
Shower	25 litres
Wash basin	9 litres
Bidet	4 litres
Sink (with washing-up bowl)	7 litres

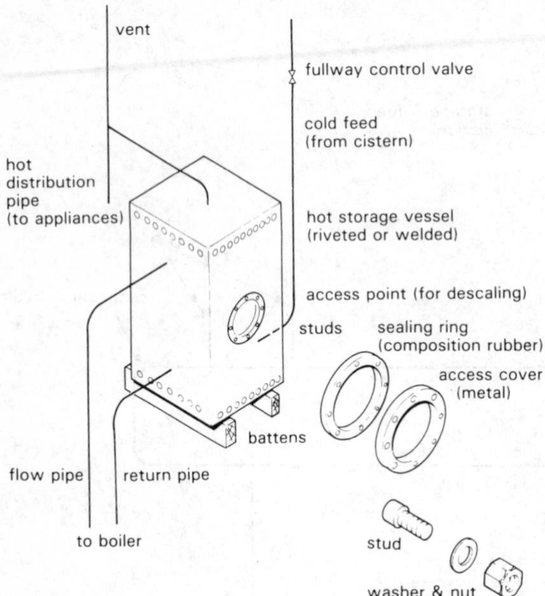

Figure 1.12 *Galvanised steel hot water storage vessel*

amount of hot water required, and consequently the size of the storage vessel, can vary considerably from one household to another, even if each has the same number and type of appliances. Therefore only an approximation can be made. It is always good practice to err on the large side.

Experience has proved that the best shape of vessel is cylindrical. This gives great strength and enables the vessel to be manufactured from thin material, which leads to an economical product. An added advantage of the cylindrical vessel is that it aids stratification, as described earlier.

Cold water is heavier per unit volume than hot water (Book 1, pp 135–6), therefore the cold feed supply pipe must always be connected at the base and the hot water distributing pipe must always be connected at the top of the hot storage vessel. The best location for the vessel is as near to the boiler as is practical, so that the primary circulation pipes are kept short and heat losses reduced to a minimum. These can be reduced still further by insulation.

Spreader tee This directs the flow of cold water in a horizontal direction and aids stratification (see Figure 1.13).
Primary flow To provide hot water quickly an extension pipe is required to deliver the water to the top of the cylinder.
Types of thread In line with metrication, the connection threads have been revised and reference to inch sizes has been discontinued. The thread which was known as 'BSP' is now termed 'series G'. The thread sizes are referred to by a range of numbers, e.g. ½, ¾, 1, 1¼. Male threads are referred to as 'external' (ext). Female threads are referred to as 'internal' (int). It should be noted that there is no physical change in the threads, the change being one of terminology only. For example:

1 A 1″ BSP female thread becomes a G. 1 int thread.
2 A 1¼″ BSP male thread becomes a G. 1¼ ext thread.

Unvented hot water systems

The supply of hot water in the United Kingdom has for many years been of the boiler/cylinder/cistern type system as described and illustrated previously. Recent changes in the model water bye-laws (1986) now permit the installation of unvented hot water systems. This type of system allows the hot water system to be connected directly to the main cold water service. The whole system is then under a controlled pressure, and a pressure reducing valve is fitted.

The advantages of this type of system over the traditional system are:

1 No cold water storage cistern, together with the relevant pipework, is required. This overcomes many problems regarding the use of storage cisterns, e.g. space, head, frost damage etc. There is also a reduction of on-site work, with an inevitable saving of cost;

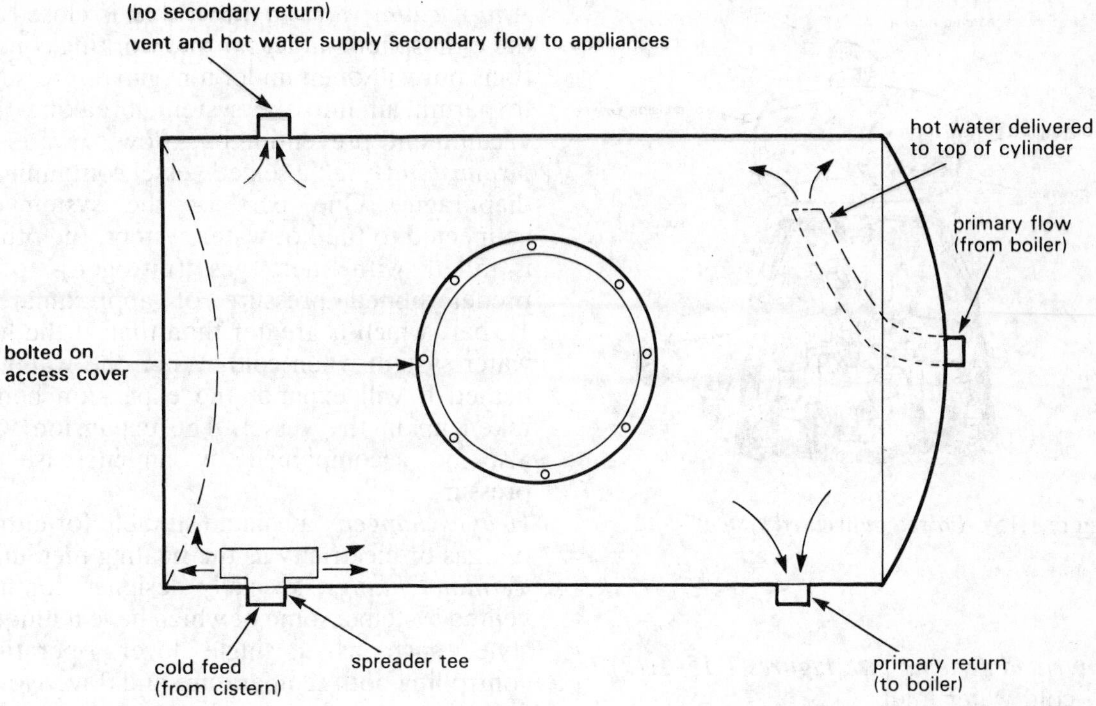

Figure 1.13 *Horizontal hot water storage vessel*

2 Because both hot and cold water supplies are at the same (equal) pressure, the use of mixer units – and in particular showers – is improved. The increased pressure also gives a more invigorating shower;
3 Due to the pressure available, smaller diameter pipes can be used, also giving a saving in the cost of installation, water and fuel;
4 It is a much quieter system in operation than the traditional one;
5 Less risk of reduced flow with simultaneous demand;
6 Less risk to health because there is no stored cold water which could become contaminated;
7 Greater flexibility is gained in siting the unit.

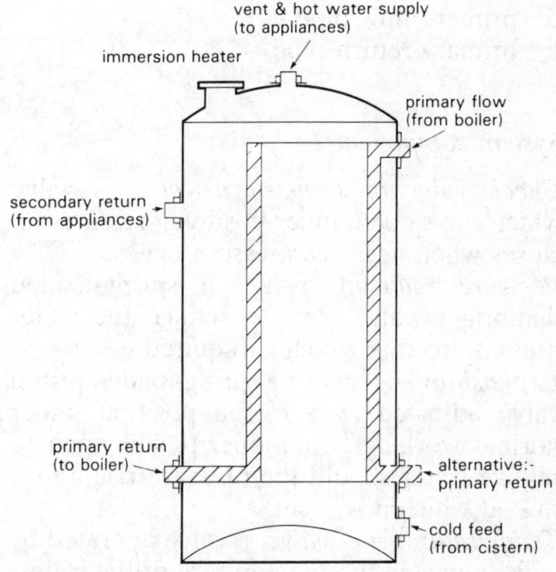

Figure 1.14 *Vertical indirect hot water storage vessel*

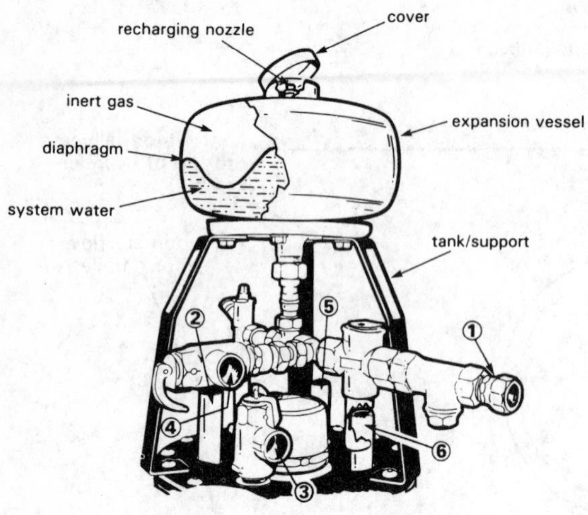

Figure 1.15 *Polystel unvented system unit*

Unvented systems (see Figures 1.15–1.17)

1 cold water inlet
2 hot water outlet
3 discharge to warning pipe (temperature relief valve)
4 discharge to warning pipe (expansion relief valve)
5 primary flow pipe
6 primary return pipe

System components

Check valve or non-return valve: a valve which stays open under positive pressure but closes when negative pressure exists.
Pressure reducing valve: a spring-loaded diaphragm valve used to reduce the water pressure to that which is required.
Expansion relief valve: a spring-loaded piston valve adjusted to a closed position under normal working conditions. It will open to relieve water should the pressure rise above that at which it was set.
Temperature relief valve: a valve operated by a difference in the temperature of the water, its function being similar to that of the expansion relief valve.
Anti-vacuum valve: a valve that is closed to the atmosphere under normal working conditions but will open under a negative pressure to permit air into the system, breaking the vacuum and preventing backflow of water.
Expansion vessel: a sealed vessel containing a diaphragm. One part of the system is connected to the hot water system, the other is filled with inert gas (nitrogen) to a predetermined pressure of approximately 1.5 bar, which is greater than that of the hot water system when cold. When the water is heated it will expand, the expansion being taken up in the vessel. The expansion will also be accompanied by an increase in pressure.
Heat exchangers: available suitable for either oil, gas or electricity as the heating medium.
Terminal fittings: specially designed for unvented systems, some of which have a unique style, such as a single lever operation controlling both temperature and flow. Noise factor is overcome by various types of nozzle outlet, and by means of aerators and laminar flow jets.

Boilers

Most homes today are fitted with domestic hot water systems, and many of them also have either full or part space heating systems. The householder of today has a wide choice of heating boiler and heating fuel. The type of boiler chosen will be governed by personal preference, whether free standing or wall mounted, cost, and availability of a particular fuel.

The choice of fuel would be between:

1 solid fuel,
2 gas,
3 oil,
4 electricity.

Size and efficiency
Boilers are described by their heat output. Heat output is the amount of useful heat

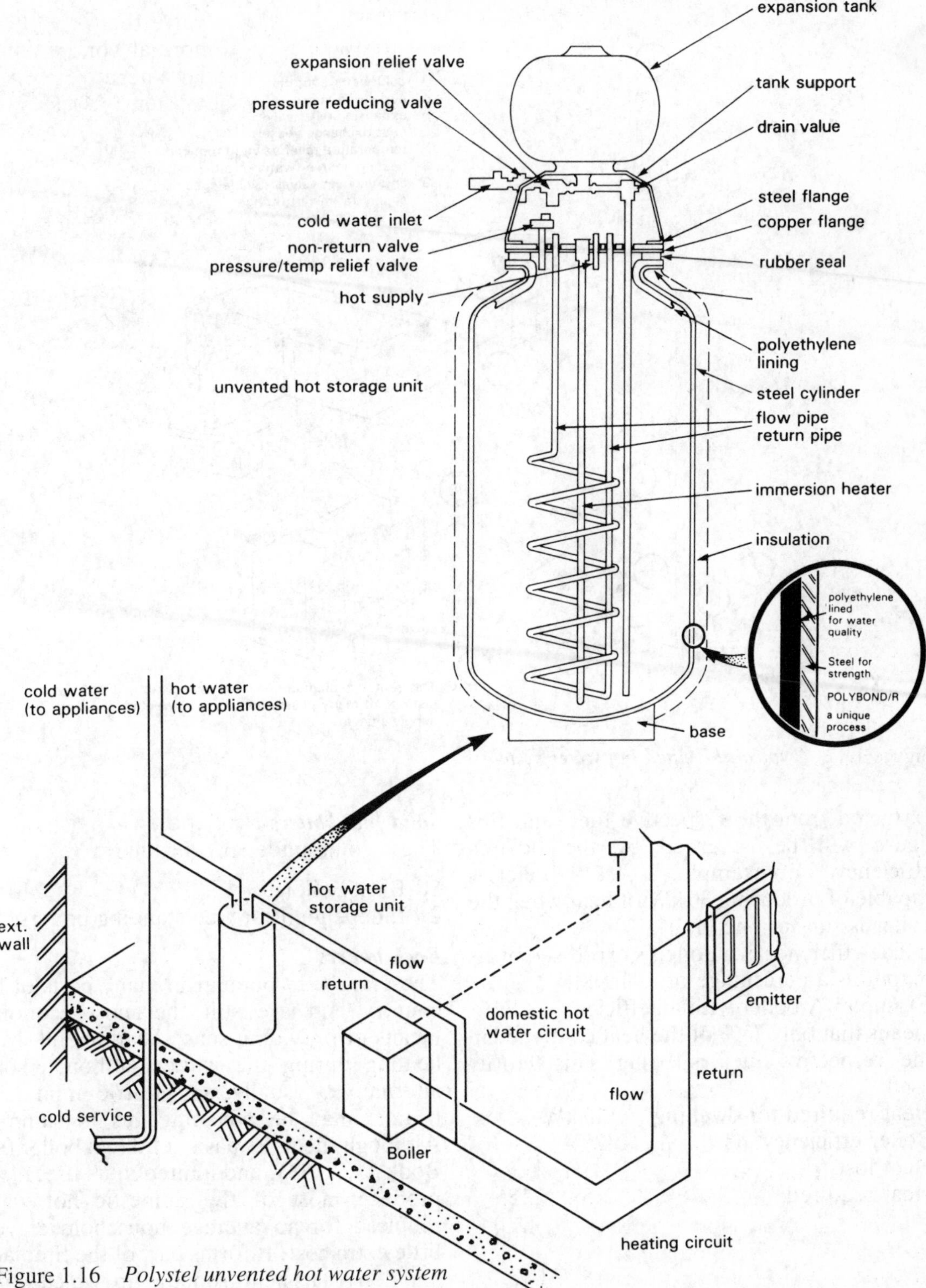

Figure 1.16 *Polystel unvented hot water system*

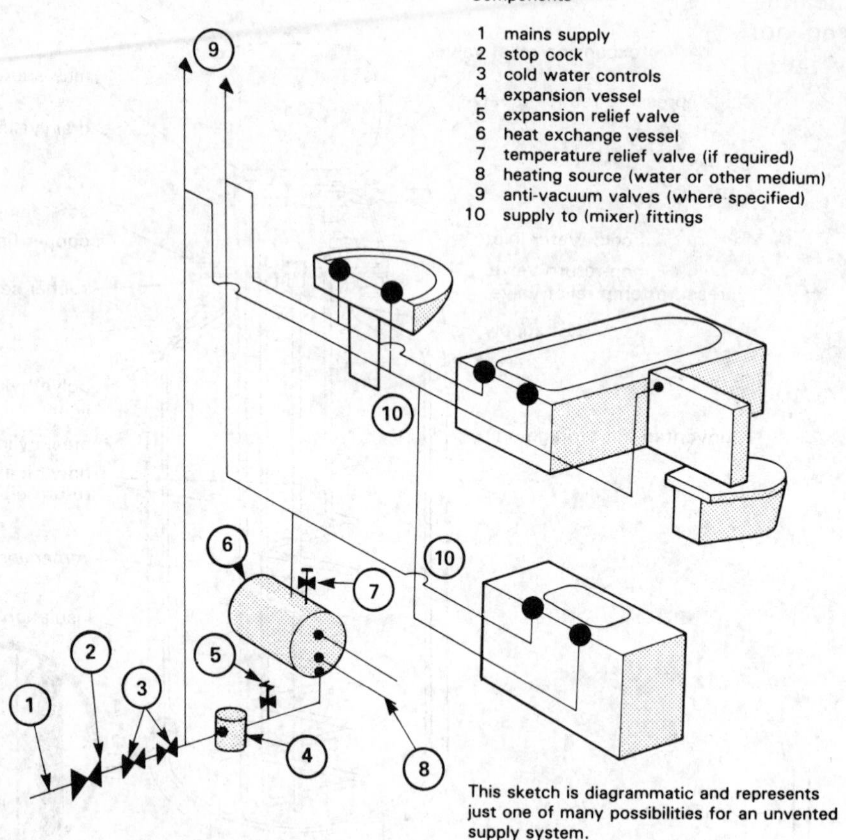

Figure 1.17 *Typical unvented hot water system*

extracted from the respective fuel, and this figure will be governed by the boiler's efficiency. For example, a 30 kW boiler is capable of producing 30 kW of heat when the boiler is run for one hour.

The efficiency of a boiler is expressed by its output as a percentage of its input.

Example: A boiler with an efficiency of 75% means that only 75% of the heat contained in the respective fuel is being satisfactorily used.

Heat required for dwelling	15 kW
Boiler efficiency	75%
Heat lost	25%
Heat required	$15 \times \frac{125}{100}$
	18.75 kW

Solid fuel boilers

These come under two headings:

1 Back boilers,
2 Independent or free standing boilers.

Back boilers

These are very popular in many parts of the country and are still the most economic means of providing space heating and water heating during the winter, although some alternative, usually an electric immersion heater, may be advantageous in summer. Although the most basic solid fuel boiler (see Book 1, p. 133), and limited in its use, it will produce most of the domestic hot water required for an average household at very little extra cost. It forms part of the fireplace, and derives its heat from the fire which would be normally heating only that particular room. Larger and more intricate back boilers are available (Figure 1.18), some with the flue forming an integral part of the boiler enabling the maximum amount of heat to be extracted from the hot flue gas as it passes up to the chimney. Although these boilers are able to impart a much greater heat value to the water, this would still not be sufficient satisfactorily to heat even a small domestic dwelling, but could supply what is known as 'background heating' in addition to the domestic hot water requirement.

High output boilers

An example of the development of this type of back boiler is the 'Esse Bramble 30', designed to produce a heat output of approximately 8.8 kW which should be sufficient for many smaller dwellings. This appliance can easily be installed in both new or existing dwellings. It will burn either smokeless fuel or household coal with maximum efficiency. It incorporates a unique damper with three settings, and a special flue brick which reduces smoke emission when burning household coal.

This boiler is of the wrap-round type, fabricated in heavy gauge low carbon steel. A central flue is controlled by the three-position damper which opens and closes the internal flueway.

1 *Closed position* Prevents hot flue gases passing through the boiler flue – water heating reduced to a minimum.
2 *Mid position* Recommended for maximum efficiency when burning smokeless fuels.
3 *Full open position* This position will give maximum draught condition and is recommended when burning household coal.

The burning rate can be controlled by adjustment of an air control cam at the centre of the fire front. The boiler has an efficiency of approximately 75%.

Solid fuel fires and high output back boilers are once more becoming very popular. No other form of heating has so many benefits, including:

1 Warm comforting feeling of an open fire,
2 Several radiators can be fed from the back boiler,
3 Cost effective,
4 Wide range of fires to choose from,
5 Local one-room heating possible.

Link-up system

Solid fuel boilers can be used as the sole heating system or in conjunction with another heating system. An existing gas or oil fired system can be linked up to an open fire or solid fuel room heater, thereby giving all the advantages of a real fire.

There are two types of 'Link-up system', both of which mean a saving can be made on oil or gas bills.

1 The solid fuel boiler is connected to the hot water storage vessel to provide domestic hot water while the gas or oil boiler continues to provide the space heating via the radiators.
2 The system is connected so that you have the choice of using either or both systems to provide heating and hot water. It must be remembered that solid fuel open fires or room heaters have a lower heat output than gas or oil boilers, therefore a less efficient system would result if using only the solid fuel appliance. It may be necessary either to reduce the temperature in some rooms by automatically controlled thermostatic valves or to turn off selected radiators when using only the solid fuel appliance.

Not all existing central heating systems can be successfully linked up to an additional boiler. Some of the factors governing this are:

1 Location and distance from solid fuel appliance to point of link-up into existing system,
2 Location of hot water storage vessel,
3 The need for each appliance to have a separate chimney.

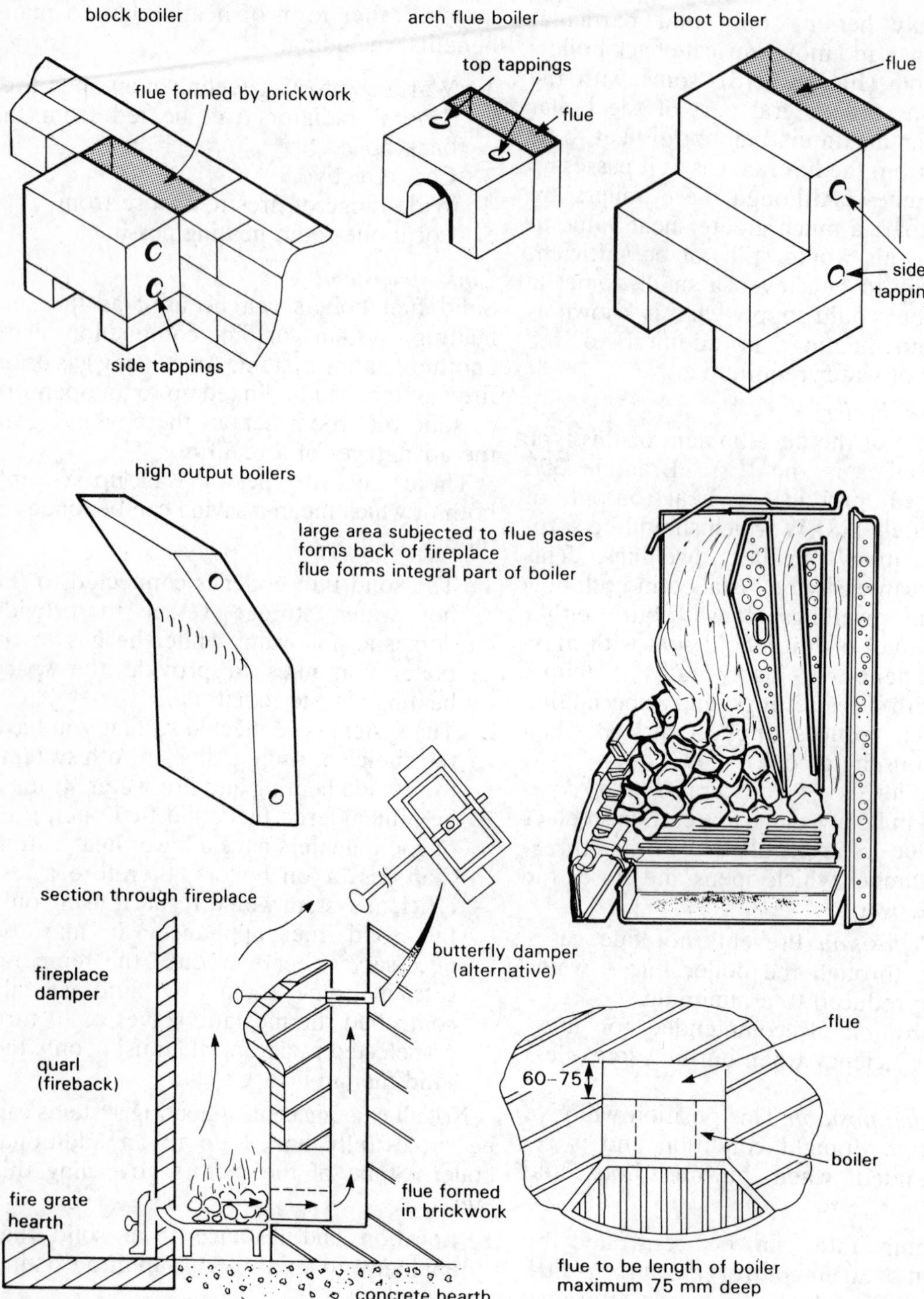

Figure 1.18 *Solid fuel boilers*

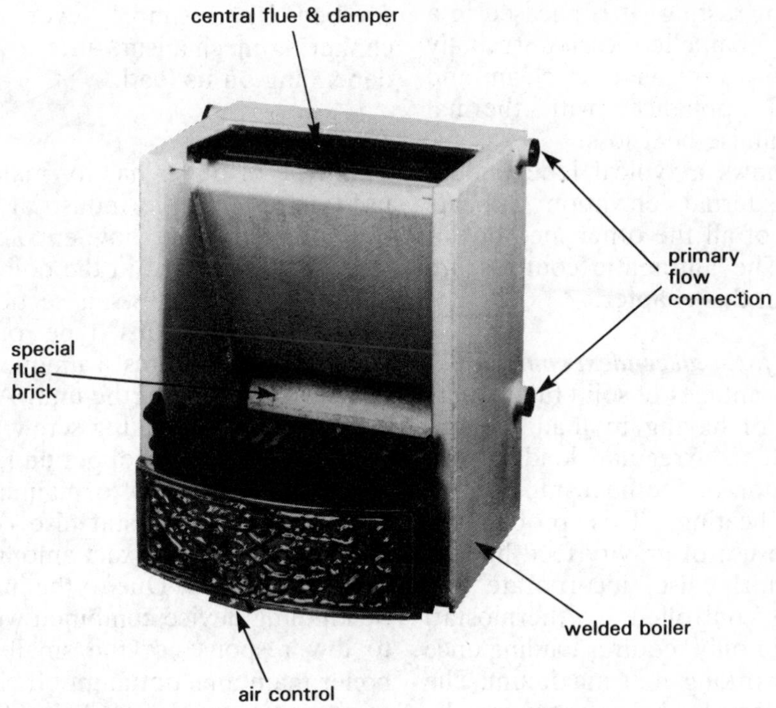

Figure 1.19 *High output boiler 'Esse Bramble 30'*

Independent or free standing solid fuel boilers

The natural progression from the back boiler with its limited use is to a more sophisticated and larger boiler capable of providing both the domestic hot water and space heating of the dwelling. Independent boilers can be fired by solid fuel, gas or oil and modern boilers are now operated automatically, being fitted with electrical and thermostatic controls.

There are two main types of independent solid fuel boilers: (i) sectional boilers; (ii) gravity feed boilers. Both come in sizes large enough to provide all the domestic hot water and satisfactorily to space-heat domestic dwellings.

These boilers are suitable for use with gravity central heating, small bore and micro bore systems, and the systems can be controlled by thermostat, time clock and individual thermostatic radiator valves for economic and automatic control.

Sectional solid fuel independent boilers

The sectional solid fuel independent boilers on the market today are capable of providing the heat required for the small domestic household, for both water and space heating. The boiler flue outlet must be connected to a chimney, independent of any other appliance. The flue and chimney must be regularly cleaned depending upon its use: it is recommended that this should be carried out at least once a year.

Once the fire is lit and the thermostat set, the boiler requires only a minimum of attention. Refuelling and ash removal is required from time to time, possibly two or three times a day when the boiler is working at its maximum. The fuel is fed through either a front or a top door. The door incorporates a fireproof gasket, ensuring an airtight seal when closed, which results in controlled burning. The boiler is constructed of heavy gauge low carbon steel, bower barff treated to prevent rusting. It is encased in a vitreous (stove) enamelled casing (usually white in colour) to provide a clean and durably finished appliance, with thermal insulation to minimise heat loss.

Figure 1.20 shows a typical independent boiler whose external elevation appears identical to that of all the other methods of heating water. The automatic controls are dealt with later in this chapter.

Gravity feed solid fuel independent boilers

One of the disadvantages of solid fuel boilers is the necessity of having to load the fire regularly with fuel. Irregular loading will result in fluctuation of the heat imparted to the water for heating. This problem is overcome by the use of gravity feed hopper type boilers which also incorporate fan assisted burning controlled by thermostat. This type of boiler only requires loading once each day, when working at its maximum. The fire burns continuously, being automatically fed with fuel from the hopper. The sensing thermostat governs the fan when heat is called for and boosts combustion. This ensures a high burning temperature, which in turn fuses the ash into a solid clinker, which may be automatically ejected from the fire-bed by a simple lever operated de-clinkering mechanism, once or twice a week depending on its load.

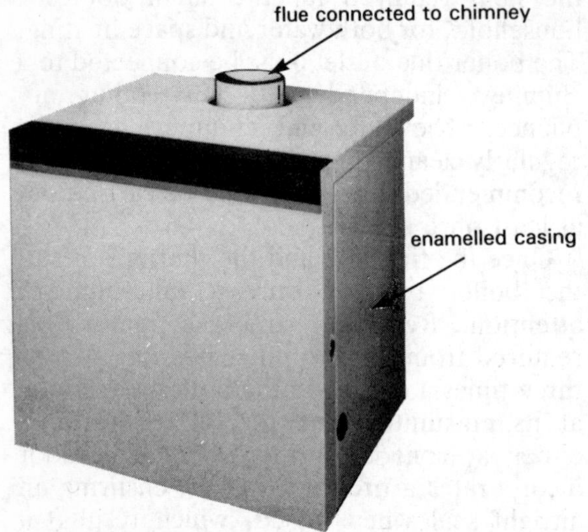

Figure 1.20 *Independent solid fuel boiler (Trianco)*

Underfeed boilers

This type of boiler has for many years been used solely in large industrial work; recent developments have now enabled a boiler of this type to be used in the domestic market.

Figure 1.21b shows a section through a burner used in this type of boiler; the 'Coalstream' features a unique coal feeding device at the base of the unit. A thermostatically controlled rotating screw automatically feeds coal from the hopper directly to the fire at the rate required to maintain the boiler output. The thermostat also controls a fan which provides the exact amount of combustion air required. Due to the inbuilt automatic kindling device combined with rapid high to low response of the small fire bed the boiler maintains optimum efficiency from its minimum output of 0.6 kW (2000 BTUs) up to 17.5 kW (60 000 BTUs) suitable for full domestic central heating and hot water.

This type of boiler has been designed to burn low cost bituminous coal (smokelessly) instead of the more expensive anthracite, this is due to the high temperature of combustion in the firebed.

These very efficient boilers require little attention, only the removal of the burnt clinker periodically, which can be either manual or automatic.

Oil-fired boilers

In their outward appearance, solid fuel and oil-fired boilers appear very similar. It is only upon opening the boiler casing and carrying out an inspection that it becomes evident which fuel is being used. Oil-fired boiler flues are usually connected to chimneys to discharge the products of combustion into the atmosphere at a safe height. It is now possible to purchase boilers which burn *kerosene* with low level, balanced flues, so overcoming the necessity of being tied to a chimney location or the expense of building

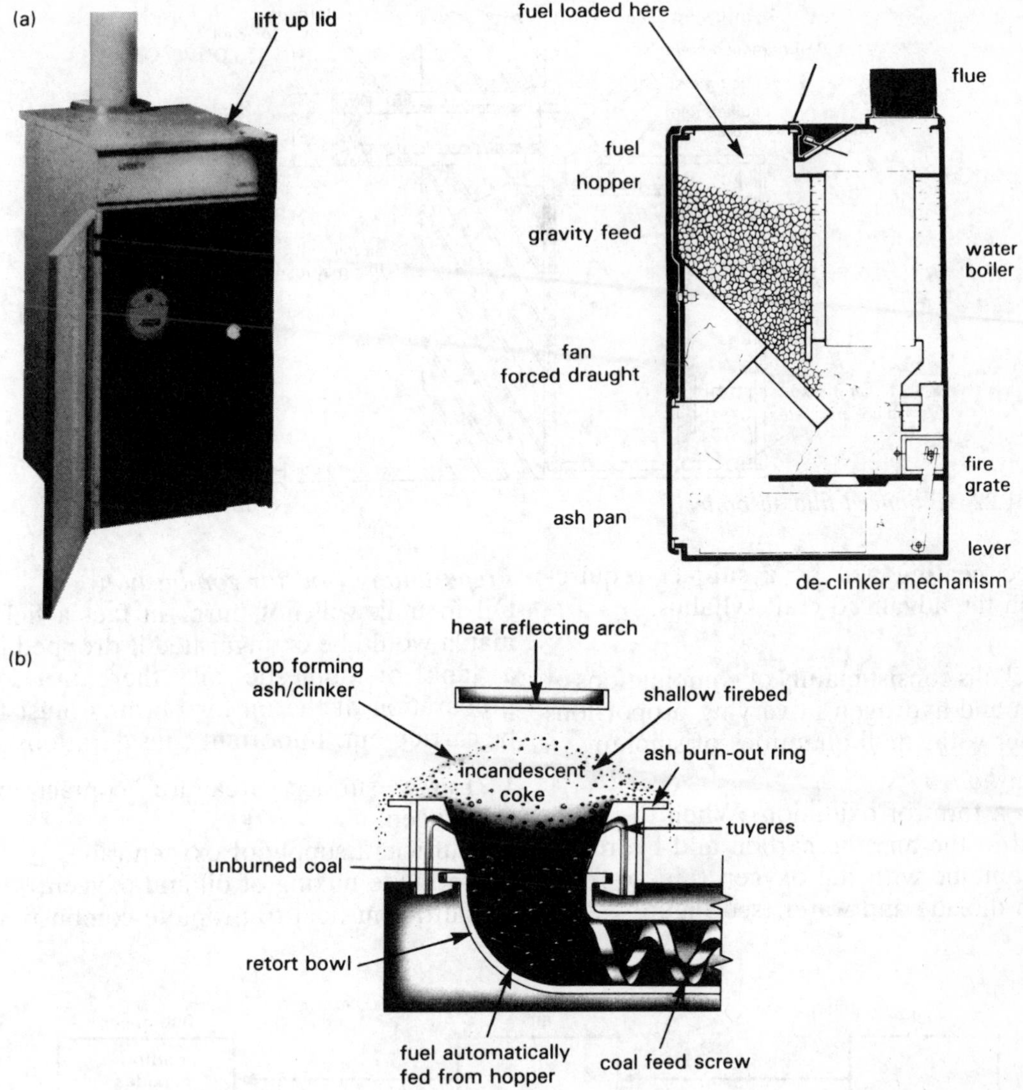

Figure 1.21 *Gravity feed solid fuel boiler and underfeed solid fuel boiler (burner)*

one. A balanced flue ensures correct combustion without the aid of the draught of a chimney. The balanced flue arrangement enables an adequate supply of fresh air for combustion and at the same time allows equalisation of pressure between the inlet and outlet under unfavourable wind conditions.

Oil-fired boilers were, a few years ago, a very popular method of providing domestic hot water and space heating. Owing to the increase in the cost of the fuel, fewer oil-fired systems are now being installed. Although few new systems may be installed, however, there are still a great many oil-fired systems in operation today and the modern plumber must be familiar with both the installation and maintenance of these. It is not intended to deal with this subject in depth here: only the basic requirements will be treated, as

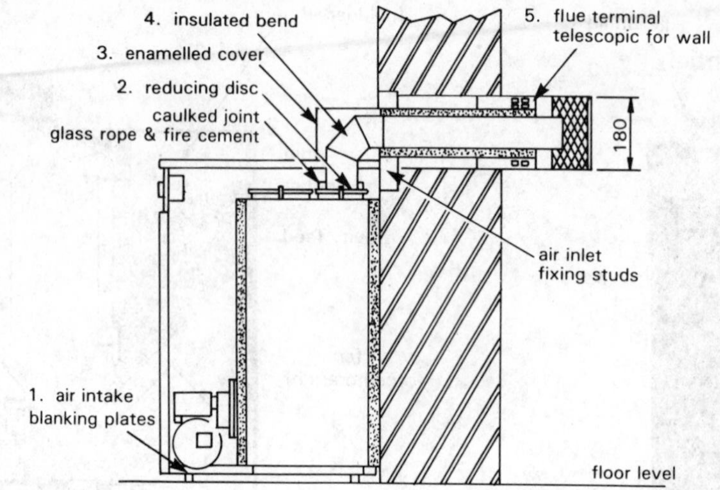

Figure 1.22 *Balanced flue oil boiler*

oil-fired heating will be a subject requirement in the advanced craft syllabus.

Fuel
All fuel oils consist mainly of combinations of carbon and hydrogen in varying proportions together with small quantities of sulphur.

Combustion
This is a form of oxidation. When the oil is burned in the air, the carbon and hydrogen (oil) combine with the oxygen (air) to form carbon dioxide and water (see Figure 1.23).

Preparation of oil for combustion
Oil in bulk will not burn, in fact a lighted match would be extinguished if dropped into a tank of domestic oil: therefore some preparation of the oil for burning must first be carried out. Important considerations are:

1 Large surface area for contact with oxygen,
2 Sufficient supply of oxygen (air),
3 Intimate mixing of oil and oxygen,
4 Sufficient heat to promote combustion.

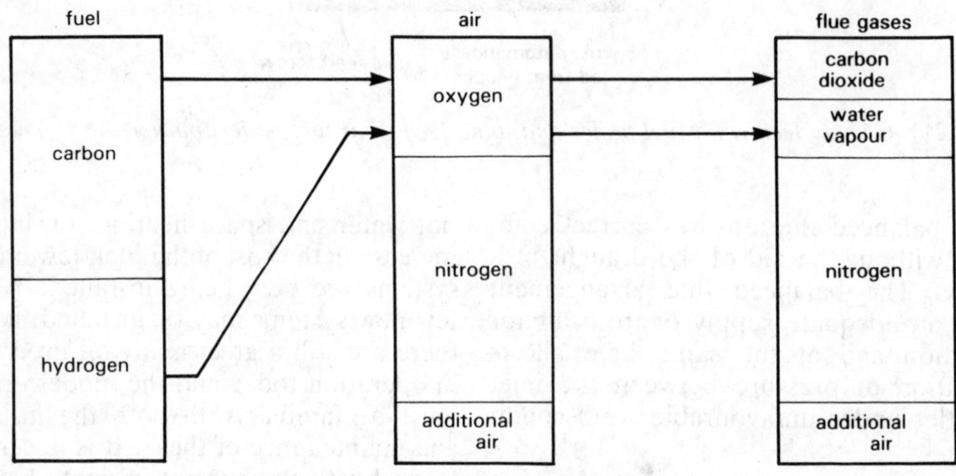

Figure 1.23 *The combustion process*

Note: it cannot be over-emphasised that adequate air is essential for correct combustion and ventilation sources should be as close to the boiler as practicable.

Burners

The function of the burner is to convert the oil into a state in which it can be satisfactorily mixed with air for combustion, and to intimately mix the oil and the air in the correct proportions to obtain maximum efficiency from the combustion process. This is achieved by one of the following processes:

1 Vaporisation (by heating the oil),
2 Atomisation (mechanically breaking up the oil into fine particles).

Vaporising burners are further classified as:

1 Sleeve or drum burners,
2 Pot burners.

Sleeve burners are only satisfactory for small boilers with rating of approximately 10–11 kW. Sleeve burners are suitable for use with kerosene-type fuel, and are silent in action.

Pot burners can be either:

1 *Natural draught burners* in which the air is drawn into the combustion chamber naturally by the pull of the chimney draught. They are satisfactory for boilers with a rating of up to approximately 17 kW; or
2 *Fan assisted burners* in which the air is blown into the burner pot or combustion zone by means of an electrically operated fan. This type of burner is reliable up to approximately 23 kW.

Rotary burners There are three different types of 'Wallflame' burner. All of them are now obsolete but of course many are still operating very satisfactorily, so the practising plumber must have knowledge of this type of appliance. They are available in sizes up to approximately 87 kW, although they are used primarily in the domestic market and rated up to 23 kW. Above this size the pressure jet burner is the type usually recommended. The fuel is partially atomised by the rotating spinner, and is then ignited by a high tension spark, or a low tension hot wire coil, which creates a hot spot where the oil is initially vaporised. Air is enabled to mix with the vapour, thus ignition then takes place. The heat from this point quickly increases the vaporisation process until the flame spreads around the entire rim of the burner.

An electric motor drives the spinner, on top of which is fitted a paddle wheel type of fan for directing the air to the point of combustion. A low but stable chimney draught is essential for satisfactory results. Kerosene fuel is used for these burners.

Note: The pot boiler and the wallflame boiler have been almost totally superseded by the 'Mini pressure jet' boiler, due to development that has led to the manufacture of very tiny nozzles.

Atomising burners

The pressure jet burner was developed primarily for very large domestic, commercial and industrial boiler applications and it has proved to be extremely efficient and reliable although somewhat noisy in operation. Fairly recent development has seen the introduction of smaller output burners of approximately 13 kW, with even greater reliability and quieter action: these boilers are the ones most commonly installed in the domestic market today. The oil is pumped under pressure (usually 6.9 bar) through a specially designed nozzle, where it is emitted as a mist of tiny droplets. The oil mist combines with the air delivered by the fan and is then ignited within the combustion chamber of the boiler. For pressure jet boilers a gravity head for the oil supply is desirable (because of cost) but the fuel pump of the burner can be adapted to enable the oil storage tank to be below the level of the burner.

Oil storage tanks

The following points should be considered when specifying an oil storage tank. It should

1 Conform to local authority regulations,
2 Be as large as possible to capitalise on bulk buying – minimum size 1140 litres,
3 Be accessible for filling, de-sludging and repair, with
   (a) a maximum distance of 30 m from the vehicle/tanker delivery point
   (b) as short a distance as possible from the boiler
   (c) a minimum head of 0.3 m, and a maximum of 3.0 m (see note on pressure jet),
4 Have adequate support (1140 litres = 1.25 tonnes approximately),
5 Be manufactured from black low carbon steel with welded joints,
6 Be protected externally by good quality oil paint,
7 Have a plate thickness dependent on size.

The oil supply lines to the boiler may be constructed from either:

1 *Copper*. This should be soft temper or half hard with flared fittings (manipulative joints), no soft-soldered joints or jointing compounds to be used. Alternative brazed or silver soldering may be used; or

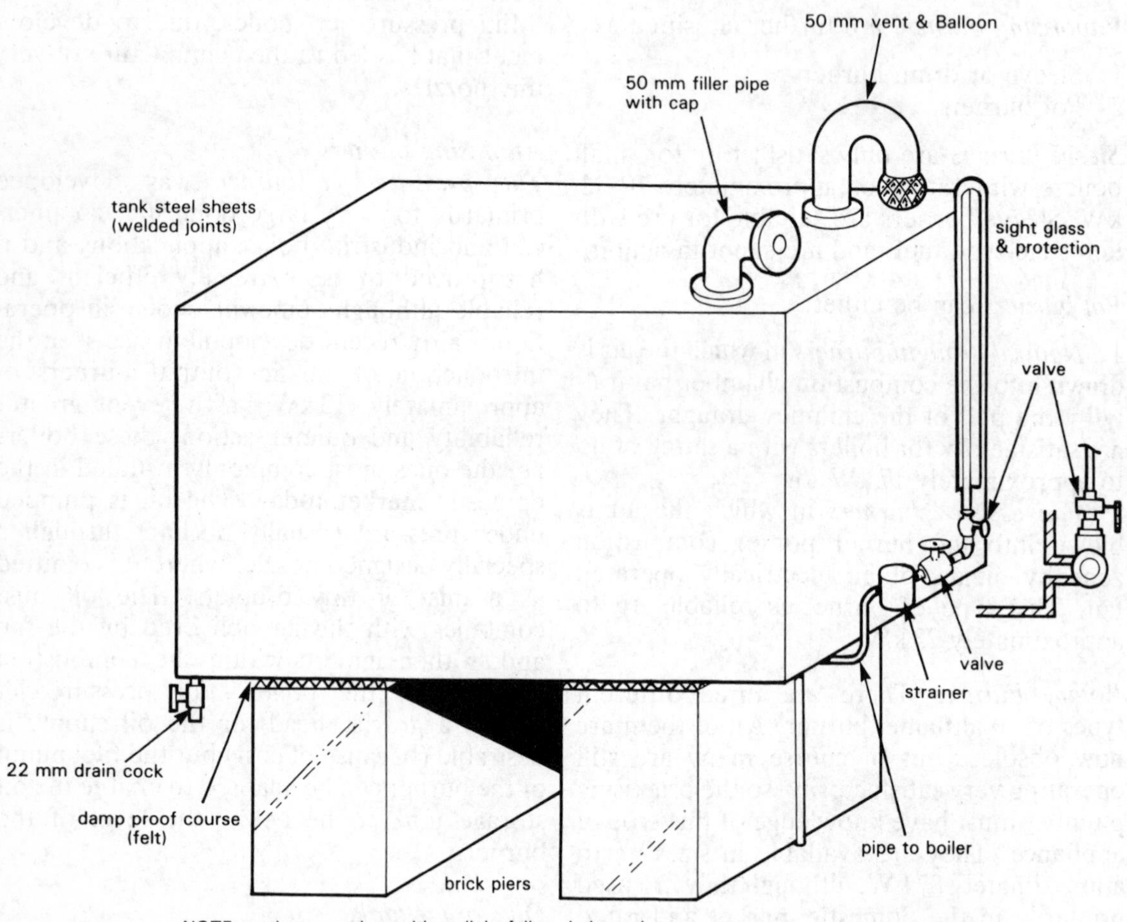

Figure 1.24 *Typical oil storage tank layout*

2 *Steel*. Black low carbon steel (not galvanised) with tapered threads only, sealed with oil resistant (shellac based) jointing compound. P.T.F.E. tape may be used.

Valves These should be of the glandless fullway pattern. *Fire valves* to shut off the oil supply automatically on an excessive rise in temperature should be fitted, and may be of either fusible link, hydraulic or electric pattern.

Filters To be of 5 meshes per millimetre and to be made from stainless steel, monel metal or phosphor bronze. Alternatively the more modern paper element filters which are of a throw-away pattern may be used.

General considerations

The layout should be as simple and economical as possible, with a minimum number of joints. Soft copper oil pipeline should be well supported, clipped at 500 mm spacing. Pipe should be laid to a steady fall (automatic venting) avoiding dips.

The oil supply should be controlled by an automatic fire valve fitted next to the point of entry to the building, the sensing element being near the boiler for safety reasons. The boiler must be set on a strong fire resistant surface unaffected by heat or oil, such as ceramic tiles or concrete. A good chimney is essential to ensure correct operation of vaporising and rotary burners, and it should develop a draught sufficient to induce the correct quantity of air for combustion.

Electric boilers

The development of electric boilers now makes them an attractive alternative to other forms of energy. They can be fixed as a replacement for an existing boiler, or as part of new systems. They make use of off-peak electricity known as 'Economy 7' by means of an array of heating elements near the base of the boiler. Further heating elements are fitted near the top of the boiler to give an extra boost which may be required due to abnormal demand or a prolonged cold spell.

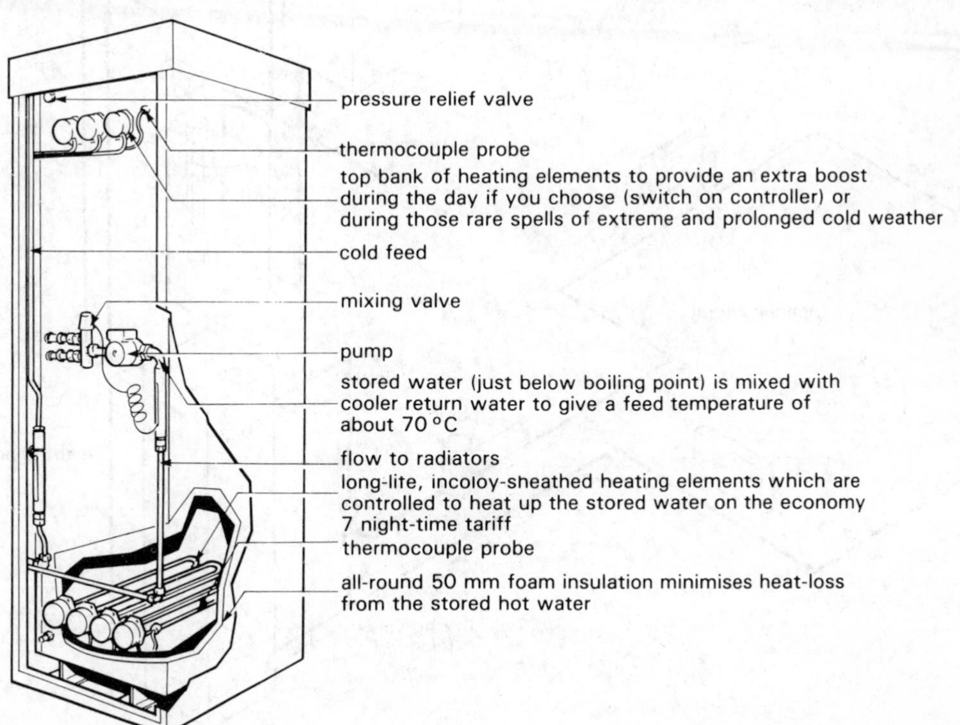

Figure 1.25 *Electric domestic hot water boiler*

Figure 1.25 shows an electric boiler which outwardly appears similar to a vertical chest freezer. Inside is a well-insulated water storage unit heated by thermostatically controlled heating elements.

The hot water is pumped around the heating circuit by a pump as in the previously described systems. There is one important difference between electric water heating and other systems: the water is heated in the electric boiler to a much higher temperature, almost to boiling point, and is too hot to be used without being passed through the mixing valve (blender). The function of the blender is to mix the near boiling water from the boiler with cooler water returning from the heating circuit, giving a blended water at the usual 70 °C for circulation to the heat emitters. The system is controlled in the normal manner with automatic control timers, room thermostats, etc.

The advantages of this type of boiler are:

1 Clean, quiet and reliable,
2 No storage of fuel required,
3 No flue required,
4 Little or no maintenance required, maintains its efficiency,
5 Pollution free,
6 Can be sited almost anywhere (under cover with electric supply),
7 Easily installed,
8 Factory assembled and tested,
9 Suitable for new or existing installations.

Figure 1.26 illustrates how an existing system

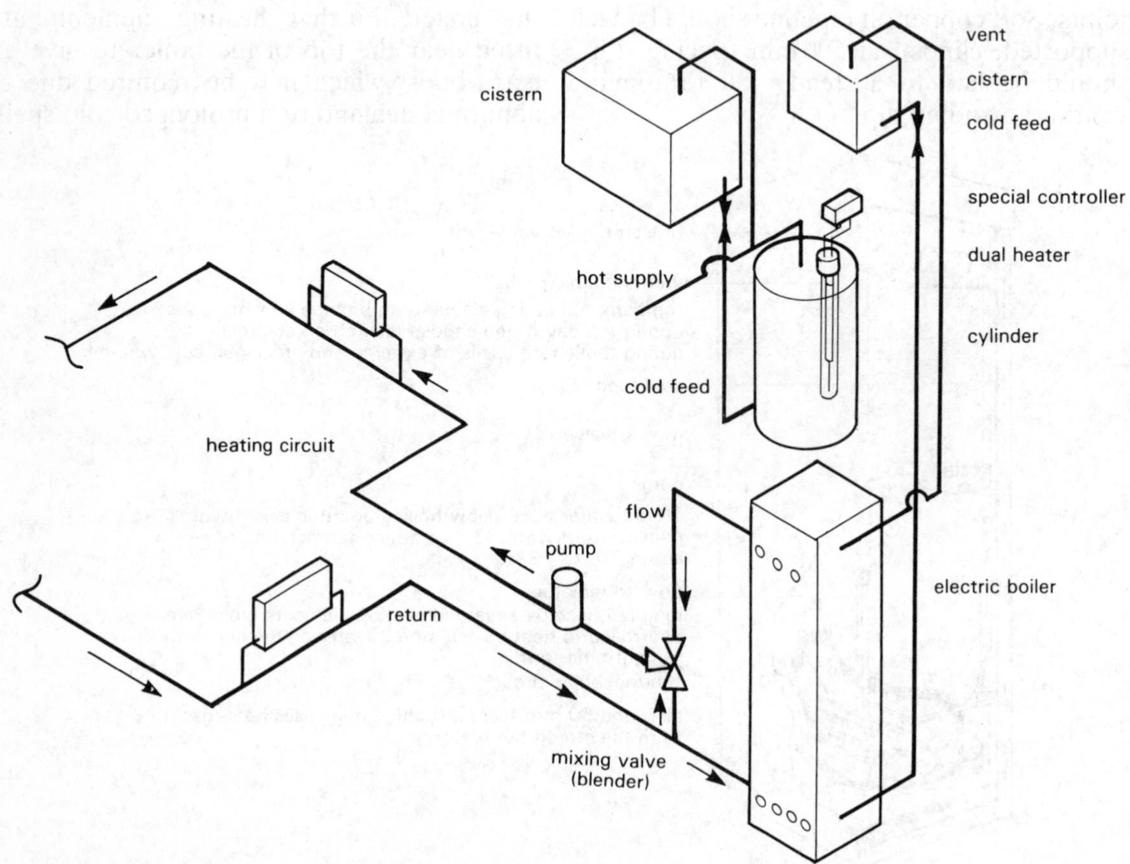

Figure 1.26 *Separate domestic hot water and heating system (electric)*

of hot water and space heating can be adapted to incorporate an electric boiler and dual immersion heater in hot storage. The cylinder should be well insulated and isolated from the heating system.
Figure 1.27 illustrates a typical arrangement of a domestic hot water and space heating system.

Domestic space heating

Some years ago many homes in the UK had only cold water on tap, but today most homes have not only hot water and bathrooms but the house is heated by hot water systems known as central heating. The name central heating comes from the fact that the source of heat is a boiler from which the hot water pipes radiate to all the rooms.

These heating systems form another part of the plumber's work and it is necessary for him or her to have knowledge of design, installation and fault finding procedures. The design work is a fairly complex operation and is not usually carried out by the installer, the calculations involving the heat loss of the building, the heat requirements of the occupants, and the sizing of boiler, radiators and piping being carried out by experts.

In this section we will be dealing with the basic principles required for the effective installation and working of the heating system.

Gravity circulation systems
This is the system by which domestic hot water systems circulate the heated water (see Book 1) and which depends on the

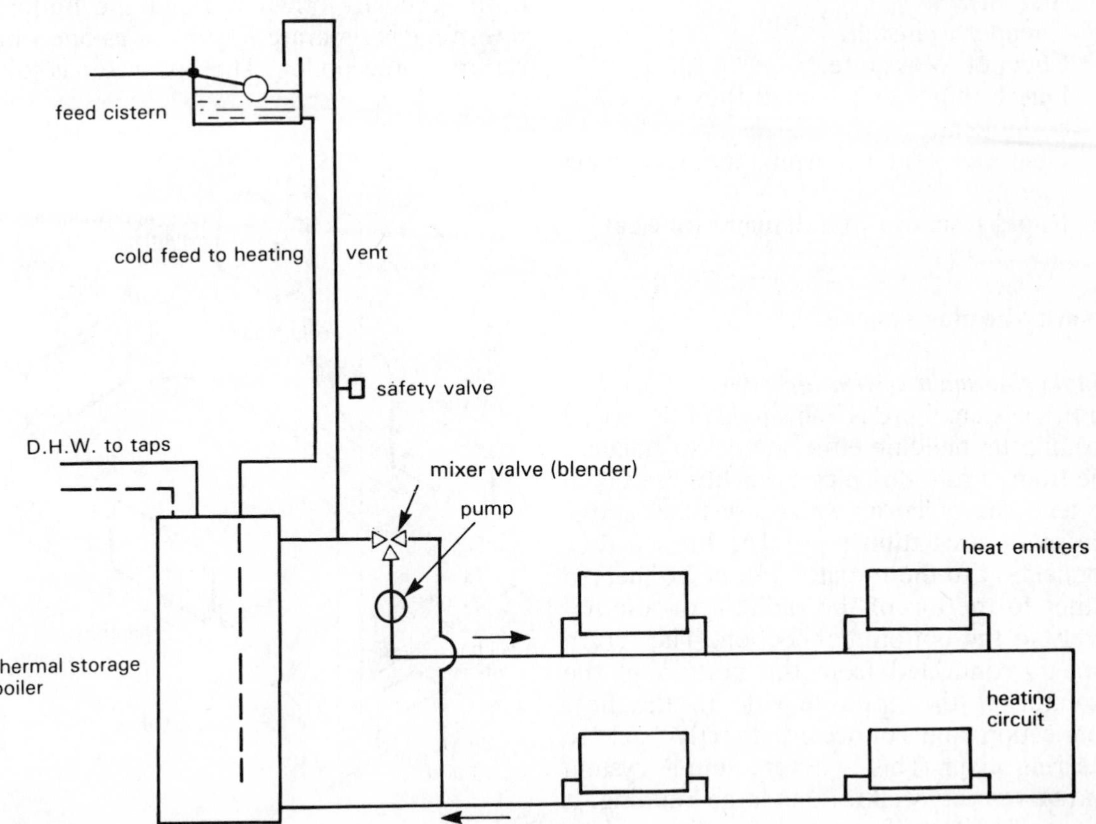

Figure 1.27 *Electric boiler combined system*

movement of a column of hot water (lighter) being displaced by a column of cold water (heavier). This movement is known as convection. The very early space heating systems worked on the gravity circulation system and required careful and accurate pipe sizing, the use of large diameter pipes and large radius bends (Figure 1.28).

Forced circulation
This is the system in which the heated water is forced or drawn around the pipework system by the operation of an impellor. This type of system has many advantages over the gravity system, including:

1 The use of small diameter pipes,
2 Frictional resistance in systems easily overcome,
3 Isolated radiators present no problem (design),
4 Cheaper to install,
5 Cheaper to operate,
6 Due to pipes being small they can easily be hidden,
7 Heat losses smaller (small diameter pipes used),
8 Rapid response to a demand for heat.

Gravity heating systems

Single ring main system (gravity)
In this system there is one single pipe carried around the building either above or beneath the floor. From this pipe branches are taken to feed the radiators which are fixed above the main circulation pipe. The branch flow connection to the radiator can be connected either to the top of the radiator or alternatively to the bottom connection. The return pipe is connected from the bottom of the radiator at the opposite side to the flow connection and connected directly back to the ring main. This is a very simple system and serves very well for one large building. It is not so suitable for a building divided into rooms (ideal for community halls, etc.).

The main disadvantage of this system is that the cooled water from each radiator is returned to mix with the hot water in the main, so cooling it down. This means that each successive radiator along the main is cooler than the previous one, and in a large system the last radiator will be appreciatively cooler. This can to some extent be minimised by accurate pipe sizing, and adjustment of lock-shield or thermostatic radiator valves.

The one pipe drop system (gravity)
This system is suitable for buildings several storeys high and consists of a main flow rising through the building and terminating as a vent above the feed and expansion cistern. The return pipes (of which there may be several) which feed the heat emitters are connected high up into the flow pipe. They drop vertically down through the building, returning as separate returns or as one single return to the boiler. The hot water is taken

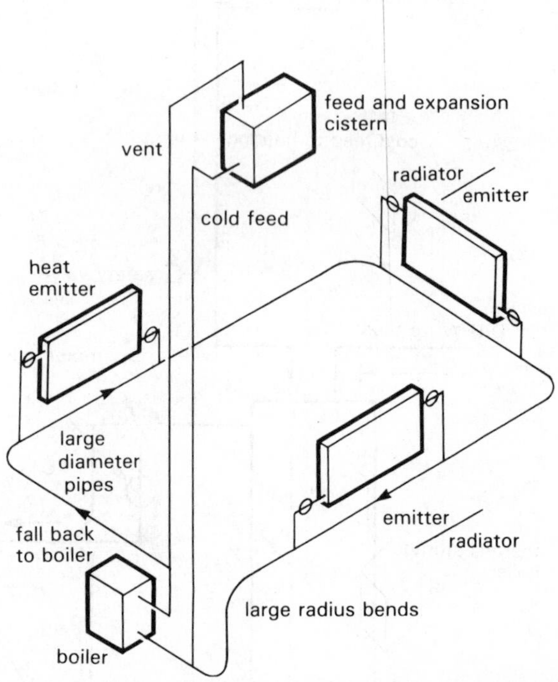

Figure 1.28 *Single ring main system (isometric projection)*

from these returns, passed through the emitter and returned back into the same return where it is taken back to the boiler for re-heating (Figure 1.29).

The two pipe return system (gravity)

The two pipe system of heating is suitable for buildings with single- or several-storey design. The cistern is located in the highest possible position to provide the required head of pressure. The flow pipe from the boiler rises up through the building to terminate over the cistern as the vent to the system. At each floor level a pipe is connected which conveys the water to the heat emitters, branch pipes being taken from this pipe and connected to the top of the emitter by means of a control valve. The used hot water leaves the emitter through a lock-shield valve on the bottom connection and is returned via a common return pipe to the boiler for re-heating (Figure 1.30). It is common practice to fit a flow pipe with a diminishing size of pipe as each emitter is passed. The return pipe is the opposite, it starts as a small diameter pipe and increases in size as it picks up the water from each successive emitter.

The gravity systems illustrated are but a few of many systems to be found in everyday work. These systems have become obsolete over a period of time for the following seven reasons:

1 They are often an unsightly system.
2 There is no possibility of hiding the pipes.
3 They are more costly to install.
4 Some isolated areas are difficult to heat.
5 There are larger heat losses.
6 They are more expensive to operate due to the large volume of water.

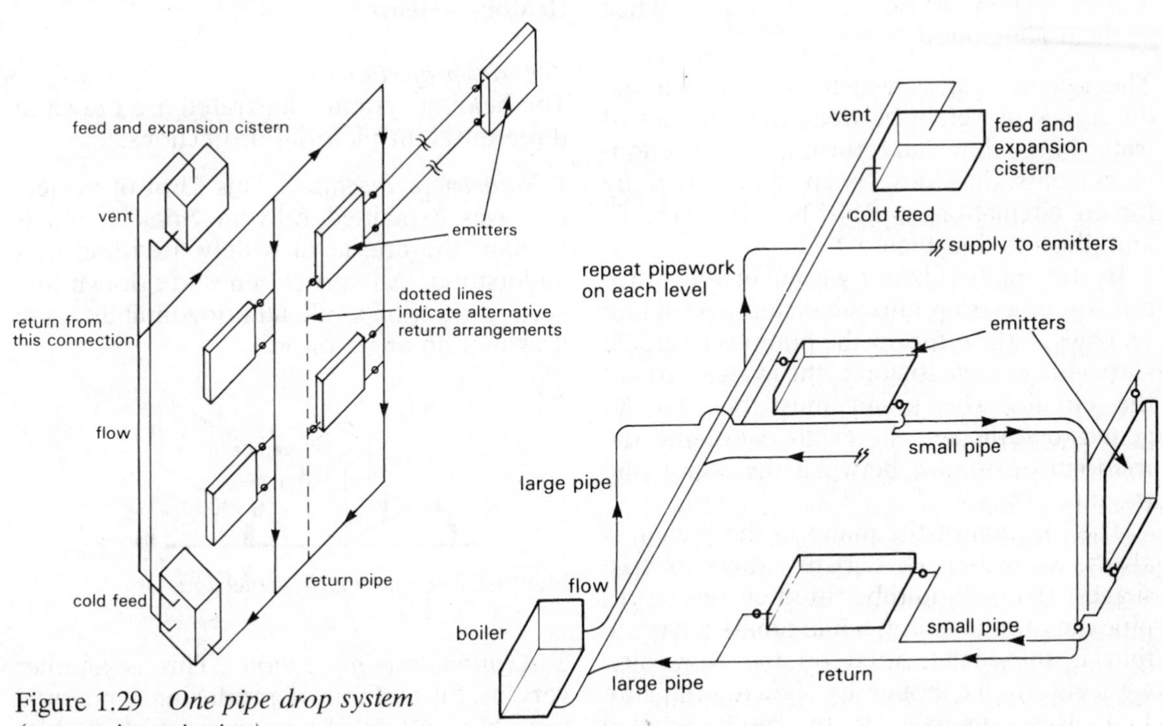

Figure 1.29 *One pipe drop system (axometric projection)*

Figure 1.30 *Two pipe return system (planometric projection)*

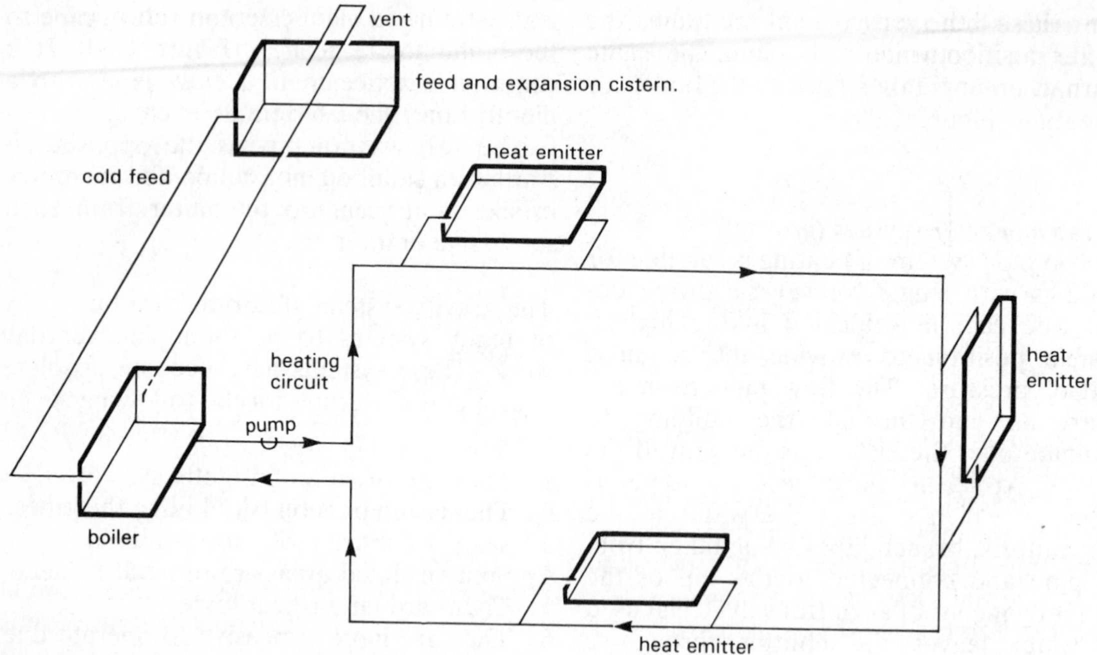

Figure 1.31 *Single ring main system (planometric)*

7 The system is slower to respond when heat is required.

The gravity system which operates on the difference in weight between the columns of water in the flow and return, i.e. convectional currents, has now been superseded by forced circulation as described on page 27 and illustrated in Figure 1.31.

By forced circulation we mean the introduction of a pump into the system. As stated on page 28 the effect of the pump is to create sufficient energy to force the water around the circuit. The pump must be able to generate sufficient energy to overcome the frictional resistance between the water and the pipework.

The location of the pump in the system is always a matter of debate. Some people suggest that it should be fitted on the return pipe near to the boiler, while others advocate that it should be fitted on the flow pipe adjacent to the boiler as shown in Figure 1.31. Refer to page 35 for more detailed information on pumps.

Heating systems

Pictorial projection

The heating systems illustrated are drawn in three different pictorial projections:

1 *Isometric projection* This form of projection was explained fully in Book 1 and is perhaps the one most widely used and best understood. All vertical lines are drawn in a vertical position while all horizontal lines are drawn at an angle of 30°.

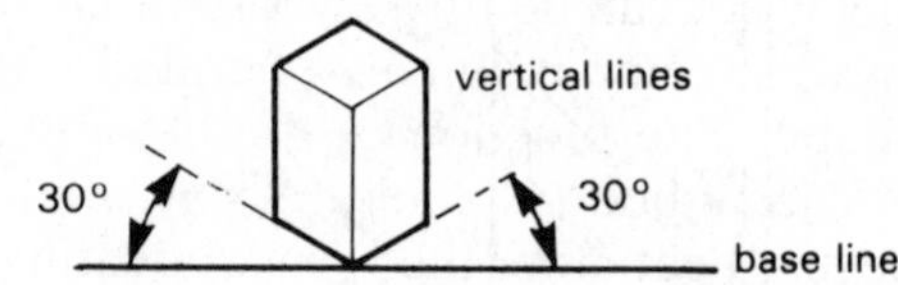

Figure 1.32a *Isometric projection*

2 *Axonometric projection* This is another very useful method of producing a pictorial view of a system of pipework which enables you to follow clearly and easily the various

pipe runs. All the vertical lines are drawn in a vertical position while the horizontal lines are drawn at an angle of 45°.

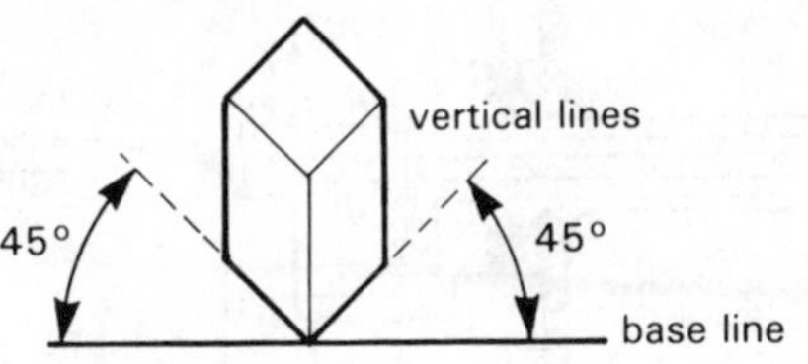

Figure 1.32b *Axonometric projection*

3 *Planometric projection* This method of projection is accepted as perhaps the best way to reproduce a plumbing or heating system in pictorial projection. This method, as the title suggests, gives both a true plan and a pictorial view at the same time. It is ideal for reference when taking off pipe lengths while estimating or preparing lists of material. The horizontal lines are drawn horizontal and all the vertical lines are drawn at an angle of 60° to the base line.

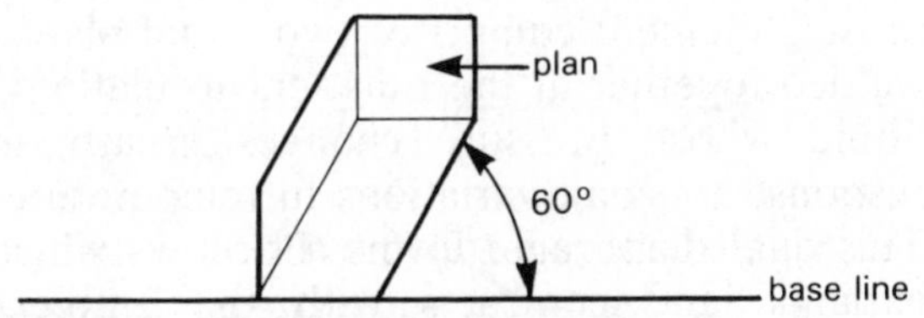

Figure 1.32c *Planometric projection*

Controls

Hot water and heating systems, regardless of the fuel used, can be fitted with various types of controls, some manually operated, others automatic. The function of these controls is to switch the boiler on and off automatically whenever heat for the space heating or domestic water heating is required, so effecting a saving of fuel by controlling the temperature of the water. The use of pumps enables mini-bore and small-bore systems of heating to be installed in most domestic dwellings almost regardless of design. To install an effective central heating and domestic hot water system in the home and to economise on the running costs it is important to have an effectively and efficiently controlled system.

The following are some of the controls available to assist in obtaining maximum efficiency at minimum operating cost:

1 Drain cock,
2 Thermostat,
3 Programmer,
4 Pump,
5 Safety valve,
6 Diverter valve,
7 Non-return or check valve,
8 Economy valve,
9 Thermostatic radiator and lockshield valve.

Drain cock This is perhaps the most simple and basic control in the whole system, yet it is one of the most important. The purpose of the drain cock is to facilitate the draining of all the water from the system, i.e. boiler, storage vessel and pipework. Drain cocks are usually made of brass with washers of fibre or nylon.

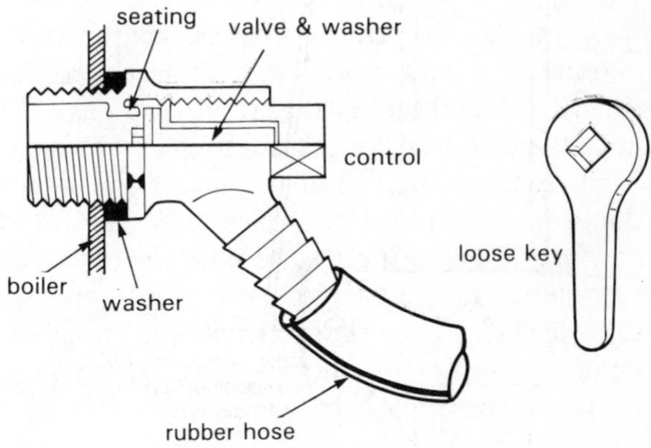

Figure 1.33 *Drain cock*

Thermostat This is a very cost-effective control, its function being to close down the appliance when the temperature of the water has reached a predetermined level set on the thermostat. The setting of the thermostat should be in the region of 60 °C for domestic

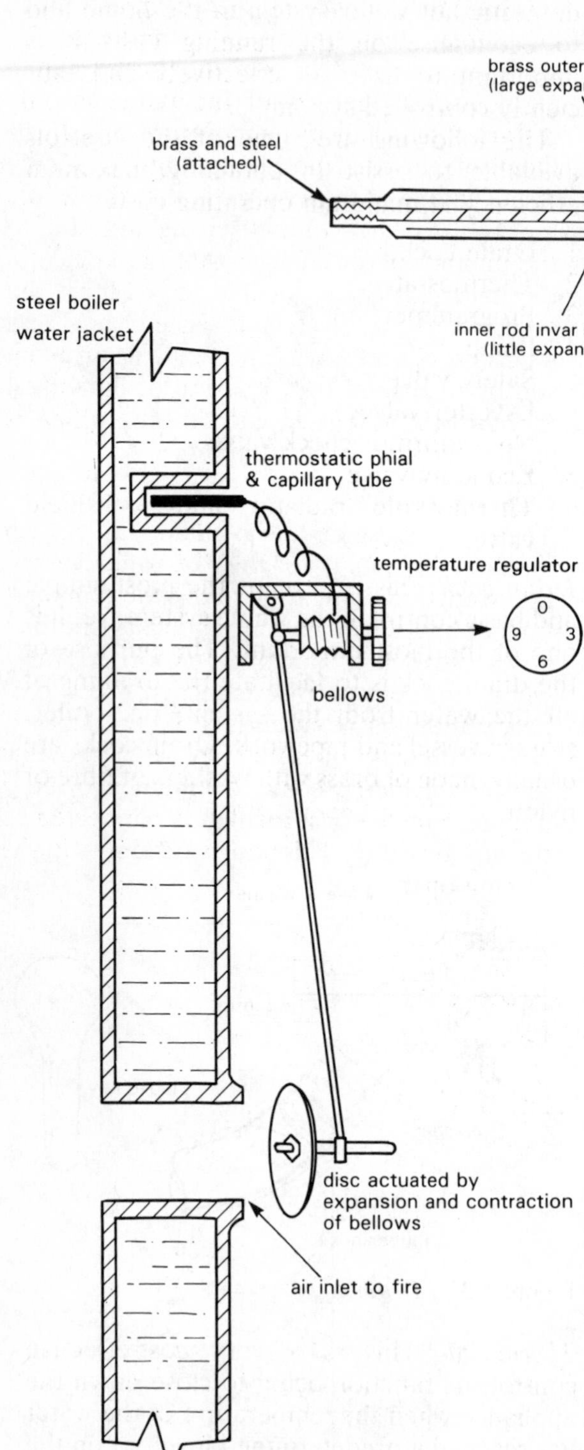

Figure 1.34 *Bi-metallic thermostat*

installations. This is governed by the nature of the water and the requirements of the user.

Operation of thermostat

Sensing element (Honeywell) The thermostat sensing element comprises two metal plates, welded together at the rims, encapsulating a liquid whose pressure changes greatly in response to small variations in temperature. This dual diaphragm forms a bellows which expands and contracts with the ambient temperature changes. This movement serves to operate a snap-action micro switch rated to control the heating circuit.

The action of this type of thermostat is brought about by the fact that metals have varying degrees of expansion over the same range of temperature. The two metals in this case being brass with a high rate of expansion and invar steel with a low rate of expansion. When the system is cold the metals will have contracted to their shortest length and the electrical contact will be made. When the water is heated and the temperature has reached that set on the control the brass outer casing will have increased in length and, due to the fact that the invar steel rod is

Figure 1.35 *Automatic air control thermostat*

secured to the brass at the end, the rod is taken with it and so the electrical contact is broken. When the water is cooling the brass tube will also cool and contract in length until the electrical contacts touch again, whereupon the boiler fuel is ignited and the whole cycle is repeated.

Room thermostat

This type of thermostat is manufactured in many different sizes, shapes and colours to fit in with most decorations. Irrespective of appearance they all perform the same function, which is to automatically close down or alternatively, operate the boiler at a pre-set temperature.

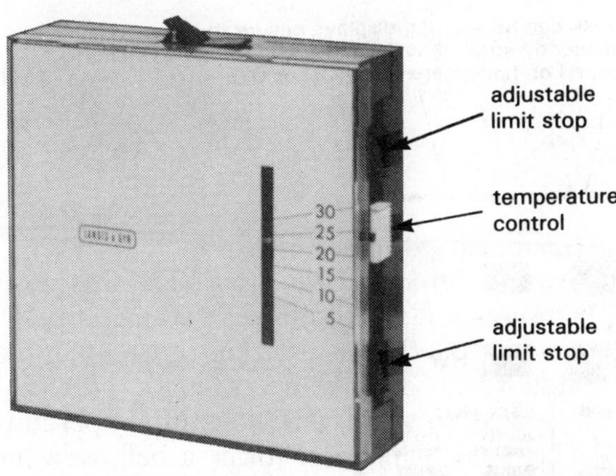

Figure 1.36 *Room thermostat*

The thermostat should be fixed on an inside wall approximately 1.5–2 m above the floor. The unit should not be exposed to direct sunlight, cold draughts, be adjacent to hot water pipes or radiant heat from appliances, as these will affect the pre-set working temperature. For wiring diagrams and other specific information always check the manufacturer's literature.

Room thermostats are also used to control the operation of the boiler and/or the pump. They are extensively used in domestic heating systems and are usually situated in the *hall*. When the air temperature of the hall has reached that at which the thermostat has been set, the electrical current to the appliance is broken and the boiler flame extinguished. Ignition of the flame is only restarted by a fall in temperature or adjustment of the thermostat setting. The only disadvantage of this control is that the whole building is heated to that setting and, due to the number of air changes that take place in the hall, it is quite possible to overheat the other rooms, resulting in a waste of heat and incurring unnecessary expense. This problem can be overcome by the fixing of individual thermostatic valves in each room, which is explained later in this chapter.

Cylinder thermostat

These thermostats are available as clip-on units to the outside of the hot water vessel, their function being to regulate the temperature of the water by automatically shutting down the boiler, pump or motorised valve when the water has reached the predetermined temperature (see Figure 1.38).

The unit is operated by a bi-metallic sensing element; some companies offer a unique remote sensor which is connected to the unit by cable. This allows flexibility in the placing of the unit.

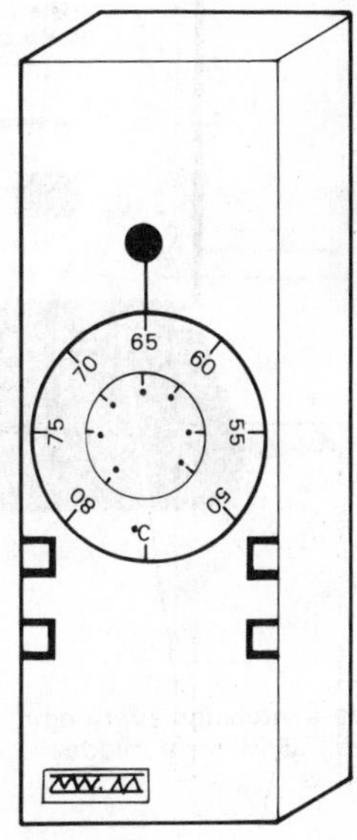

Figure 1.37 *Cylinder thermostat*

vent
vent
motorised valve
hot water supply
cylinder thermostat
flow pipe
cold feed
boiler
return pipe

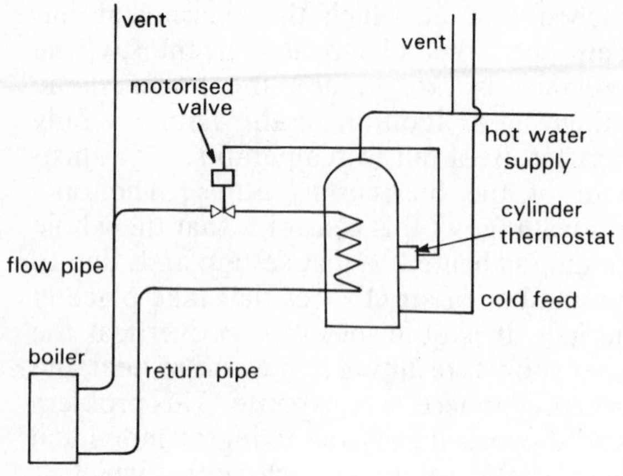

Figure 1.38 *Cylinder thermostat located in system*

Programmer

In addition to being able to control the temperature of the water as described, i.e. with the thermostat, it is also necessary to be able to set the boiler to start and shut off automatically at predetermined times of the day or night. Figure 1.39 shows one of the many programmers available to the installer. This one is a Landis & Gyr electronic model, incorporating many features.

Heating timeswitch

It may be that the system only requires a control to switch the heating circuit on or off. In this case a *timeswitch* has been designed

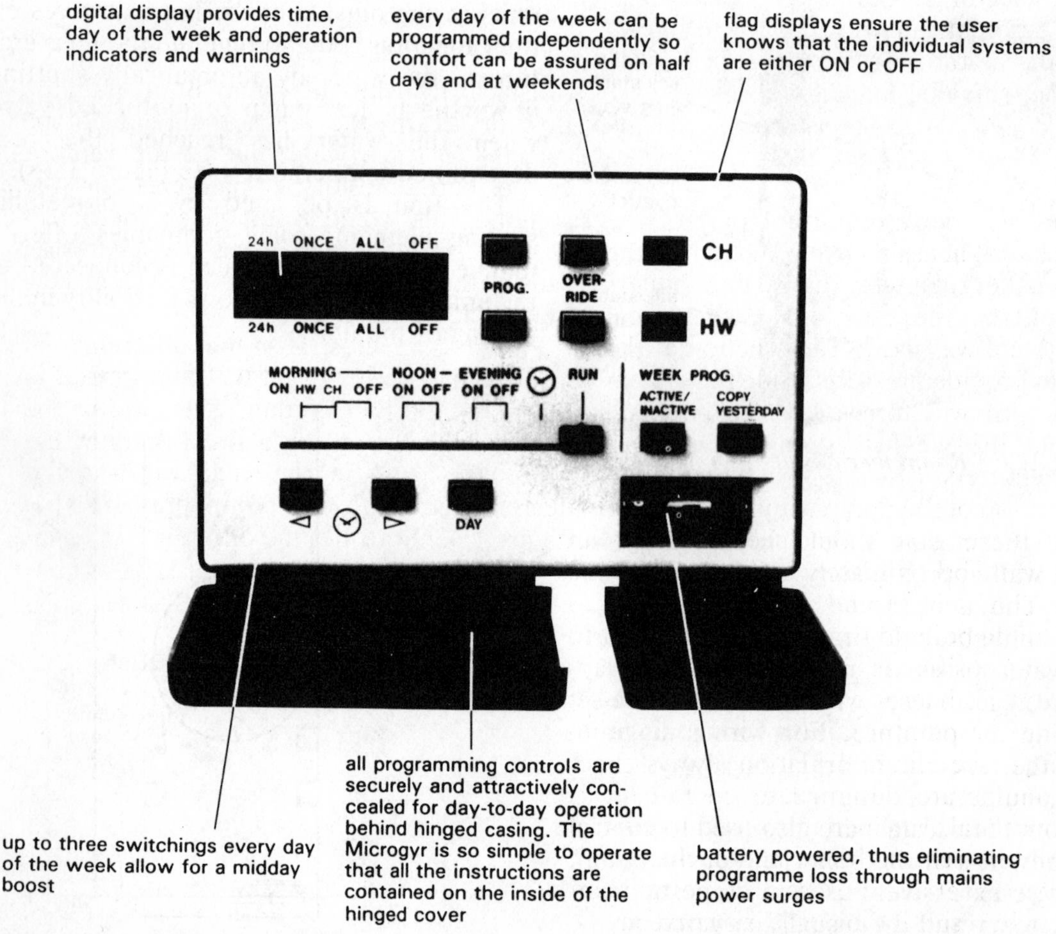

digital display provides time, day of the week and operation indicators and warnings
every day of the week can be programmed independently so comfort can be assured on half days and at weekends
flag displays ensure the user knows that the individual systems are either ON or OFF
24h ONCE ALL OFF
PROG.
OVER-RIDE
CH
24h ONCE ALL OFF
HW
MORNING —— NOON — EVENING
ON HW CH OFF ON OFF ON OFF
RUN
WEEK PROG.
ACTIVE/ INACTIVE
COPY YESTERDAY
DAY
up to three switchings every day of the week allow for a midday boost
all programming controls are securely and attractively concealed for day-to-day operation behind hinged casing. The Microgyr is so simple to operate that all the instructions are contained on the inside of the hinged cover
battery powered, thus eliminating programme loss through mains power surges

Figure 1.39 *System programmer*

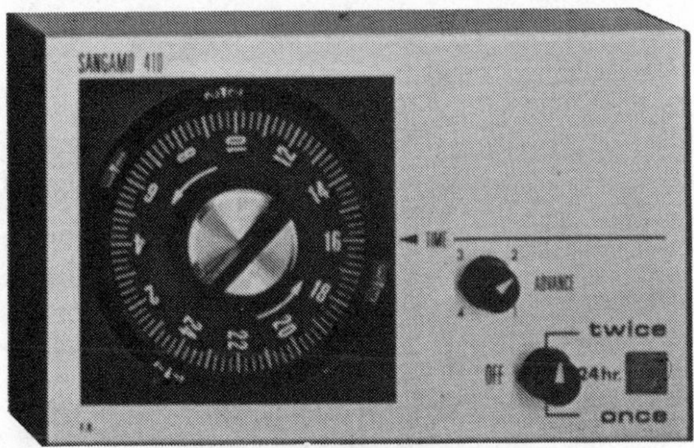

Figure 1.40 *Timeswitch*

which can be pre-set to operate the circuit once or twice a day. The selected programme can be overridden by the operation of the override selector on the switch. Some time-switches automatically revert back to the set programme at the start of the next period, while others may be manually reset.

Pumps

The pump is a very important part of the modern central heating system. It enables the use of smaller diameter pipes and boilers than would be the case if a conventional gravity system was used. The function of the pump is to provide pressure inside the system which in turn will force water to circulate throughout the whole system of pipework and heat emitters.

The location of the pump within the system can be on the *flow pipe* (now generally accepted as the best position) or on the *return pipe*. In some instances it is connected adjacent to the boiler inside the boiler casing. One popular position is in the horizontal line with isolating valves at each side. These are to facilitate the pump removal for repair or replacement (see Figure 1.41).

Some pumps are designed to operate in either horizontal, inclined or vertical positions. Always check the manufacturers' information sheet as regards this point as damage could result if the pump is incorrectly located. Although the system may function

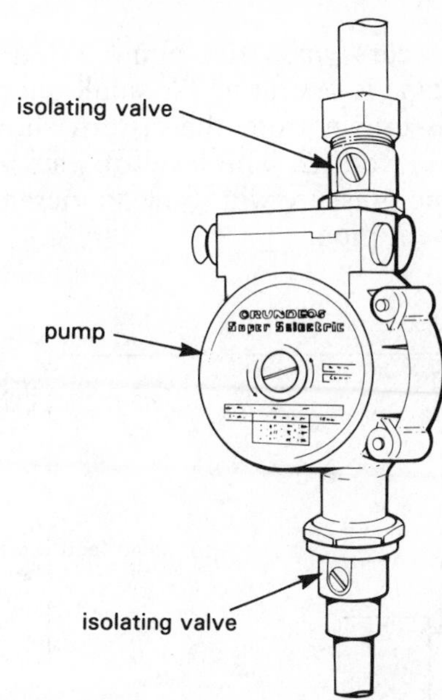

Figure 1.41 *Pump and isolating valves*

satisfactorily with the pump fitted into the flow or return, the pressure should not be such as to force water out of the vent or create sub-atmospheric pressures in the system. This would result in an inefficient heating system and excessive corrosion caused by oxygen entering the system (see Figures 1.42 and 1.43).

Figure 1.42 illustrates the difference in levels which occurs when the pump fitted to the flow pipe is operating. The level in the cistern will increase by a small margin, h2, while the level in the vent will show a decrease equal to the pump pressure, h1. Figure 1.43 illustrates the difference in levels

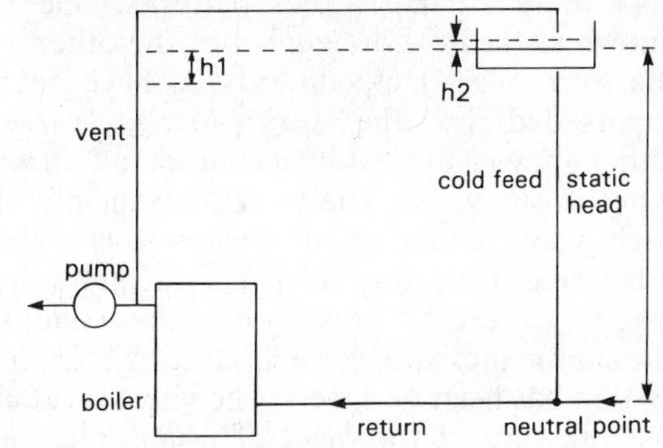

Figure 1.42 *Pump located on flow*

which occurs when the pump fitted in the return pipe is operating. A small quantity of water is drawn from the cistern showing a reduced level, h2. The level of water due to the pump pressure will show an increase, h1, in the vent pipe.

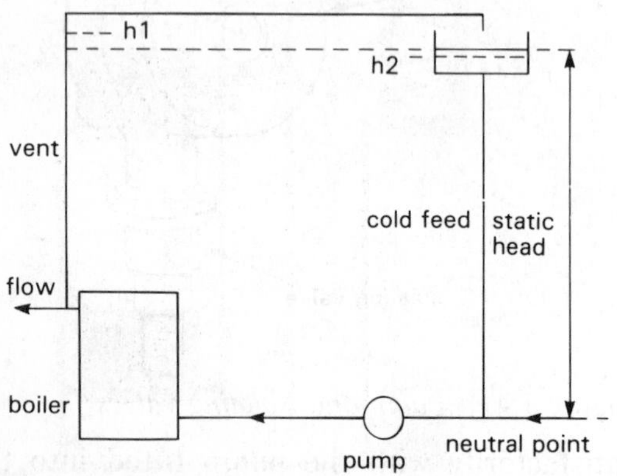

Figure 1.43 *Pump located on return*

Safety valves

Although all systems are not fitted with safety valves, they are still an important consideration when dealing with boiler controls and should be fitted wherever possible. The most commonly used safety valve is of the type illustrated in Figure 1.44 which is of a spring-loaded control.

In addition to the valve shown in Figure 1.44 there are two other patterns: one is known as the *dead-weight valve*, the other as the *lever valve*. Both these types have been superseded by the *spring-loaded valve*, although you may still encounter all three types in site work. The working principle of each valve is shown in Figure 1.45. The function of the safety valve is to facilitate the escape of excess pressure in the system, should the pressure exceed that for which the system has been designed. The valve is set at the pressure of the *head of water*, plus an approved safety factor.

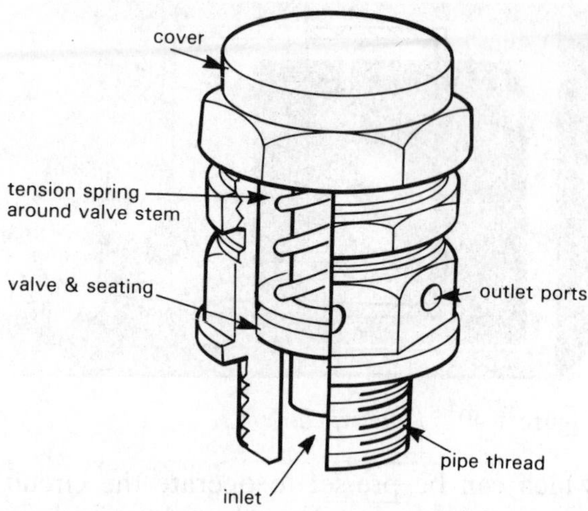

Figure 1.44 *Spring loaded safety valve*

The recommended fixing position of the valve always causes some debate. If the water is *hard*, the type of hardness and the degree of hardness may have a bearing on the position chosen. In the case of temporary hard water, with the possibility of furring taking place in both the boiler and the flow pipe, serious consideration must be given to the position of the safety valve so that it is protected as far as possible from the risk of becoming sealed with the lime incrustation and so becoming useless. In this situation it may be advantageous to fix the valve in the return pipe where the least amount of furring will take place. It is of course argued that the valve in this position is now subject to fouling by sediment and possible corrosion but this is the lesser of the two problems. It is also believed that the pressure is greater at the top of the boiler, but this is not in fact the case as the pressure will be transmitted equally in all directions.

To sum up, if water is neutral or permanently hard, fix the valve either on or near the top of the boiler or flow pipe; if the water is temporarily hard (lime deposit), fix the valve on the return pipe.

Regular checking and maintenance of all safety valves are strongly recommended.

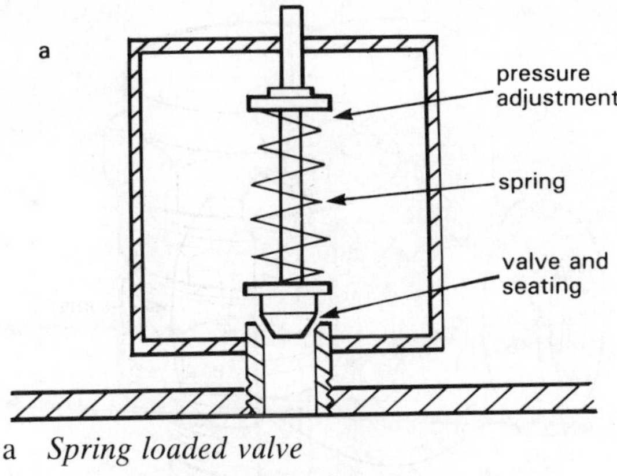

a *Spring loaded valve*

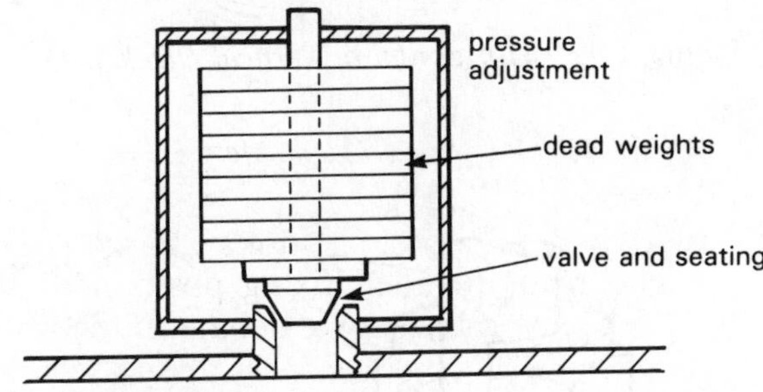

b *Dead weight valve*

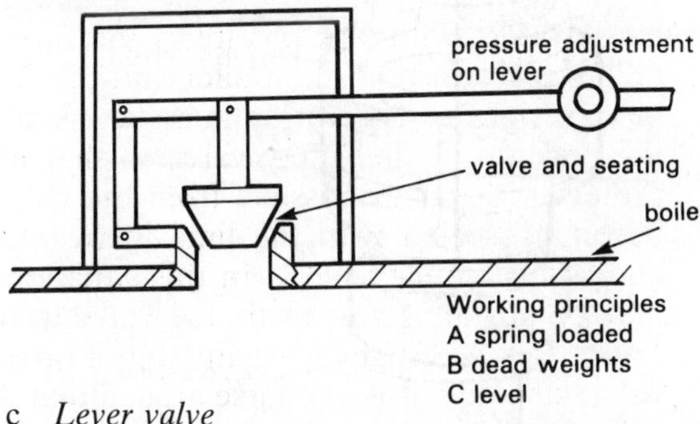

c *Lever valve*

Figure 1.45 *Working principles of safety valves*

Motorised valves

These valves are manufactured as two port and/or three port controls with a motorised unit as an integral part. Both types are suitable for use in the fully pumped combined central heating and domestic hot water systems. The function of the valve is to control the flow of water through the system. To achieve full temperature control, room and cylinder thermostats should be used in conjunction with these motorised valves.

Two port valves

Figure 1.46 shows how both the domestic hot water and the central heating circuits can be controlled by the two port valves.

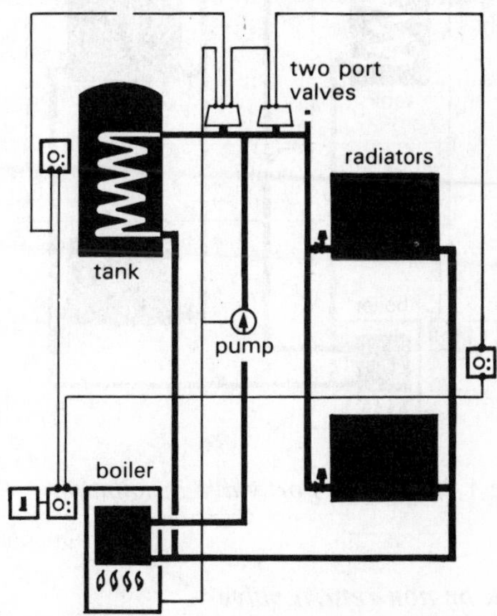

Figure 1.46 *Two port valve installation*

Three port valves

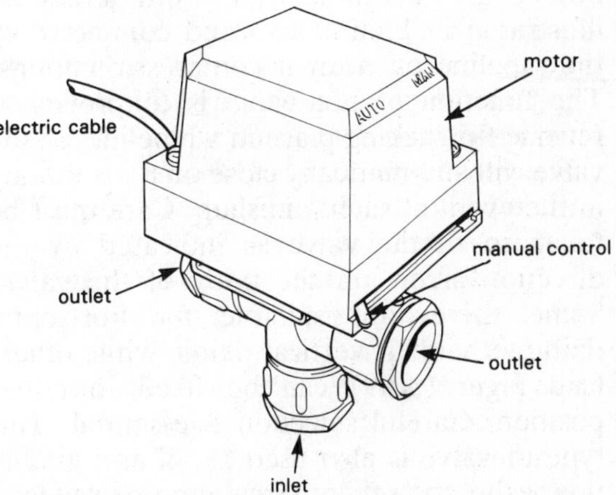

Figure 1.47 *Three port valve*

Figure 1.48 (overleaf) shows the positioning of a three port valve, also known as a diverter or mid-position valve. This valve will control both the domestic hot water and the central heating system simultaneously.

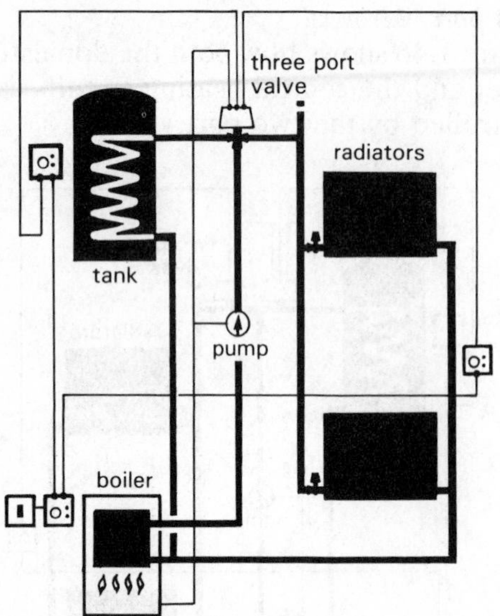

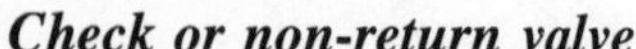
Figure 1.48 *Three port valve installation*

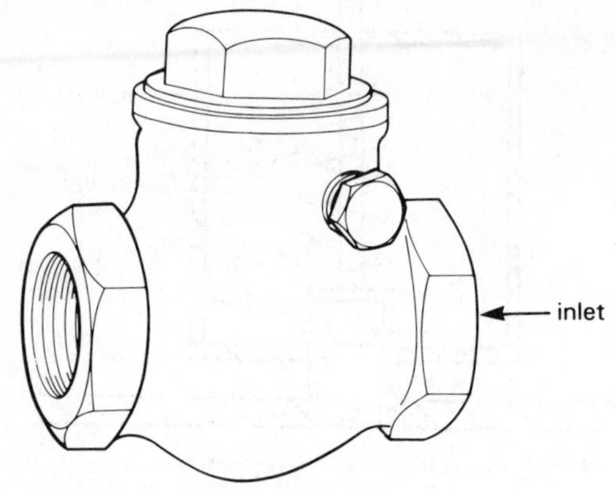

Figure 1.49 *Horizontal or vertical check valve*

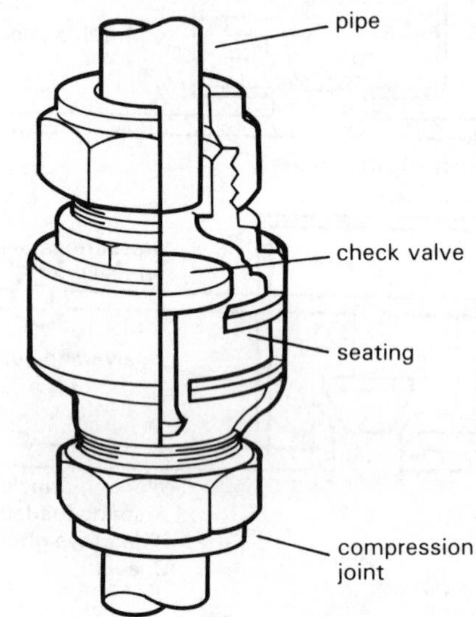

Figure 1.50 *Check valve (prestex)*

Check or non-return valve

These valves are manufactured in various shapes and sizes. Some are made from bronze with flanged plates; some with internal threads as shown in Figure 1.49, while others are manufactured from brass as illustrated in Figure 1.50 and connected to the pipeline by normal compression joints. The function of the valve is to prevent a reverse flow taking place in a pipeline, as the valve will automatically close on to its seating in the event of such a mishap. Care must be taken to fit the valve as indicated by the direction arrow on the body of the valve. Some valves are suitable for horizontal fixing, others for vertical fixing, while others (see Figure 1.49) can be fixed in either position. Careful selection is essential. This type of valve is also used as an anti-gravity flow valve to prevent circulation of water in the heating system when the pump is switched off.

Economy valve

This is a form of three-way valve fitted into the return pipe from the cylinder to the boiler or heater. The function of this type of valve is to facilitate the heating of only part of the contents of the hot storage vessel when this is required.

Location of economy valve in system

Figure 1.52 shows the location of the valve in the system. Under normal working conditions with the valve open as in Figure 1.51 the whole of the contents of the hot storage vessel will be circulating through the heater.

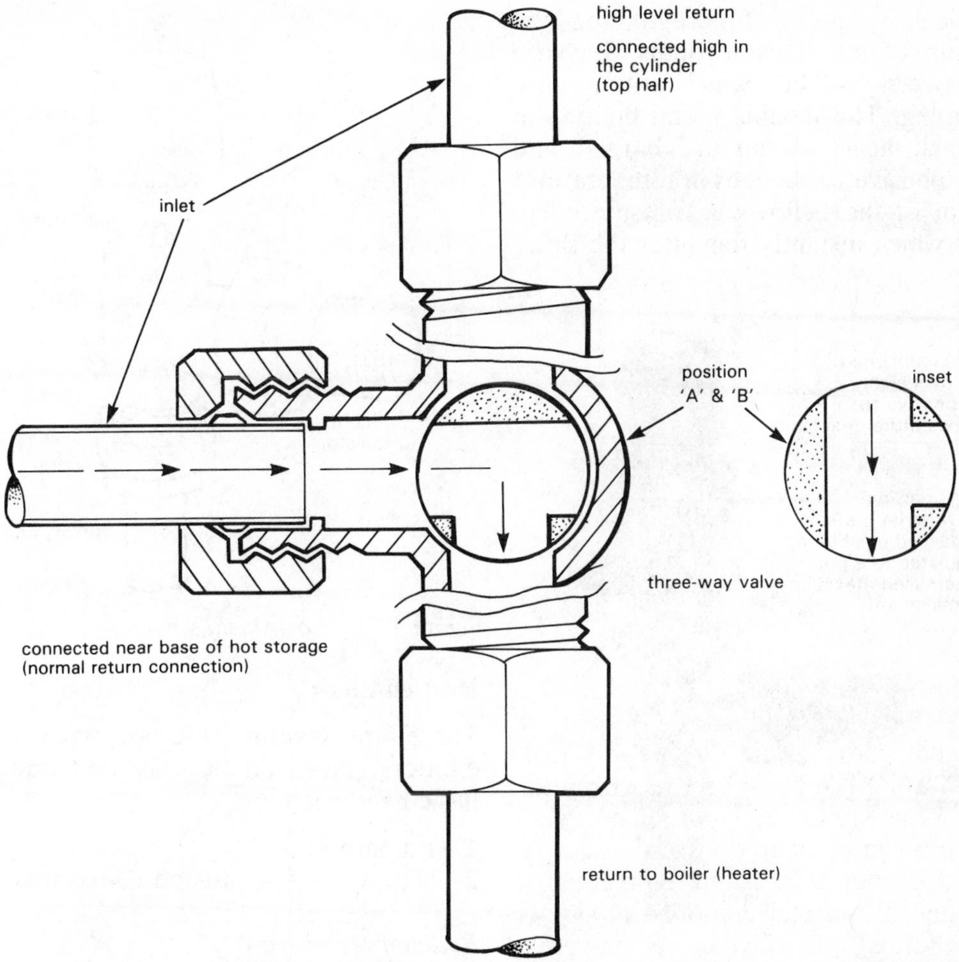

Figure 1.51 *Economy valve*

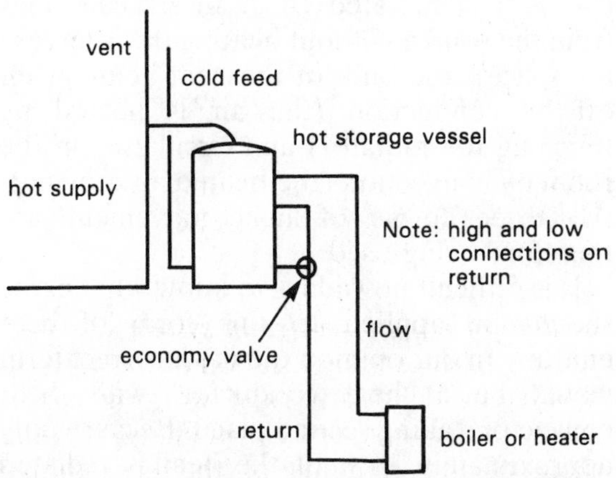

Figure 1.52 *Economy valve located in system*

On occasions when only a small amount of hot water is required then by a simple quarter-turn of the valve head the waterways would be positioned as *inset*. This would then allow circulation of the water in the top half of the cylinder only, effecting a considerable saving in heating: hence its name, *economy valve*.

Thermostatic valves

This valve is usually calibrated from 1 to 5 or 1 to 7 to give controlled temperatures from approximately 10 °C to 28 °C. The valve senses the air temperature in the actual room and will respond, allowing hot water to flow through the radiator until the air temperature reaches that at which the valve is set. They are particularly useful for regulating the temperature of bedrooms or rooms which are not in general use but which still require some heating. The flexible metal bellows in the thermal head is vapour charged and highly responsive to changes in temperature. Movement of the bellows is transmitted to the valve which instantly regulates the flow.

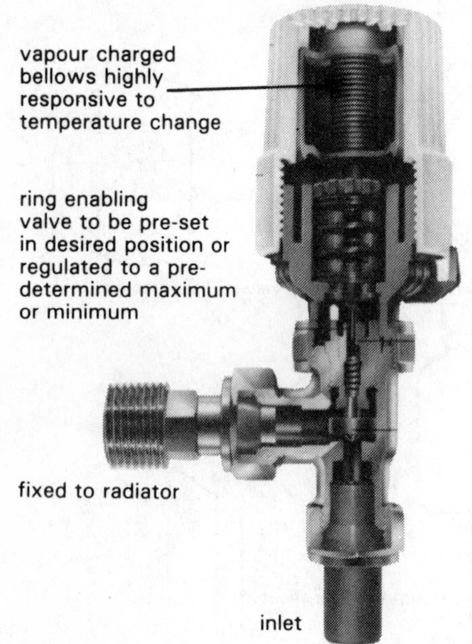

Figure 1.53 *Thermostatic radiator valve*

Lockshield valves
This type of valve is fitted to the outlet end of the radiator, its function being:

1 By adjustment of the regulator the flow of water through a radiator can be accurately controlled. This may be necessary in the case of some radiators being starved of hot water. This is known as *balancing* the system.
2 The valve can be used as an 'off' control and, in conjunction with the other control valve, it is possible to remove the radiator for repairs or renewal or to allow re-decoration of the wall behind the radiator to be carried out without draining down the whole system.

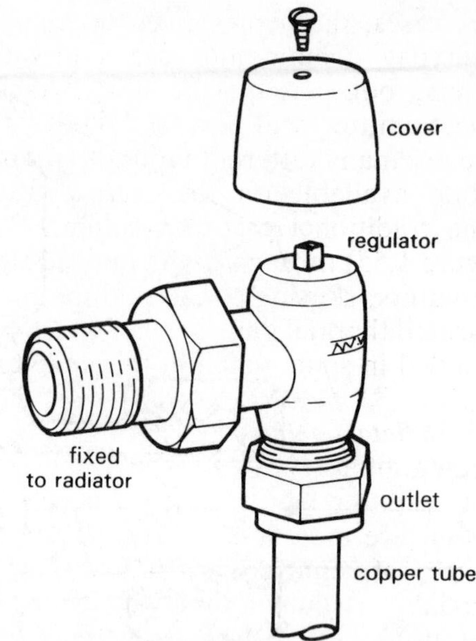

Figure 1.54 *Lockshield valve*

Heat emitters

There are several different types of heat emitters, they can be classified under two basic headings:

1 Radiators,
2 Natural or fan assisted convectors.

Radiators (emitters)
The name given to this appliance would indicate that the heat given off is *radiant heat*, but this is not the case. Only a small amount (5–10%) is radiated (heat in straight lines from the source without heating the intervening space), the bulk of the heat being given off by conduction (the air is heated by touching the radiator) and circulating in the room by convection (the heated air moving). All three forms of heat movement are therefore being used.

It is difficult nowadays to know what name should be applied to this form of heat emitter. In our opinion the most correct term would be a 'heat conductor' with 'heat convector' taking second place. Because only approximately 10% of the heat is radiated heat the term 'radiator' would come a very poor third. It can be readily seen that, as in so many cases, the names given to some of our everyday commodities are sometimes misnomers, but perhaps the all-embracing term 'heat emitter' will best fit the bill.

There are many different types of emitters (radiators) available on the market today from the traditional cast-iron column type (see Figure 1.55) through to the pressed steel panel emitters, skirting panels and the more sophisticated thermal panels, all of which will be explained in some detail in this chapter.

Column (radiator) emitters

The open column types are generally more efficient as they present the maximum heating surface area to the air in the room. This type of emitter is the one used commercially, its main disadvantage being that because of its open construction it contains many dust traps. A variation of the column emitter is shown in Figure 1.56. The columns are of a solid or plain elevation, still presenting a fairly large surface area for heating purposes but not having the problem of the dust traps. This type of emitter is very hygienic and is therefore known as the hospital pattern. It is recommended where hygiene is listed on the requirements.

Figure 1.55 illustrates the column type emitter: the flow pipe can be connected either at the top or bottom of the emitter by means of a control valve. The return pipe is connected at the opposite end of the emitter to the bottom connection by means of a lockshield valve. Open columns present maximum surface area to the surrounding air which is heated by conduction (particles touching) and circulating by convection (particles moving), together with a small amount of radiated heat (direct rays). This type of emitter was originally made from cast iron. Today it is available in both cast iron or steel, the latter being more prone to corrosion than the former.

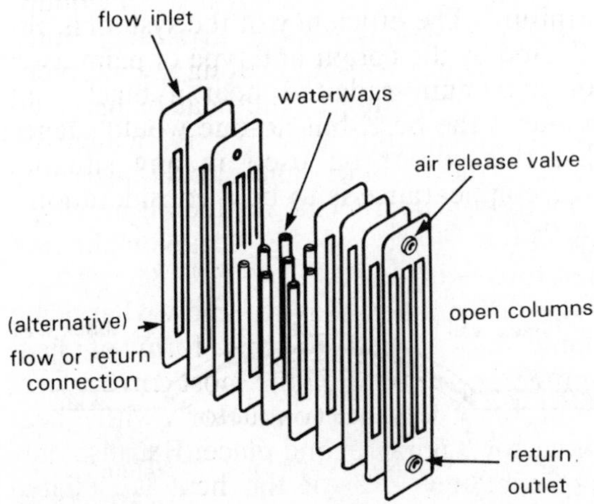

Figure 1.55 *Cast-iron column emitter*

Figure 1.56 illustrates the similarity between the two types of heat emitters. The hospital pattern emitter would be fixed and connected to the heating circuit in the same manner. By virtue of the closed or plain columns dust traps are obviated.

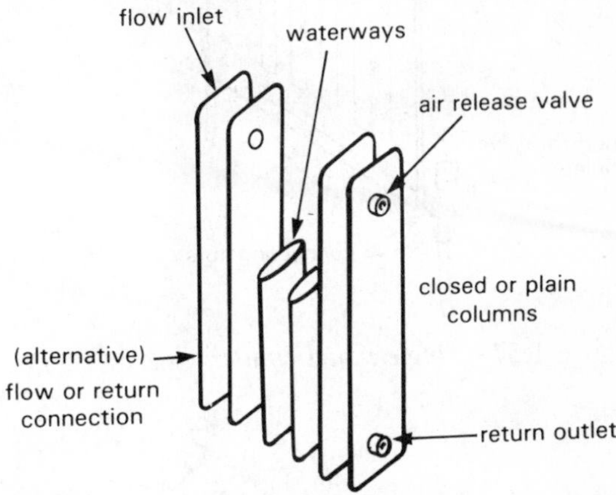

Figure 1.56 *Hospital pattern emitter*

There are on the market today a multiplicity of heat emitters (radiators) from the basic single panel (see Figure 1.57) to the double-panelled type as shown in Figure 1.58, and various modifications designed to increase the warmed surface area of the emitter as in Figure 1.58, which in turn will transfer more heat to the air surrounding it.

Siting of heat emitters

This is a very important factor and requires much consideration. Most draughts should as far as possible be eliminated but it is both impossible and undesirable to seal a room completely: all rooms must have some ventilation and air change. The siting of the heat emitter should be as near as possible to the point of entry of the fresh air, i.e. near the door or at the most important place beneath the *window*. In the case of large areas it may be advisable to fit two emitters, one at each end of the room. To enable the emitters to function properly air movement (convectional currents) must not be restricted: the emitters should be fitted approximately 100 mm from the floor and 40 mm from the wall, which will allow cleaning access as well as air circulation. It is good practice to fit the emitter along the full length of the window and, in the case of a bow-window, at an extra cost the emitter can be curved to fit neatly into the curvature of the window.

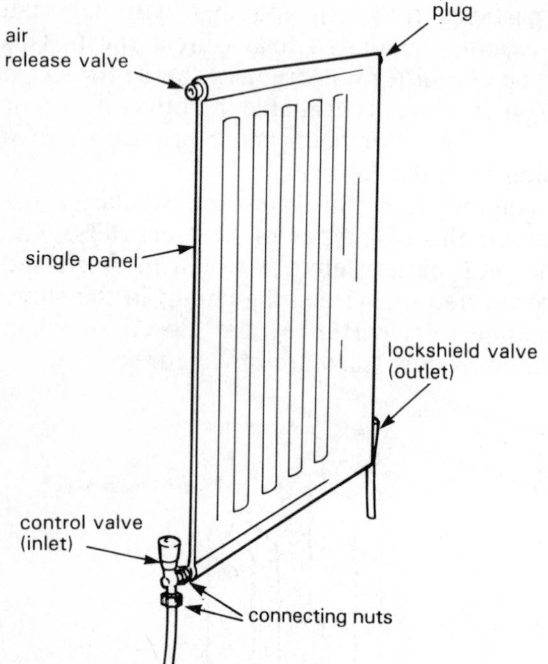

Figure 1.57 *Single panel emitter*

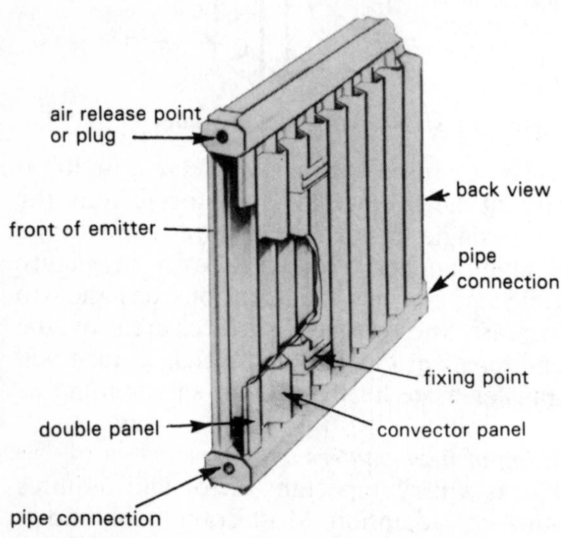

Figure 1.58 *Double panel convector emitter*

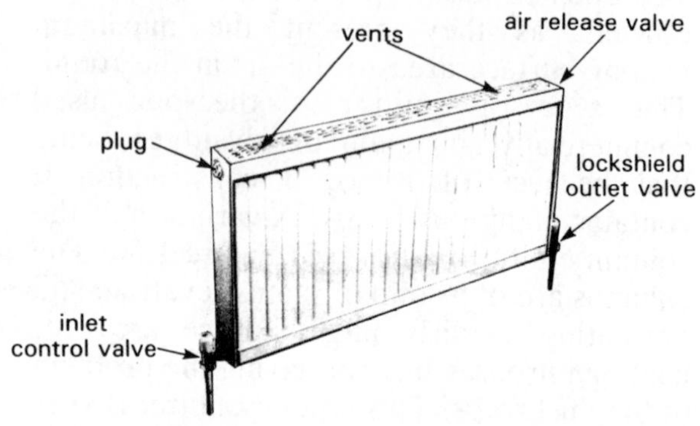

Figure 1.59 *Convector emitter*

The emitters should not be covered by curtains or fitted behind furniture as both would have an effect on efficiency. In addition, the heat and airborne dust can have a detrimental effect on both fabrics and the furniture. The efficiency of the system is also affected by the colour and type of paint used: for maximum radiated heat a black matt colour is the best, but no-one would suggest that this colour be used in any situation where appearance is to be a consideration.

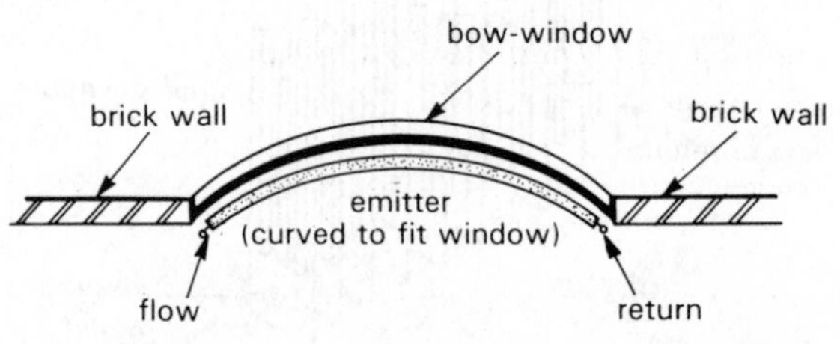

Figure 1.60 *Curved emitter*

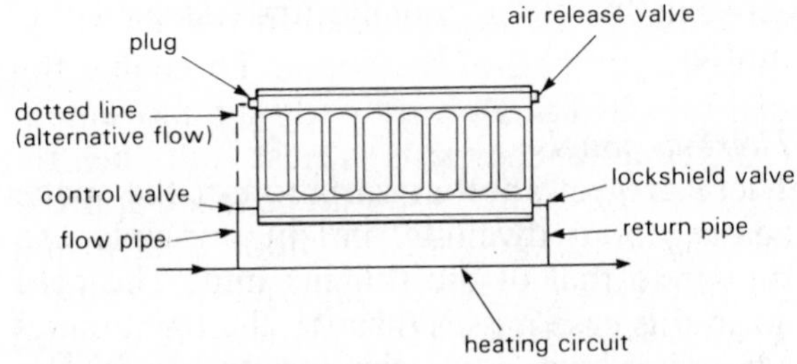

Figure 1.61 *Emitter controls and connections*

In general the same points apply to all emitters, including the convector type, although it must be pointed out that in the case of the fan-assisted convector the situation is not so critical.

The emitters (radiators) normally have four tappings, two at the top and two at the bottom. It is possible to fit the flow connection either at the top or the bottom, with the return fitted at the bottom at the opposite end.

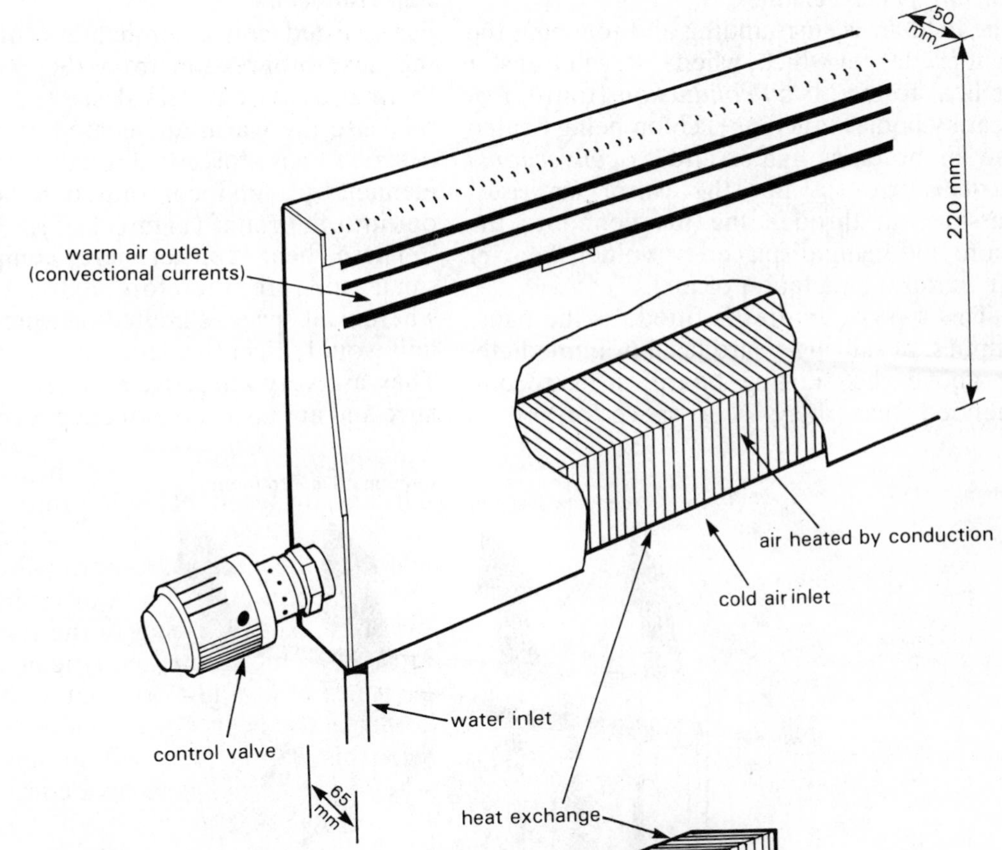

Figure 1.62 *Thermal skirting panel*

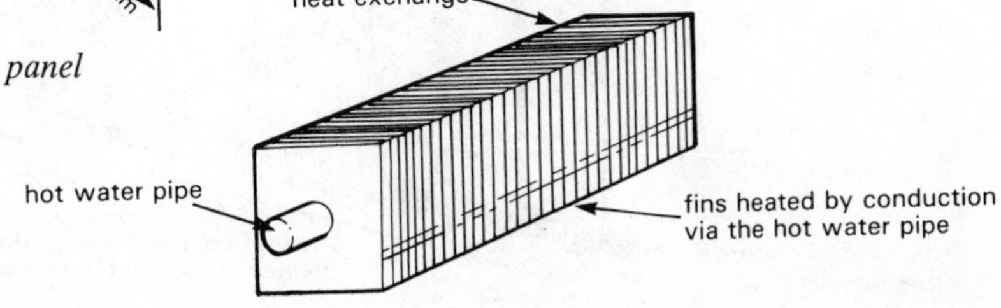

Figure 1.63 *Finned heat exchange unit*

The flow pipe is connected to the emitter by means of a control valve (sometimes thermostatically controlled); the return pipe is connected by means of a lockshield valve (sometimes known as a balancing valve). An air release valve is fitted in one of the two top connections, its function being to bleed the system of air during the filling operation and it is sometimes required for the same purpose during the working of the system. The remaining spare tapping is simply plugged off.

Skirting pattern emitter
The cold air is surrounding and touching the heat exchanger which, when hot, will transfer its heat to the air by *conduction* (transfer of heat by bodies touching). Upon being heated the air becomes lighter and so *convectional currents* are set up: the warm air rises, passing out through the top vent into the room and being displaced by colder, heavier air (natural circulation of air).

This type of emitter is fitted, as the name implies, at skirting-board level or immediately above the skirting board; it is recommended that there is a space of 60 mm between the floor and the underside of the emitter.

Thermal panels
These are yet another variation of the space heating of a dwelling and in principle are similar to that of the skirting unit. The cold air in this case passes through the front panel where it is heated by being in contact with the hot-water pipes. The heated air then passes out of the top of the unit into the room by convectional currents (Figure 1.64).

Fan convectors
Fan-assisted convector heaters are perhaps the next progression from the skirting and thermal panel emitters described above. In this case the warm air, heated as previously described, is forced through the heating element by a fan incorporated in the unit and out into the room (Figure 1.65).

These heat emitters are comparatively small and are therefore extremely useful where wall space is limited or where a larger unit would affect the decoration of the room. They are very attractive and neat in appearance and are also very effective: because they

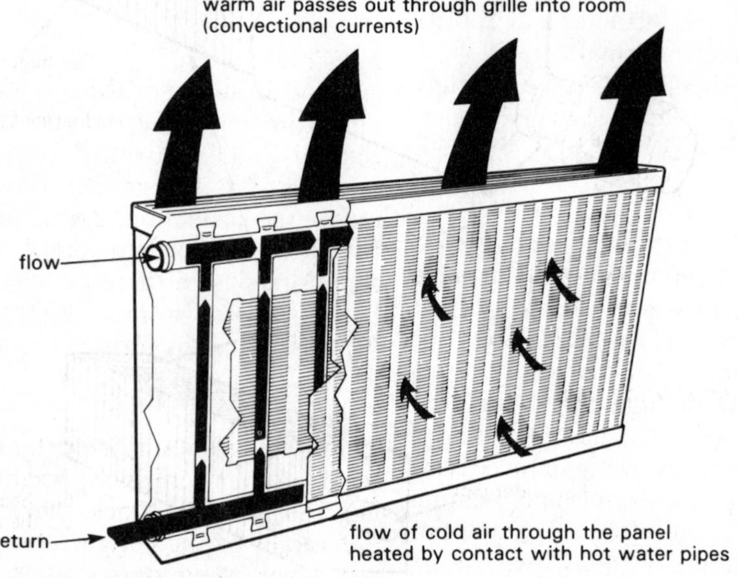

Figure 1.64 *Thermal panel*

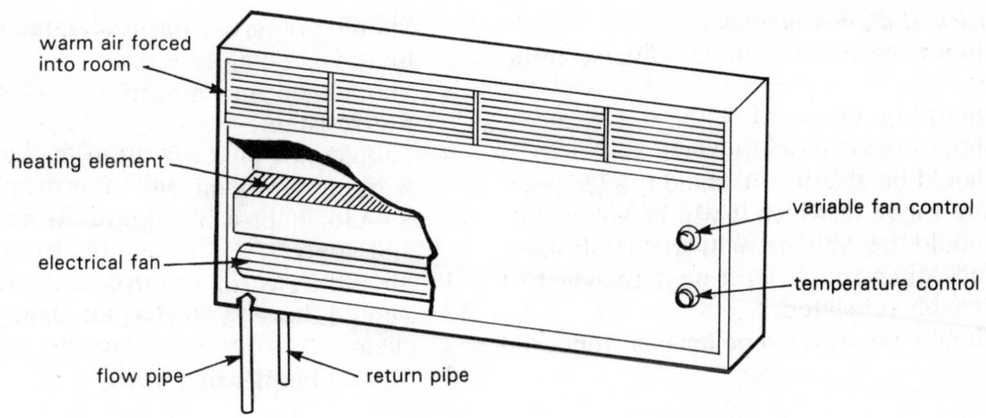

Figure 1.65 *Fan assisted heater*

are fitted with a variable fan they are able to boost the heat of the room quite quickly. They are manufactured in a number of sizes, some capable of heating the largest room in a house, or in offices or shops. The one disadvantage is that when in operation there is the continual hum of the motor driving the fan, particularly when on high speed. However, the fact that the air in the room is subjected to a slight raise in pressure tends to prevent cold draughts entering that room. A further advantage over the ordinary radiator (emitter) is that the surface temperature is lower, no-one can be burned, and fabrics adjacent to the heater are unaffected.

In the summer the unit can be usefully employed as a cooling fan.

Insulation

For a number of years the insulation of systems was carried out to prevent damage and possible waste of water caused by frost action. Today with the vast majority of homes having hot water and heating systems, and bearing in mind the high cost of fuel, the term insulation takes on a new meaning – that of 'saving' by preventing unnecessary heat loss from the hot water and heating system. Insulation is therefore of paramount importance in the home of today to assist in:

1 Prevention of loss of heat,
2 Prevention of frost damage,
3 Prevention of condensation.

Function of an insulator

It is well to realise that all insulators are of a spongy nature, containing minute air pockets, and by surrounding the appliance with this type of material, and due to the fact that still air is a bad conductor, the passage of heat through the material is very small. Broadly speaking, all insulators function in the same way, that is, to encircle the pipe or appliances with *still air*.

Efficiency

A good insulator will have an efficiency of approximately 80%. It must be appreciated that this will be governed to a large extent by its thickness: the greater the thickness the better the insulation. To maintain its effectiveness the insulator must be kept dry; therefore should there be a possibility of its becoming wet it must be protected by a water-resisting material.

Colour

Colour is also an important factor with regard to heat emission: bright colours emit less heat than dull ones; therefore insulators are usually painted bright, shiny colours, e.g. red, whereas emitters (radiators) should be painted dull matt colours.

Properties of a good insulator

1 Should be durable under diverse influences.
2 Should be rot-proof.
3 Should resist mould growth.
4 Should be able to withstand the temperature range to which it will be subjected.
5 Should be able to withstand vibration and other rough treatment to which it may be subjected.
6 Should be water-repellent or rendered so.
7 Should not have a corrosive effect on the materials being covered.
8 Should be incombustible or at least fire-resistant.
9 Should not give off offensive smells at working temperature.
10 Should, if possible, be clean to handle and apply.
11 Should be vermin-proof.
12 Should have a surface finish easy to clean.
13 Should be pleasing to the eye.

Table 1.3 *Table of insulators*

Insulator	*Use*	*Application*
Fossil meal This is made up of various clays and fibres prepared by mixing with water and applied while in a plastic stage.	Hot water and heating systems. Boilers, pipes, etc.	Applied in three thin coats while the system is hot (often sprayed on).
Plastics composition (alternative to the above) Now very widely used.	As above.	As above.
Magnesium Contains fossil meal and 85% magnesium made up in sectional form.	Hot and cold water installations.	Purpose-made sections, clipped on.
Glass fibre Made up in blankets or strip form, sometimes fabricated on to wire net backing; also available in sections.	Hot and cold water installations.	Wrap-around pipes etc. may be covered with canvas, clip-on sections.
Aluminium This is made up in strip form and corrugated sheet. Efficiency depends principally upon its reflective quality.	Hot water and heating systems.	Wrap-around pipes and appliances. Reflector sheets.
Slag wool Suitable for packing into casings or being made up into jackets.	Hot and cold water systems.	Packed into casings.
Cork Available in sections, sheet, or granular form.	Hot and cold water systems.	Wrapped around pipes or clipped on; also packed in casings.
Foam, polystyrene, and plastics These more recent insulators are made up in either sections, sheet, panels or granules. *Note*: Some plastics and polystyrene insulators should not be used on hot surfaces (i.e. approx. 80 °C); always check the manufacturers' literature.	Hot and cold water systems.	Clipped on, or packed into casings.
Vermiculite A very efficient in-fill material.	Hot and cold water systems.	Packed into casings.

Thermal insulators

The insulating materials used today have improved tremendously over the past few years. There is now a wide selection of insulators available ranging from modern spun glass to foams and plastics. Insulators come under two basic headings:

1 Cellular,
2 Fibrous.

They are supplied for use in three main ways:

1 Pre-formed sections,
2 Loose in-fill,
3 Plastics powder.

The insulators named in Table 1.3 are but some of the many available today. Cylinder jackets in washable plastic finish can be obtained in a choice of colours.

Safety

Great care must be taken when handling and working with some of the materials. Face masks should be worn when working with powders, and no skin exposed to contact with some of the fibrous materials, e.g. glass fibre.

Pipe insulation

Pipe insulation is made in various wall thicknesses for each pipe size and in both flexible and rigid sections. It is manufactured to fit standard copper and steel pipe sizes.

Flexible insulation

Foamed polyurethane provides one such flexible insulator which is both simple and easy to fit. It is available in lengths up to 3 m for pipes up to 76 mm in diameter.

Several types of flexible polyurethane insulators are marketed. Figure 1.66 shows one type which splits horizontally for ease of fitting; it is then slipped on to the pipe and sealed with waterproof adhesive tape. Another type has a nib along its full length to lock the insulation in place and is finished with a protective PVC sheeting. Flexible sections of glass fibre and mineral wool covered with felt or plastics and galvanised

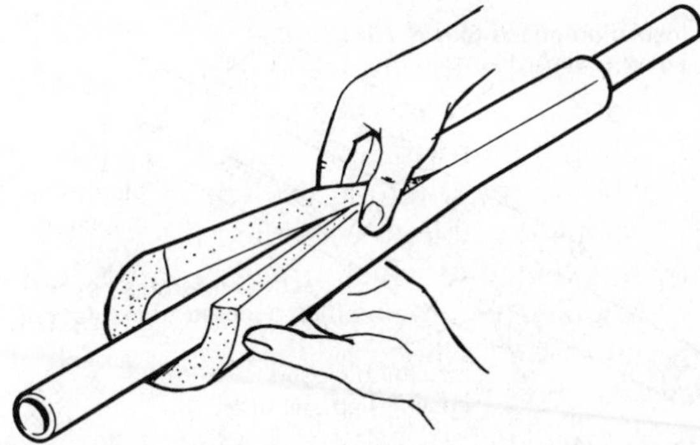

Figure 1.66 *Flexible insulation*

wire netting are recommended when the pipe is fixed externally, and when it may be subjected to adverse conditions. The insulating material is protected from the weather by felt which is secured by wire; it is then reinforced with wire netting. The whole is then painted with a bitumen solution as illustrated in Figure 1.67.

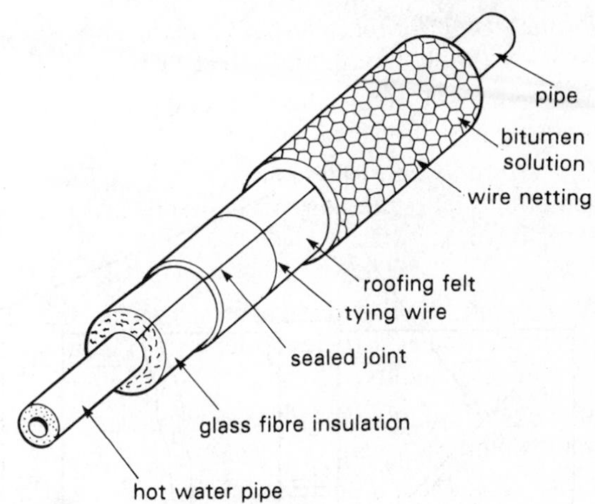

Figure 1.67 *Built-up insulation*

Rigid insulation

There are several types of rigid insulation available such as calcium silicate, cellular glass, glass fibre, cork, mineral wool or expanded plastics. They are usually supplied in half sections in 900 mm lengths. Some of the insulation is jointed and simply snaps on and locks around the pipe. The usual method

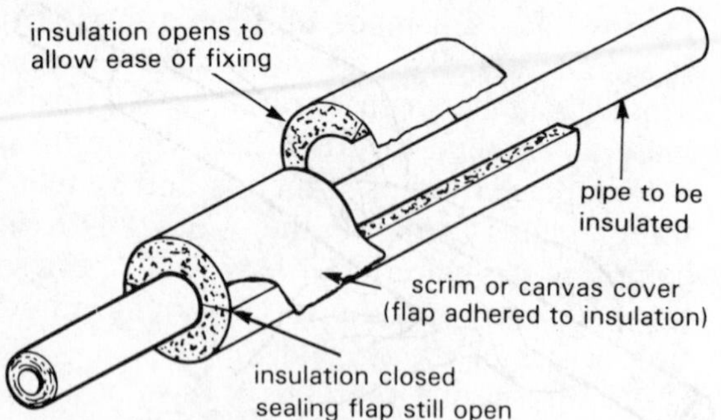

Figure 1.68 *Rigid insulation*

is to secure the sections with fixing bands every 450 mm, with additional bands at elbows and tee pieces; alternatively the sections may be stapled or fabric covers may be used.

The mitres are cut with a fine-toothed saw (hacksaw); joints may be sealed with adhesive tape.

Loose in-fill insulation

As already mentioned there are several very good loose in-fill insulators such as cork, vermiculite, plastics, mineral and glass wool. These are very useful for insulating pipes running between joists as indicated in Figure 1.70.

1 The space between the joists housing the pipes is simply filled by the chosen loose in-fill insulating material. It is advisable to seal the space with a suitable cover. Gently tamp the loose in-fill material but under no circumstances should the material be rammed in hard or compressed. See Figure 1.70(A).
2 An alternative method would be to insulate the space between the joists in the normal way, then insulate the pipes individually with one of the approved

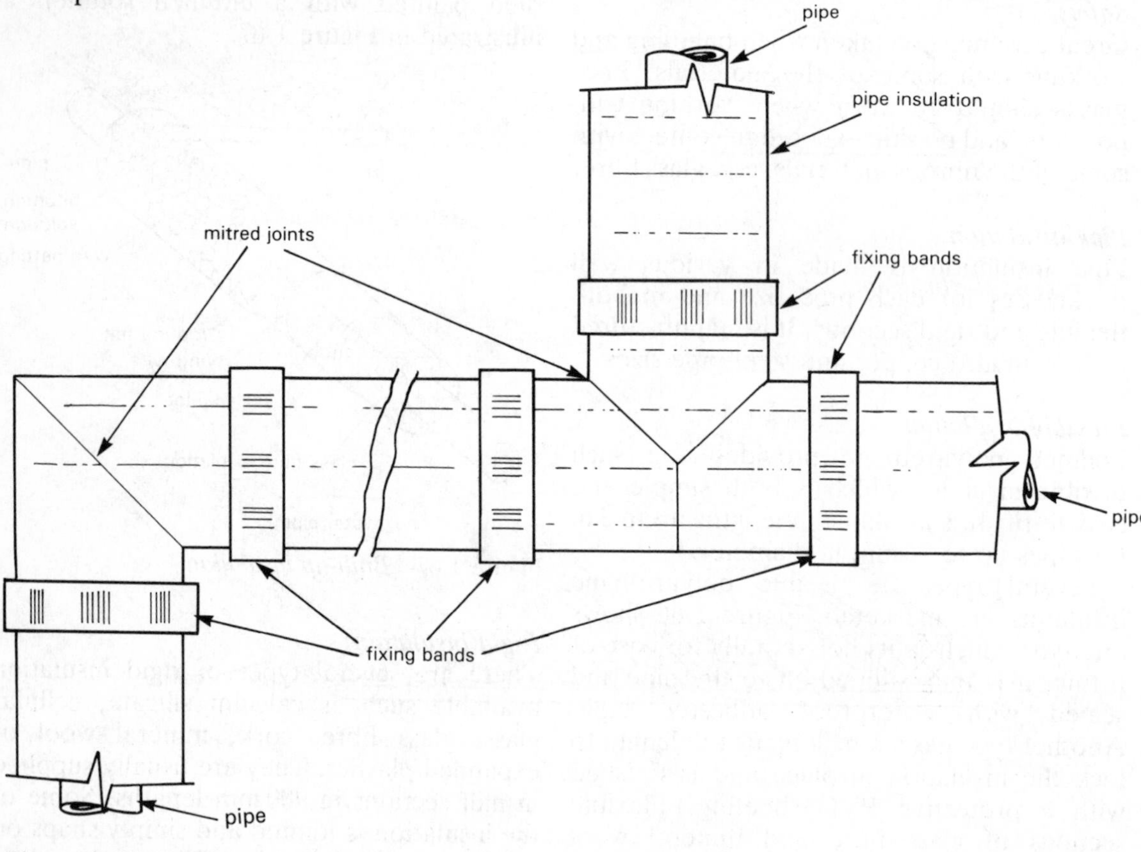

Figure 1.69 *Rigid insulation*

pipe insulators and lay them on top of the roof insulation as shown in Figure 1.70(B).

3 Another alternative method is to lay the pipes on the ceiling and to cover them with the normal roof insulation as in Figure 1.71.

Insulation of cylinders (hot storage vessels)

The most commonly accepted method of insulating a hot storage vessel is shown in Figure 1.72. It is made up from a number of segments of glass fibre or mineral wool, sandwiched between two sheets of flexible material stitched together. The segments are held in place by clips at the top and by two or three bands around the sides. Do not compress the material or leave gaps between the segments, and do not cover the immersion heater or the cables.

An alternative method of insulating the hot storage vessel is to use one of the loose in-fill materials as illustrated in Figure 1.73.

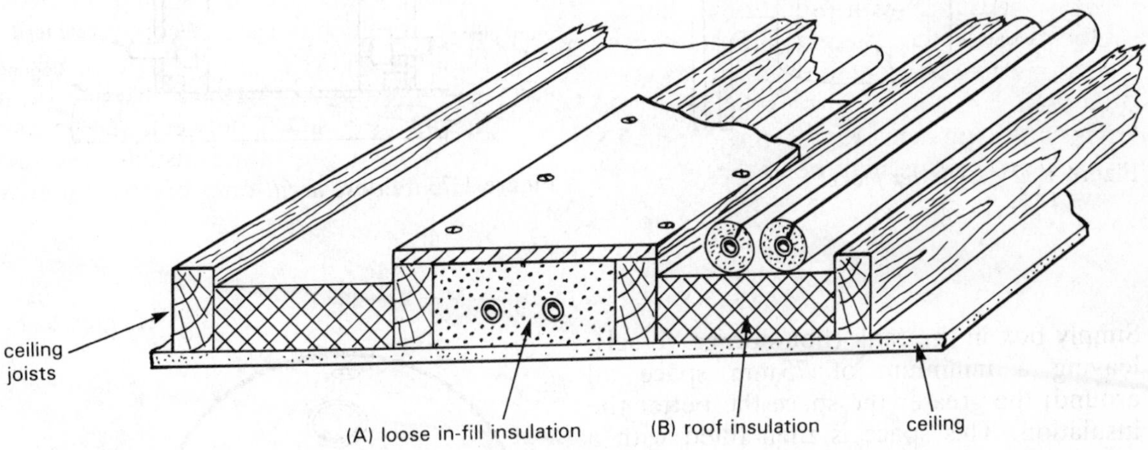

Figure 1.70 *Insulating pipework in roof space*

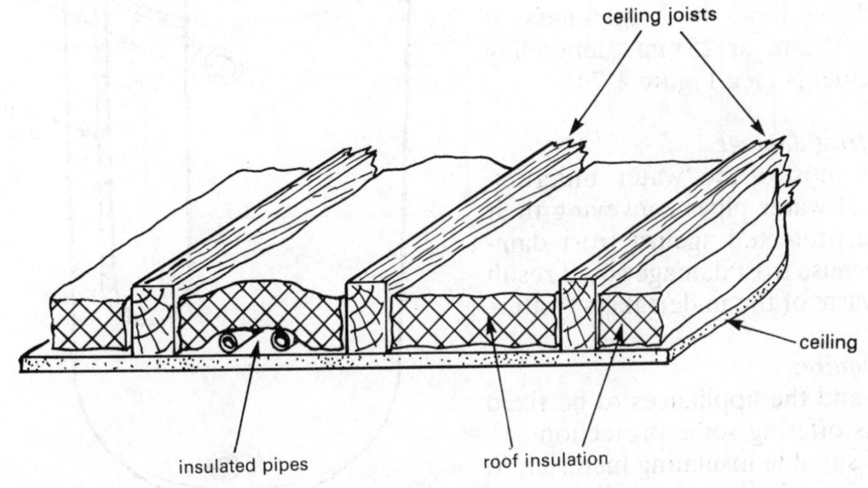

Figure 1.71 *Insulating pipework in roof space*

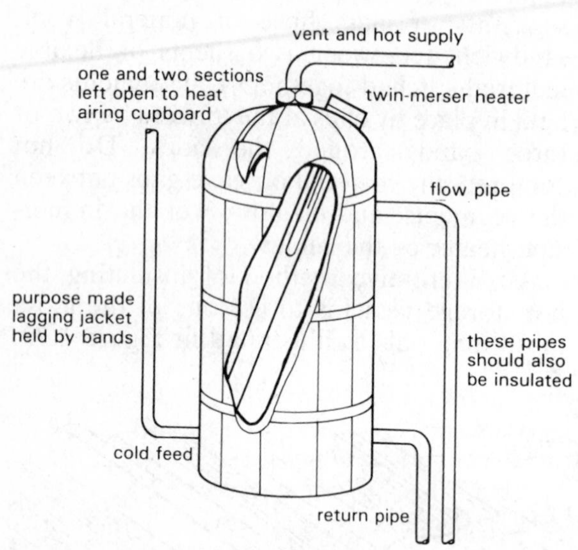

Figure 1.72 *Cylinder jacket*

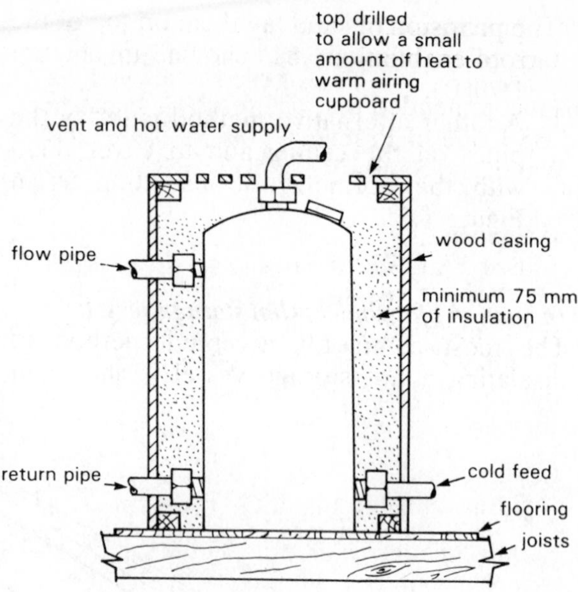

Figure 1.73 *Loose in-fill*

Simply box in the tank (hot storage vessel), leaving a minimum of 75 mm space all around; the greater the space the better the insulation. This space is then filled with a suitable granular in-fill from the list already given.

The cylinders may be purchased covered with polyurethane foam to a thickness of approximately 16 mm or 25 mm, depending on the requirements (see Figure 1.74).

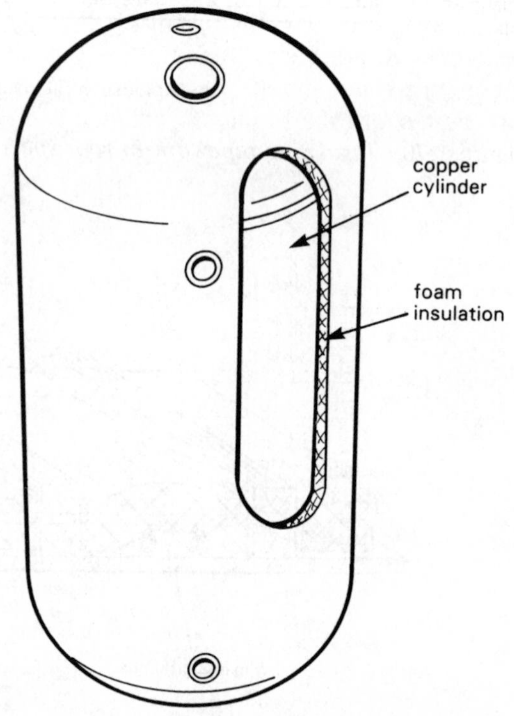

Figure 1.74 *Cylinder with foam insulation cover*

Prevention of frost damage

The by-laws of most of the 'water undertakers' require that water pipes conveying their water shall be protected against frost damage. This is because frost damage could result in a massive waste of the undertakers' water.

The by-laws require:

1 The pipes and the appliances to be fixed in positions offering some protection,
2 The use of suitable insulating material, in particular for pipes and appliances in vulnerable positions,

3 The provision of stop valves and drain-off taps to enable the system to be drained.

It is recommended that reference be made to the BS Code of Practice 99 Frost Protection of Water and Sanitary Services in Buildings.

Service pipes

1 Must be laid in the ground to a minimum of 750 mm or to whatever depth is recommended by the by-laws of your locality.
2 Should be suitably insulated wherever they are:

   (a) Exposed,
   (b) Buried to an insufficient depth.

3 Should rise as far as practicable from the external walls (see Figures 1.75 and 1.76).
4 To be controlled by stoptap and fitted with a drain-off tap.
5 Pipes laid to enable water to drain.
6 Pipes not to run through places where they are subjected to draughts. Avoid positions near windows, airbricks, external doors etc. If this is not possible protect with adequate insulation.
7 Pipes must not be positioned near the eaves or roof (see Figure 1.77).

Insulation of roof space

One of the outcomes of proper and adequate insulation of the ceiling and roof space is that very little heat can now pass from the house into the roof space (loft). As a result the cold storage cistern with its relevant pipework will undoubtedly become vulnerable to freezing. It is therefore very important to fix the cistern and the pipework in such a position as to minimise the possibility of their becoming frozen (see Figures 1.77 and 1.78).

Insulation of cistern

Figure 1.79a illustrates the use of expanded polystyrene sheets. These are manufactured ready to fit most common sizes of cistern. They should be a minimum of 25 mm in thickness. If maintenance work is required, the polystyrene parts are easy to fit or remove: they are simply held in place with tape, staples or wire. In the case of the unusual-size cistern, large sheets of polystyrene are available. These are easily cut to size and fitted as described. Purpose-made matting manufactured from mineral or glass fibre is made into a form of quilting. This is then wrapped around the vessel and secured with bands, tape or wire as illustrated in Figure 1.79b.

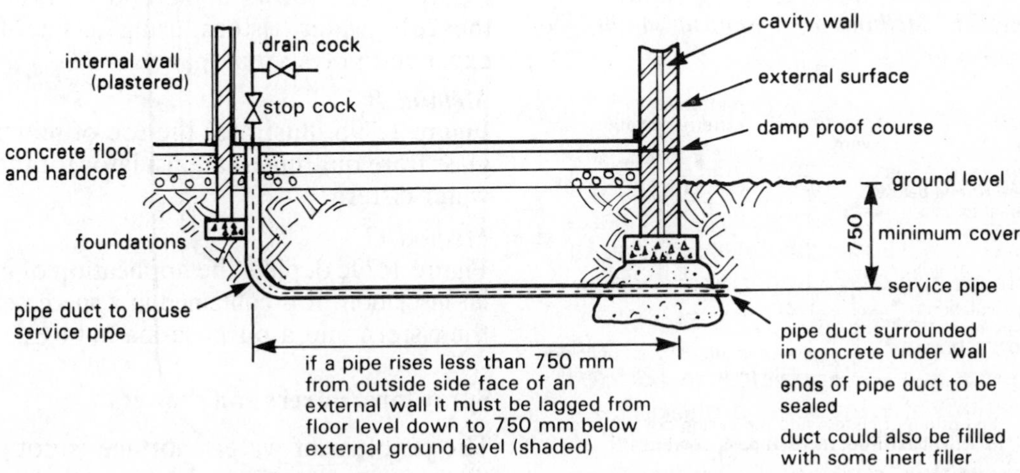

Figure 1.75 *Solid floor*

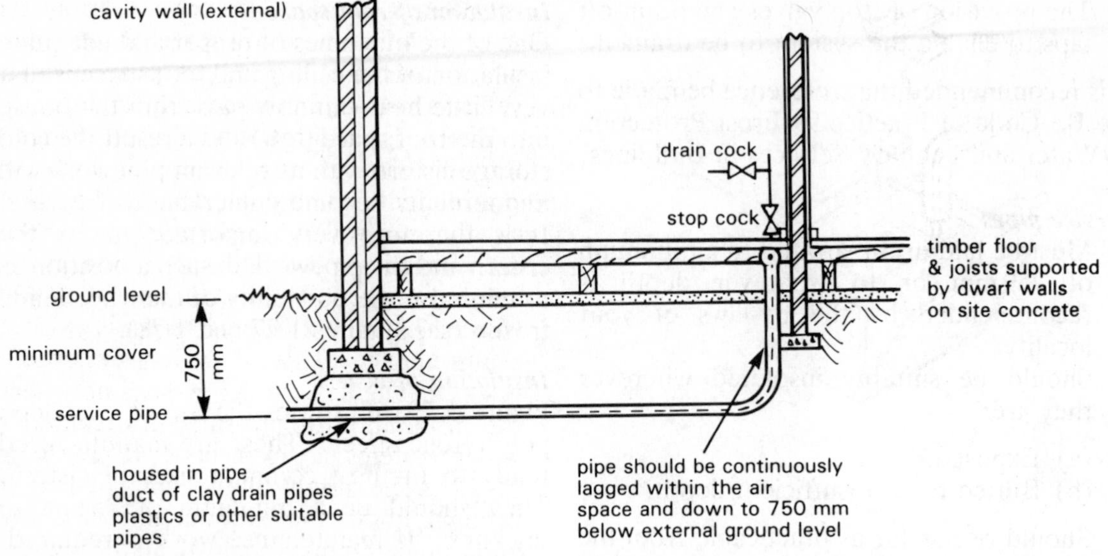

Figure 1.76 *Suspended floor*

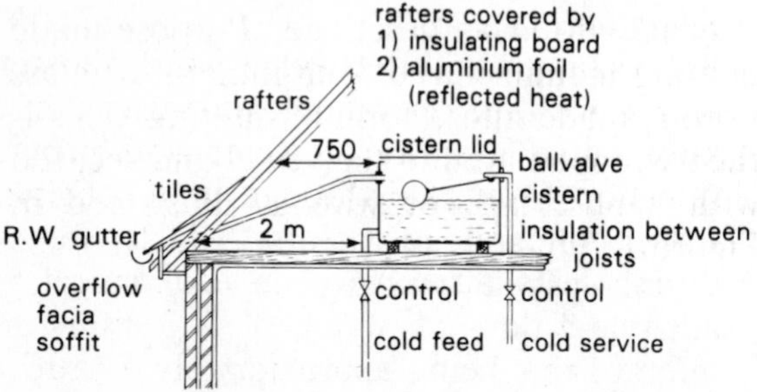

Figure 1.77 *Methods of treatment in the roof space*

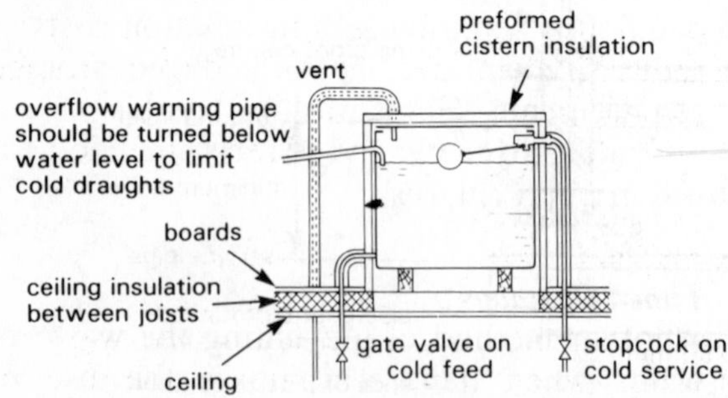

Figure 1.78 *Methods of treatment in the roof space*

A third method of carrying out this very important requirement is shown in Figure 1.79c. In this the four sides of the cistern are boxed, creating a 75 mm space around the vessel, into which some suitable granular in-fill is placed. A purpose-made insulated cover is fitted, but it is not recommended to insulate either the base of the cistern or the ceiling immediately below the cistern. This is to enable a small amount of heat to rise from the habitable room below.

Method A

Figure 1.79a shows a method of insulating the cold water cistern using a set of five expanded polystyrene panels.

Method B

Figure 1.79b illustrates the use of mineral or glass fibre quilting as an insulator for the cold water cistern.

Method C

Figure 1.79c depicts the application of granular insulation. It is contained in a space between the cistern and a purpose-made boxing.

Spray taps, mixers and showers

The problem of water shortage is not a real threat in our country. Nevertheless the careful use and conservation of water is in

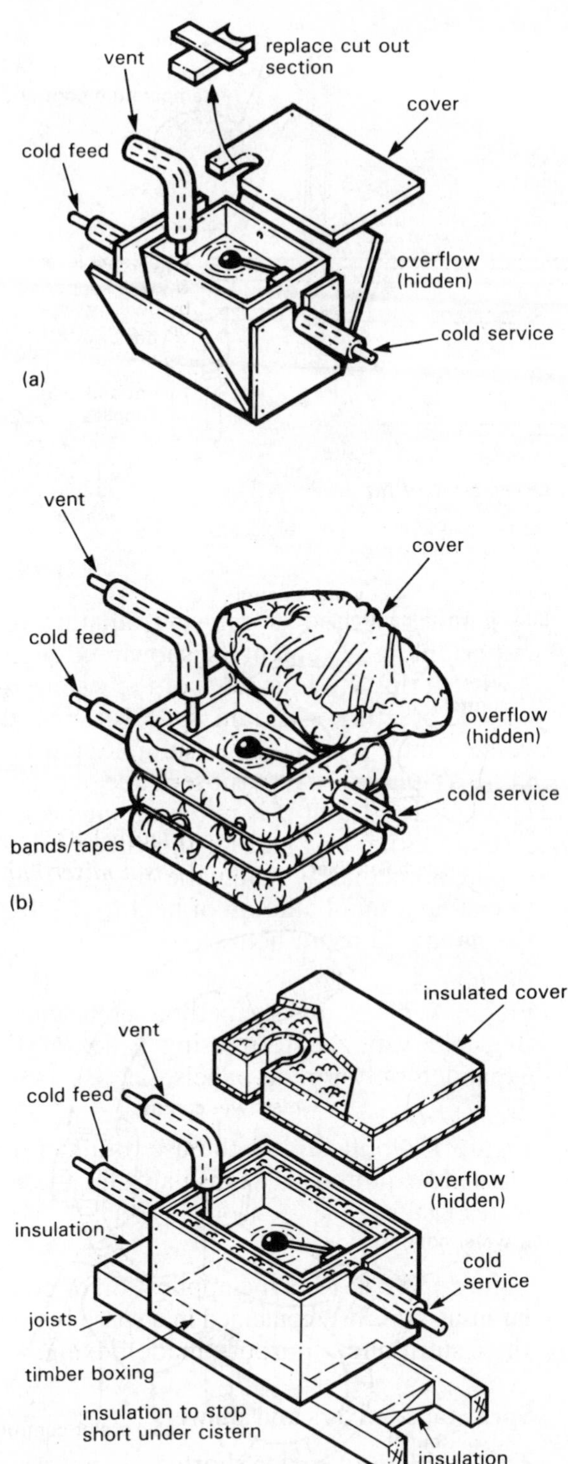

Figure 1.79 *Insulation of cold water feed cisterns*

everyone's interest. The use of spray taps, mixers and showers are some of the ways in which thorough cleaning can be achieved and yet bring about a considerable saving.

There is a large and varied range of terminal fittings now available, some with special features, but all designed to give a steady, controlled temperature flow. Some are manually controlled, others thermostatically controlled. The installation of all water (fittings) appliances must comply with the requirements of local water undertakers' by-laws. The new model by-laws now permit the installation of pressurised unvented hot water systems as explained previously (see p. 12). The result of the new by-laws has brought about the introduction of a wider and much more sophisticated range of fittings, such as single lever fittings, aerated discharge, pulsating massage jets and more precise temperature control.

Spray taps

Spray taps are effectively 'hand showers' which, with the operation of a single lever or knob, give a spray of water and provide selection from cold through to hot (Figure 1.80). Integral restrictors, which are set during installation, give an economical yet adequate flow of water of 2–3 litres per minute, this being approximately a third of that of ordinary taps. In addition to the obvious saving of water, there is also a saving in fuel required to heat the water and there could also be a saving in installation costs as reduced quantities means reduced storage requirements. Hand washing in this way is also more hygienic and is recommended for use in public places.

Timed flow taps

Another method of preventing the waste of water when hand washing is the use of non-concussive taps. These taps (see Figure 1.81) have a push-button action. They deliver water for approximately 15 seconds, after which time they automatically close off the

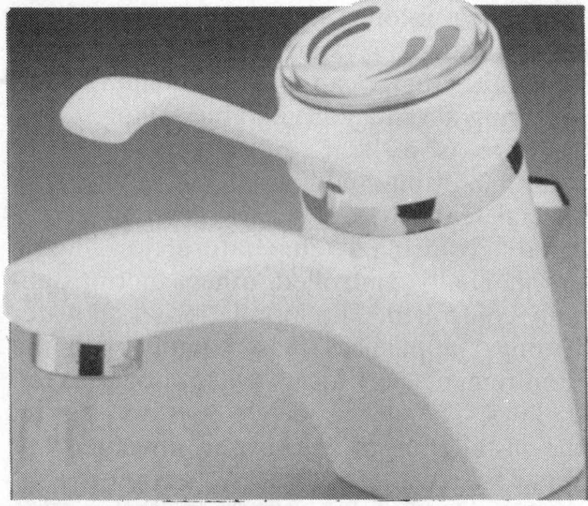
Figure 1.80 *Spray tap*

Figure 1.81 *Non-concussive tap (self-closing)*

supply. Should more water be required a further push on the button will recommence the cycle.

Mixer taps

Figures 1.82 and 1.83 illustrate just two of the many variations available. These compact one-hole fittings either have the traditional waste with plug and chain or alternatively a pop-up type. They have the advantages of simplicity of design, are simple to operate, leave an uncluttered basin, and are easy to clean. The higher balanced pressures have enabled the design of *lever valves* with precision mechanisms which give accurate and easy-to-control operation. This lever type of valve is particularly suitable for people who

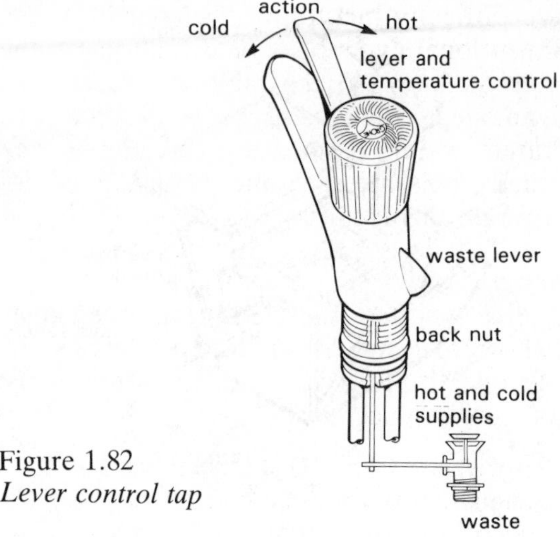

Figure 1.82
Lever control tap

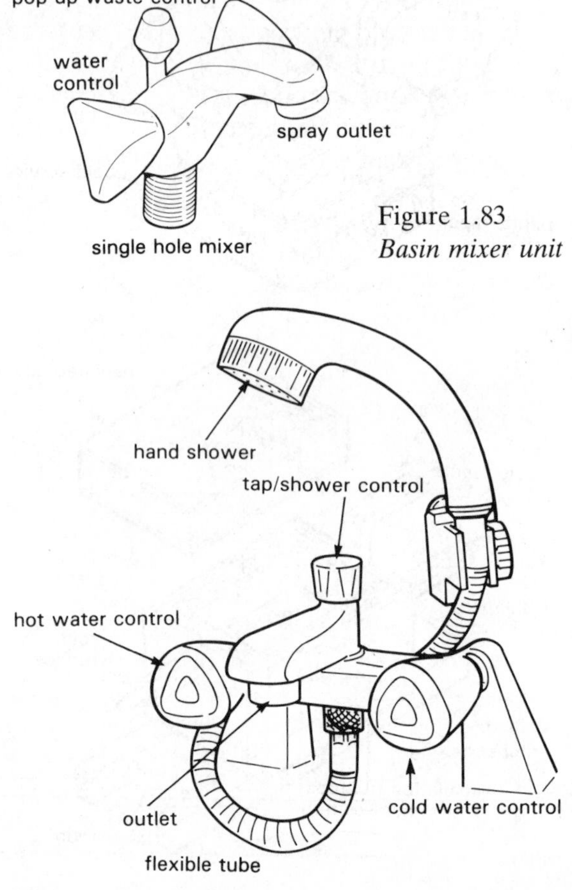

Figure 1.83
Basin mixer unit

Figure 1.84 *Bath mixer unit*

have difficulty in turning the control of the conventional type of tap. Aeration of the water is now a desirable attribute. Its advantage is that it ensures a flow of soft, aerated water, gentle to the touch and virtually splashless. Some fittings now incorporate this facility.

Showers

As already stated, there is a very large choice of shower fittings available. In selecting the right one the following considerations must be taken into account.

1 *Water pressure* This is perhaps the most important point as an inefficient shower is possibly worse than none at all. The pressure of the hot water system will be the pressure of the shower because they are subjected to the same pressure head from the cold-water cistern. This pressure should have a minimum head of water of 1 m when measured from the shower rose to the bottom of the cold water cistern (see Figure 1.85).

Suggested fixing heights

A Height of mixer approximately 1.5 m from the floor to be readily accessible.

B Approximately 2 m from floor.

C Minimum 1 m for reliable effective operation.

2 *Flow rate* Check the length of time it takes to fill a 10 litre container when placed in the position of the proposed shower, i.e. the bath. A satisfactory flow would fill the 10 litre container in approximately 16 seconds; anything longer than this would indicate low pressure.

Location

Having established the effectiveness of the hot water system, the location of the shower must now be considered. The actual choice is usually narrowed down to two positions: (a) fixed over the bath; (b) fixed in a separate shower cubicle.

(a) The shower fixed over the bath is by far the most commonly used, having the advantages of the availability of hot and cold water supply and the waste water outlet (Figure 1.86). The use of the bath as the shower tray means there is considerable saving.

All that is required is a curtain or some form of glass partition. This

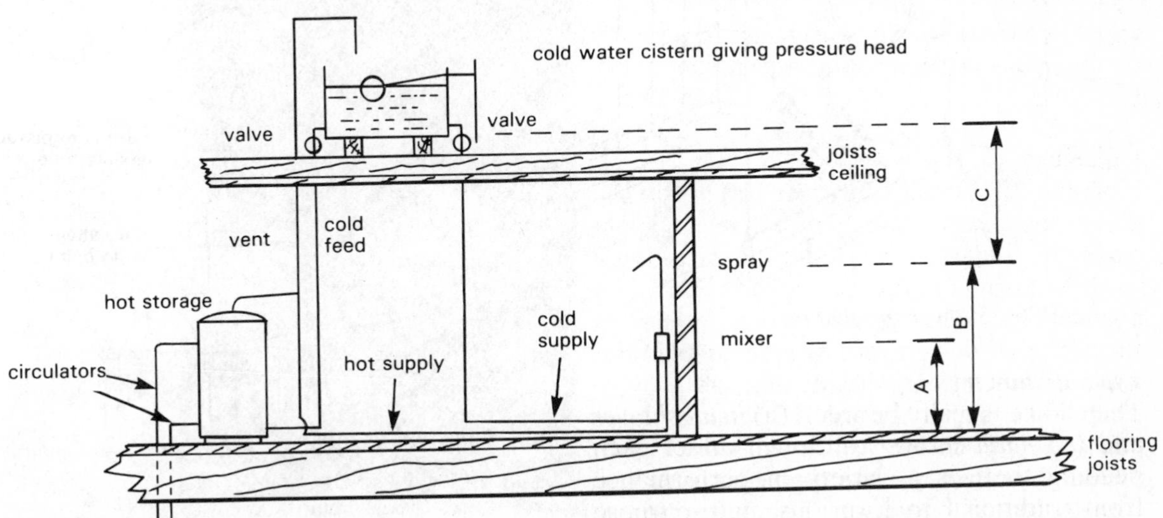

Figure 1.85 *Shower installation*

arrangement has a further advantage where there is limited space.

(b) Where space permits and where cost will allow, the making and installation of a shower in a separate cubicle is to be recommended. The advantages are:

1 Larger and more varied choice of appliance,
2 More space for occupant of cubicle,
3 Bathroom left free for use,
4 Less problem from splashing water,
5 No problem with cold curtain touching user.

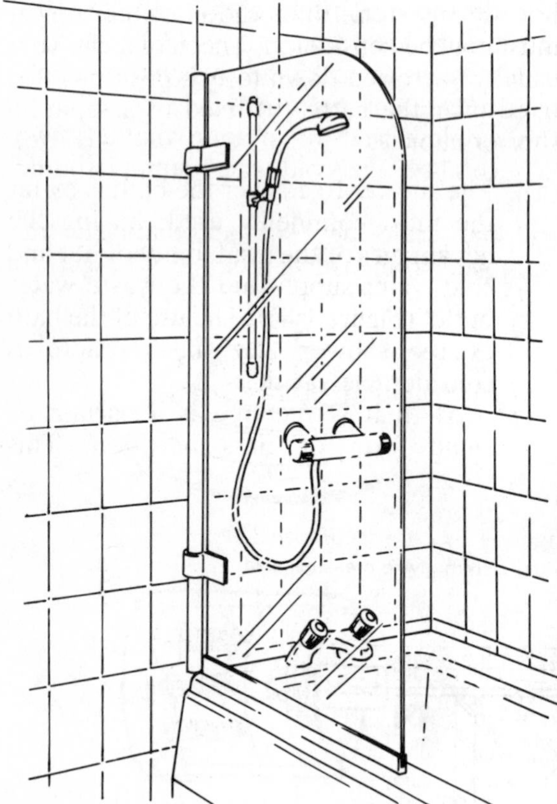

Figure 1.86 *Shower installation*

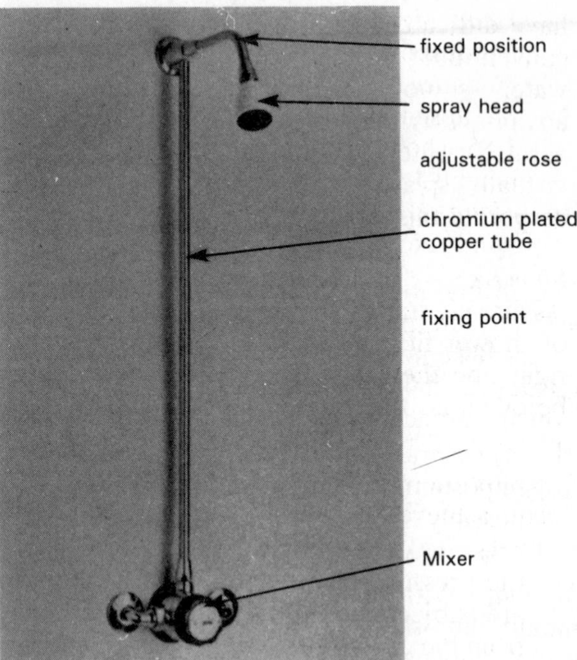

Figure 1.87 *Fixed point shower*

Type of shower

The choice is really between (a) *head shower* and (b) *hand shower*, although under each heading there is a variety of performance from traditional to luxurious multi-purpose sprays.

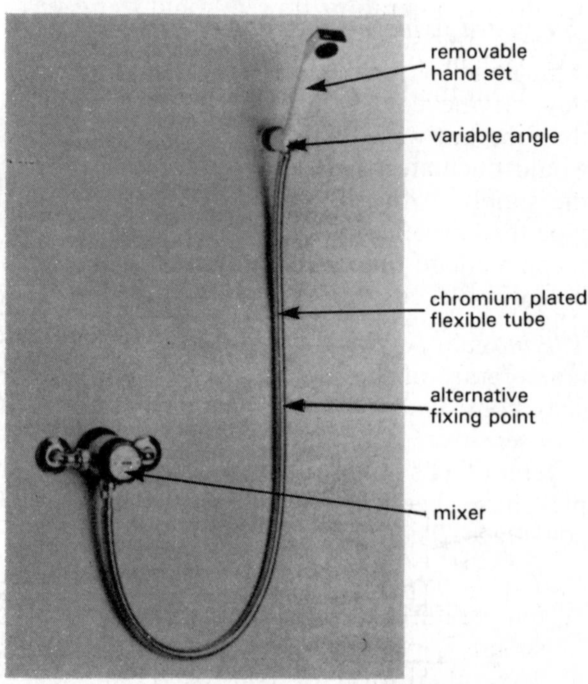

Figure 1.88 *Adjustable point shower*

(a) *Head shower* The rose height of a fixed shower is vitally important and guidance from the user is advised. In the absence of this, reference to Figure 1.85 shows the recommended fixing heights.

(b) *Hand shower* For sheer versatility the hand set is to be recommended.

The all important height can be varied very easily to suit the height of the user. It can also be easily removed from its fixing to allow a greater flexibility of use, i.e. for washing hair over the adjacent basin.

Temperature control

This is achieved by one of two methods: (a) manual or (b) thermostatic.

Manual With manual control it is of paramount importance that both hot and cold water supplies are at the same pressure, i.e. head. This is known as *balanced pressures*. This is because the flow of shower water is obtained simply by the operation of the valves, adjusting them *manually* to give the required temperature flow. Should there be unequal pressures it becomes increasingly difficult to set an even steady temperature flow. If the cold water pressure is greater than the hot water, the flow of mixed water would fluctuate: this situation could arise if the supply to the shower is taken from the pipe feeding other appliances. It is therefore recommended that the supply to the shower be a direct feed.

Thermostatic With thermostatic control the temperature of the outlet water is automatically controlled by some form of temperature-sensitive device. Until recently two different types of temperature sensing principles have been used, each with its own advantage. There are now three types:

1 *The bi-metallic coil* is very sensitive to temperature change and therefore responds very quickly (see Figure 1.89).
2 The other form of control is the wax capsule thermostat which generates a powerful force to control the proportioning mechanism of the mixer (see Figure 1.90).
3 By combining the two principles a completely new type of thermostat has been developed employing an advanced thermoscopic principle.

This new type of control claims to have the following advantages:

1 Temperature control maintained to within 1 °C of the selected temperature,
2 Immediate response to fluctuation in pressure of water supply,
3 Full thermal performance up to 10:1 pressure loss ratio,
4 Maintains accuracy even when one supply is reduced by up to 50%,
5 Complete shut-off in approximately two seconds if the cold water supply should fail,
6 Lime scale deposits automatically removed from critical areas.

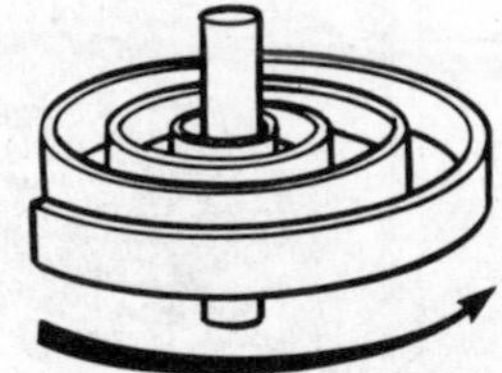

Figure 1.89 *Bi-metal principle*

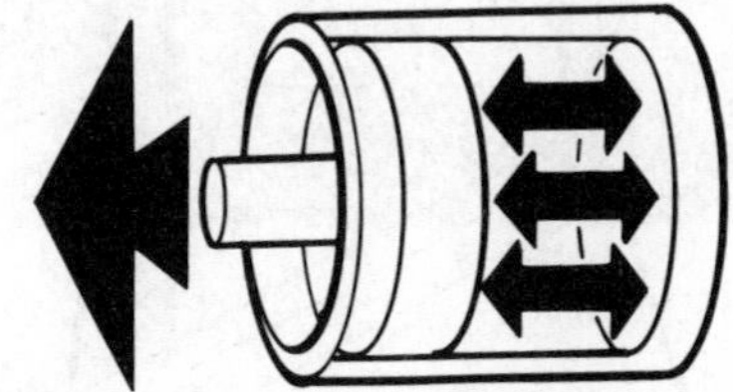

Figure 1.90 *Wax capsule principle*

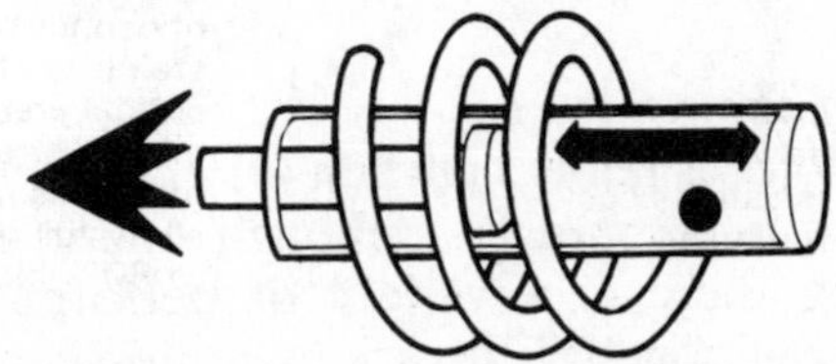

Figure 1.91 *Thermoscopic principle*

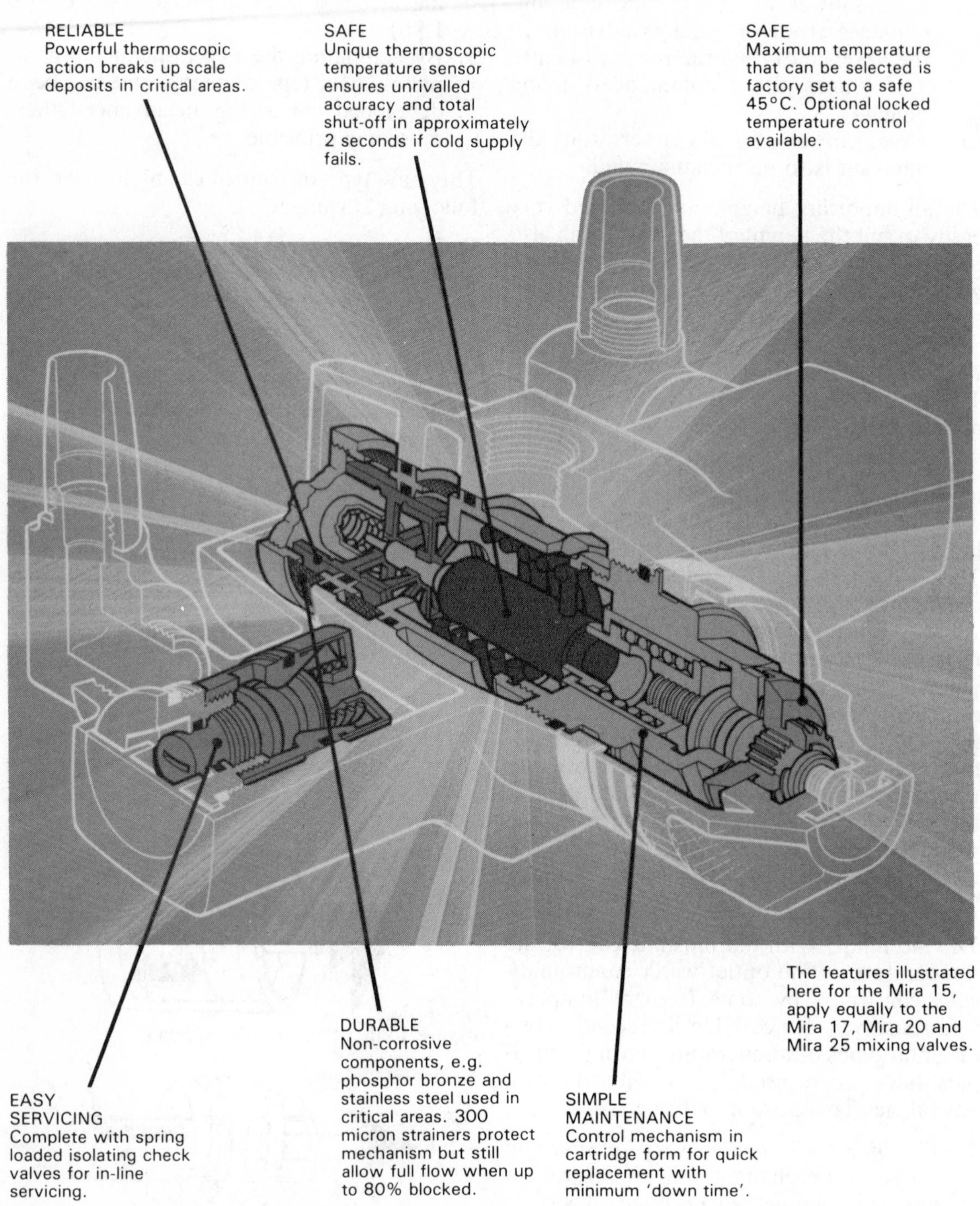

Figure 1.92 *Thermostatic control (Mira 15)*

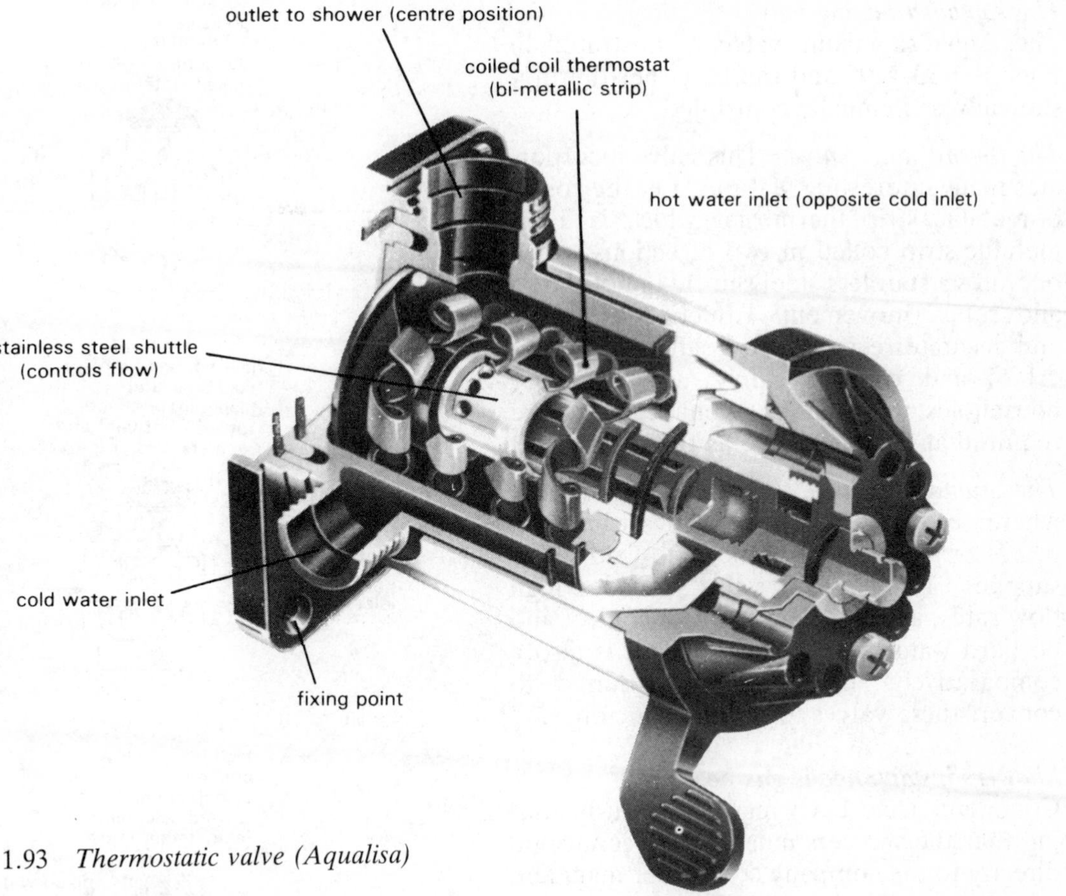

Figure 1.93 *Thermostatic valve (Aqualisa)*

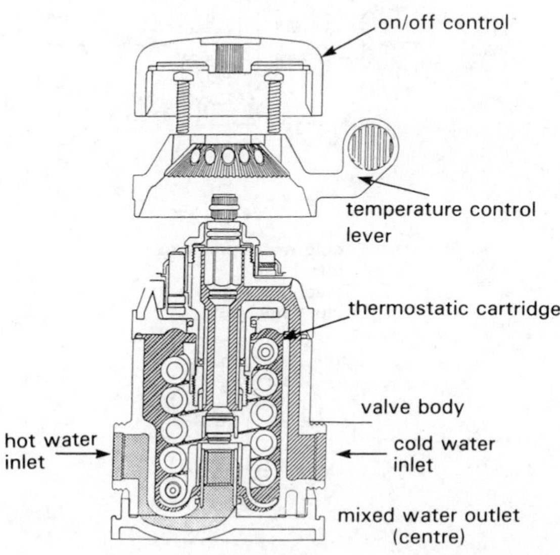

Figure 1.94 *Aqualisa thermostatic mixing valve*

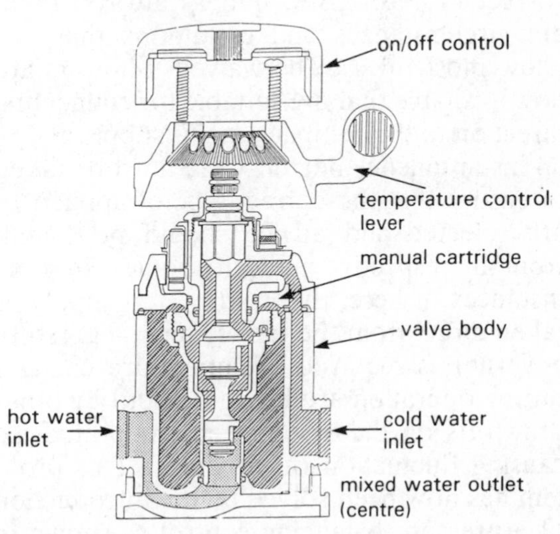

Figure 1.95 *Aqualisa manual mixing shower valve*

The Aqualisa mixing valve

The Aqualisa mixing valve is illustrated in Figures 1.93–1.95 and can be either thermostatically or manually controlled.

The thermostatic valve This valve incorporates many interesting features, i.e. the coiled bi-metallic strip thermostat which is a bi-metallic strip coiled in two directions which operates a stainless steel shuttle in both axial and rotary movement. Effecting a speedy and accurate response, it is safe in action to ±1 °C and can be safely used in busy households when other appliances may be required at the same time as the shower.

The manual valve This valve is for use where balanced pressures (described on p. 57) exist between the hot and cold water supplies. This type of valve provides high flow rates, accurate temperature control and no hard water scaling problems. It is also a comparatively simple task if required to convert these valves to thermostatic ones.

Showers/instantaneous gas heaters

Comments have been made previously stating that the showers must not be connected directly to the company cold water main and to instantaneous heaters. That is of course correct in *most cases*, but, as always, there are circumstances and conditions that will allow most rules to be waived. Showers are now available that are suitable for connecting direct on to the company mains supply and to an instantaneous heater. Care must be taken to ensure that the correct type of appliances are selected and advice should be sought from the respective authority bodies. In some instances, where the cold water supply is taken direct from the service pipe, a pressure governor is required. Temperature changes during operation are brought about by other draw-offs on the system being operated, so causing fluctuation of pressures. This problem has now been solved by the introduction of a pressure balancing control as shown in Figure 1.96. This control therefore makes the showers ideally suited to most gas water

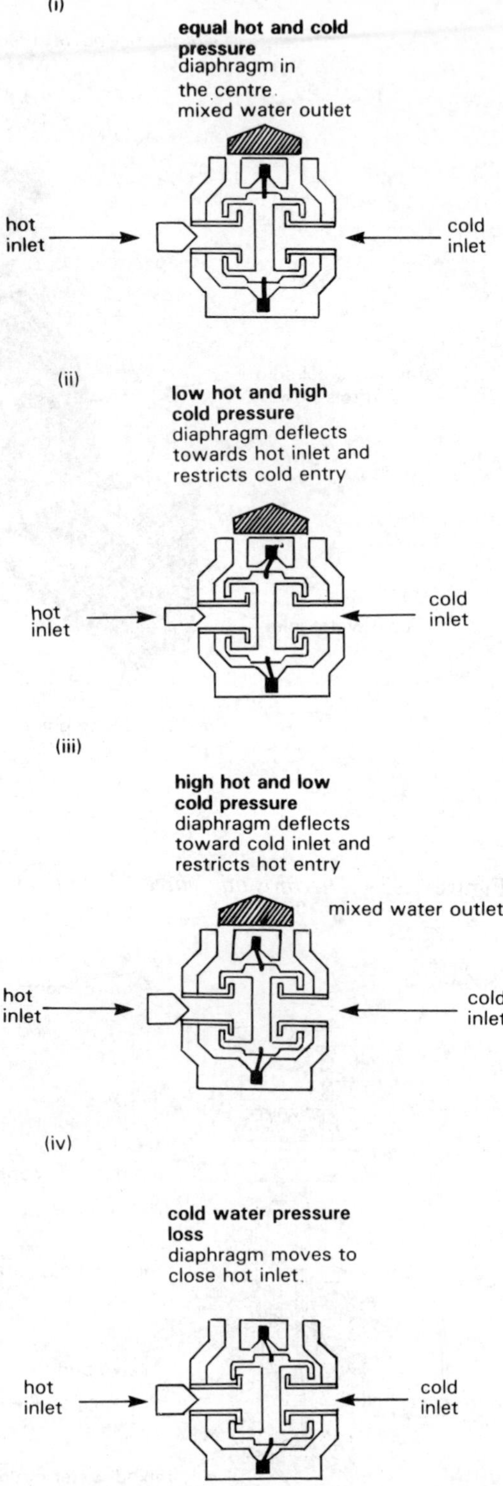

Figure 1.96 *Pressure balancing controls*

heaters and combination boiler installations. Figure 1.97 illustrates a typical piping arrangement for a shower mixer unit connected to the mains supply and an instantaneous gas heater without a water pressure governor.

Where a shower is fixed in a system incorporating a water pressure governor, it is advisable to fix the governor in the cold supply to both heater and mixer valve, and ideally there should be no other draw-off on this pipe, particularly when the shower is in use. It is also recommended that the heater be fixed as near to the shower as possible to reduce length of pipework and also to assist in temperature control. All gas water heater/shower installations must comply with the local gas regulations and water authority by-laws.

Figure 1.98 shows a typical multi-point heater/shower installation with water pressure governor.

It should be pointed out that some mixing valves may work satisfactorily without a water pressure governor, but the pressure loss ratio may make temperature control rather difficult for all but pressure balanced type mixing valves.

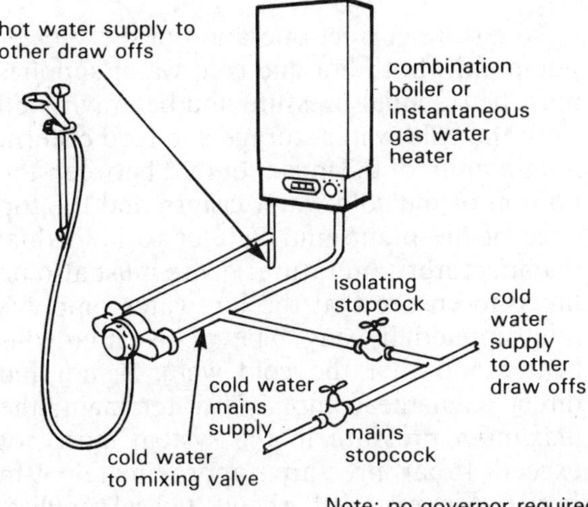

Figure 1.97 *Shower supplied by gas water heater*

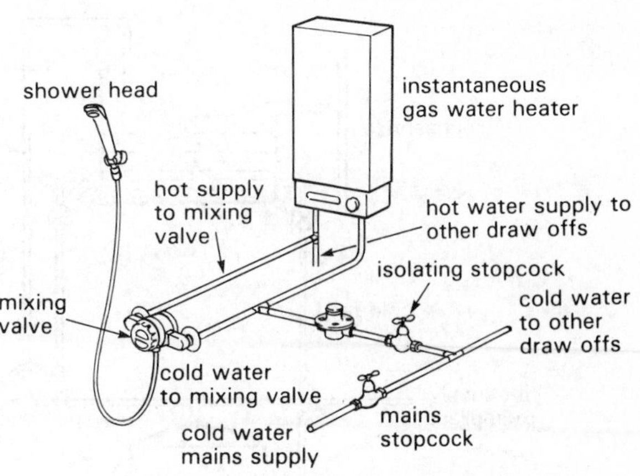

Figure 1.98 *Shower supplied by gas water heater*

Shower pumps

These are available to suit different circumstances, from a single shower where the outlet flow is inadequate and requires a boost to make it into a satisfactory unit, through to single showers to convert them into pulsating showers or for boosting the pressure to a multiple showering system with up to five or more spray heads. The object of the pump is to improve the spray force of a showering system or to compensate for lack of head (pressure). The pumps may be operated by a remote switch or automatically by a flow switch or when the shower control is operated. Installation must comply with IEE (Institution of Electrical Engineers) regulations and water authority by-laws. Very little maintenance is required – they are protected by an automatically re-setting thermal cut-out switch – but periodic cleaning of in-line strainers, where fitted, will be required. Figure 1.99 shows a possible layout of a domestic hot water system with shower and booster pump.

Electrically operated pumped shower

As stated previously there are many different variations of showers. Perhaps the most useful development has been that of the *pumped shower*. The majority of existing domestic hot water systems were installed without thought of showers and the necessary

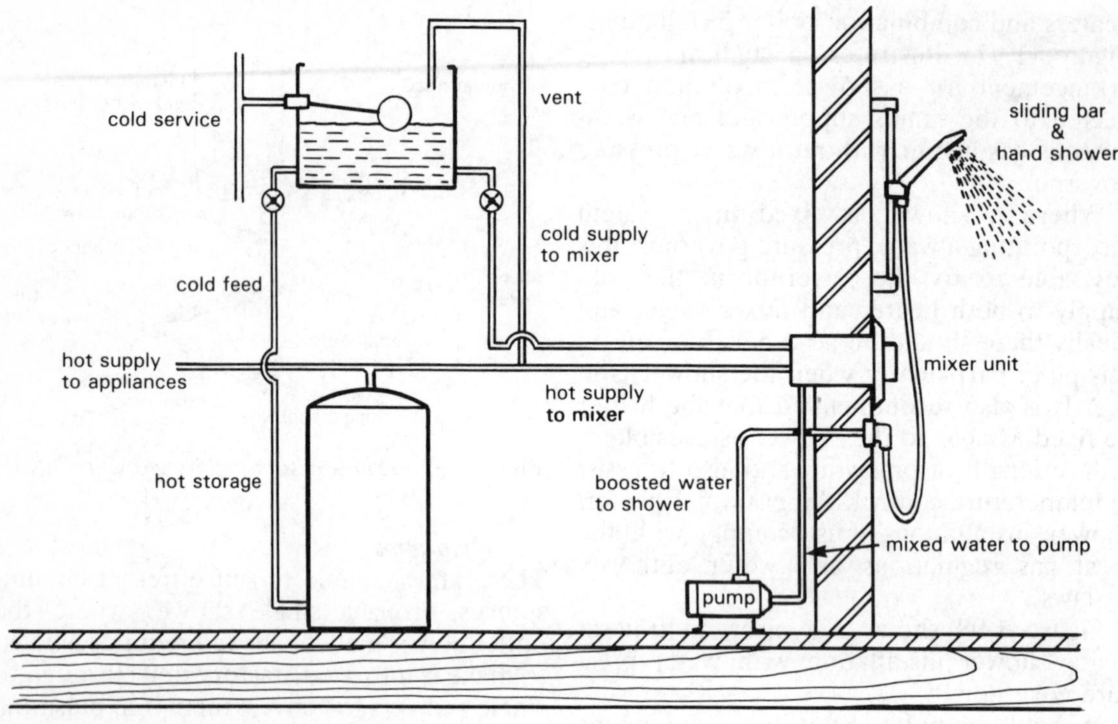

Figure 1.99 *Pumped shower*

head of water required to give an adequate flow of water at the nozzle outlet (see Figure 1.100). The problem, of course, can be solved by the re-siting of the cold water cistern in an elevated position, possibly in the loft. This can be a fairly expensive, messy and time-involving operation, including the additional insulation against frost damage that must be carried out. The fairly recent development of an electric pump capable of boosting the existing low flow of water to an acceptable flow has made a comparatively expensive job into a fairly simple operation. The pump is fixed in a convenient position near the existing shower or bath taps, the plumber's work being that of a simple connection of the flexible tube from taps to pump unit inlet and from pump outlet to spray.

It has already been stated that the plumber's work is relatively simple in this installance, but it cannot be over-emphasised that the electrical work must be carried out by a qualified electrician and to the requirements of the current IEE regulations.

To ensure correct operation of the shower pump unit, both hot and cold water supplies must be of equal pressure and be gravity fed from the cold water storage and feed cistern. A minimum of 0.3 m is required between the bottom of the cold water cistern and the top edge of the pump unit. (Refer to individual manufacturer's literature.) Care must also be taken to ensure that the hot water supply is not supplied by any type of instantaneous heater and that the cold water is not fed direct from the company's water main; the maximum pressure in the system must not exceed 1 bar pressure (approximately 9 m head). Figure 1.101 shows typical applications.

Figure 1.100 *Boosted shower unit*

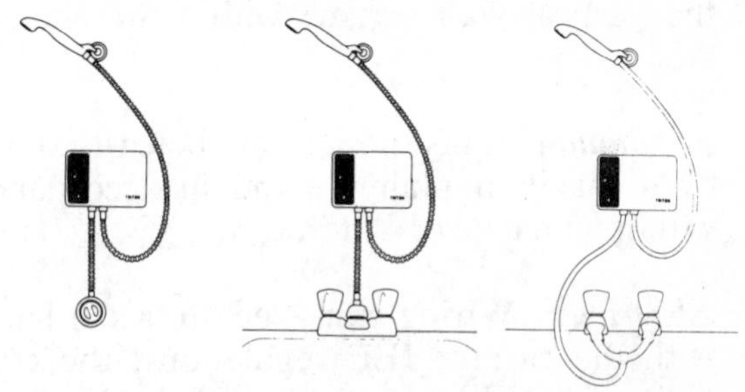

Figure 1.101 *Boosted shower unit*

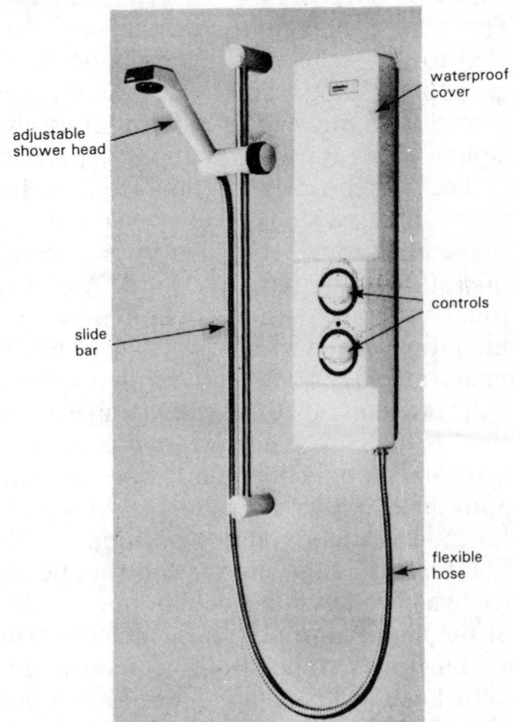

Figure 1.102 *Instantaneous electric shower unit*

Electric showers

After the electrically operated pumped shower, the next advance must be the instantaneous electrically heated shower unit (see Figure 1.102). Here again there is a very wide choice. It is important that the appliance is manufactured by the approved British manufacturers and that it carries the BEAB label (British Electrotechnical Approvals Board). This means that it has been tested and approved for electrical safety. In addition it must have the approval of the National Water Council.

Fixing One of the advantages is that it can be fixed anywhere so long as there is a mains water supply and a waste water outlet, e.g. over the bath, or any spare space can be converted into a shower cubicle. It is fixed to the vertical wall surface within the shower area.

Installation This should only be carried out by a qualified plumber and in accordance with manufacturers' literature.

Electricity Wiring must be in accordance with the current IEE regulations; the contractor should be a member of the National Inspection Council for Electrical Installation Contractors (NICEIC).

The heater should be permanently connected to an exclusively fused minimum 35 amp supply (32 amp at 220 V). A double pole pull-cord isolating switch is essential to give complete safety for the conditions appertaining. The switch-operated flow control has three settings: cold and two power settings. The temperature control gives precise temperature at the power settings. A cycling thermal cut-out switch operates if the temperature reaches 54 °C (normal shower temperature 40 °C) and an overriding cut-out switch operates if excessive temperatures occur. A pressure-sensitive switch ensures that the heater only operates if water is in the heating chamber and a system of coloured lights indicates its operating position.

The turning of the shower control opens a valve which allows the cold water to flow over the high-powered heating element, thus providing instantaneous hot water. The electric heating elements can be either 6, 7 or 8 kW. The higher the rating the larger the volume of water you can expect through it, so an 8 kW unit would give a more luxurious shower than a 6 kW unit.

Water To ensure the unit works correctly it should be connected directly to the cold service pipe by means of a 15 mm tube. The inlet and outlet of the unit has 15 mm external threads. Units are available which can operate on very low pressure and which can be fed from a cold water cistern, but it is recommended that this type of unit is best connected to the cold service pipe.

Maintenance This is restricted to cleaning out scale from the shower head which may be required once or twice a year, depending on the hardness of the water. In soft water areas it may be no more than once a year. A stiff brush or proprietary de-scalent is recommended for this purpose. Acid or abrasive materials must not be used.

In addition to domestic showers as illustrated, there are a variety of other uses and reference should be made to the manufacturers' literature for advice and guidance, i.e. multi-showers for camp-sites, factories, schools and other such places (see Figure 1.103). They include such things as:

1 Coin operated,
2 Master temperature control,
3 Timed control,
4 Electric instant shower.

Towel rails and heating coils

Connection

Perhaps the first question you should ask yourself is: what is to be the function of the towel rail and/or heating coil? No doubt the answer will tell you where they should be

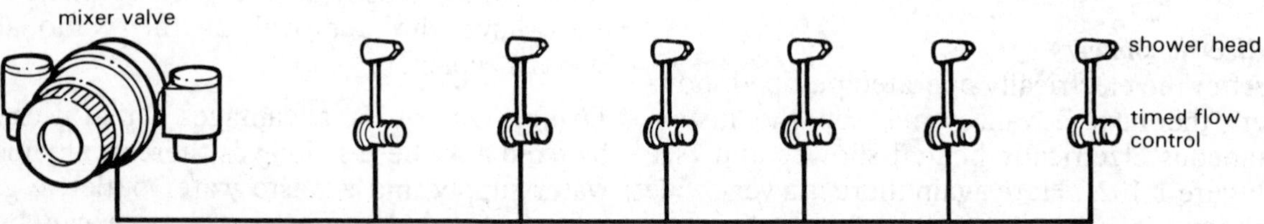

Figure 1.103 *Shower range*

connected to the system. The need to provide a hot surface *throughout the whole year* for the drying and airing of towels and clothing will dictate the position on the system where it should be connected.

Let us consider a domestic hot water and heating system (refer to Figure 235, Book 1). You will see that you have in reality two separate systems, one supplying hot water to the domestic appliances, the other supplying hot water to the heat emitters (radiators). At first it may appear that the towel rail, acting as it does as a heat emitter, should be fixed on the heating circuit. This is not the case for, as stated above, it must function *all the year* to carry out its function of drying the towels. The position must therefore be on that part of the system which is hot throughout the *whole year* and that is the domestic hot water side of the system.

Towel rail connected to direct domestic hot water system

There are several methods of connecting the towel rail to a direct hot water system, each with some minor problem. The method illustrated in Figure 1.104 is perhaps the best. In this case, the towel rail would be sited as near as possible to the primary circulators, the branch flow to the towel rail being taken from the side of the main flow to hot storage and then returning via the main return to the boiler.

Water contains air which is liberated during the heating process. It then passes along the flow pipe into the cylinder and up the vent, and it is this liberated air which could prove to be a problem unless precautions are taken. There are again two or three ways to solve this problem, by means of different ways of connecting the flow pipe, or

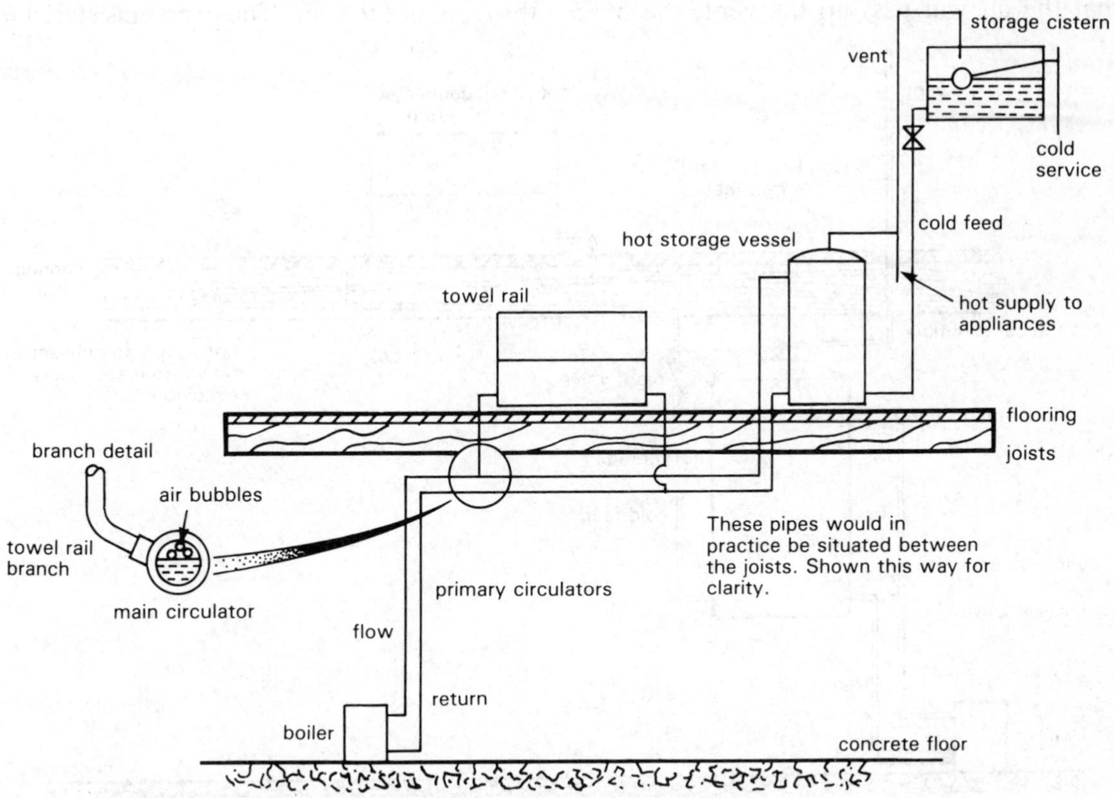

Figure 1.104 *A method of connecting towel rail to primary circuit*

manually releasing the air (but this is tedious and not recommended) or the fixing of a permanent small diameter vent pipe from the air vent connection. In my opinion the best method is to connect the branch flow into the side of the main circulating pipe (see branch detail in Figure 1.104). It will be noted that the liberated air floats along the top underside of the pipe and it therefore passes over the branch pipe connection, provided that the branch is not too vertical. It will be found in practice that little or no troublesome air will find its way into the towel rail. An advantage of this method of connection is that the towel rail becomes heated almost immediately after the boiler is ignited and does not suffer from fluctuation.

An alternative method is to connect the towel rail into a secondary circulation as shown in Figure 1.105. Although this method will overcome the air problem due to the fact that the air will pass up the vent, the main problem is that it is ineffective until much of the water in the hot storage is heated. You therefore experience a longer period of time before the whole system and then the secondary circulation becomes really hot. The temperature of the towel rail is also affected when all or most of the hot water is drawn for domestic use.

In the case where the hot storage vessel is on the same level as the towel rail, the secondary circulation return can be connected into the primary return but it *must* be fitted with a *non-return valve* in addition to the night valve.

Heating coils

A heating coil would probably be required to serve the same function as that of the towel rail, or it may be required to heat an airing cupboard. Whatever its requirements, it will most certainly need to be in operation throughout the year. The problems stated for

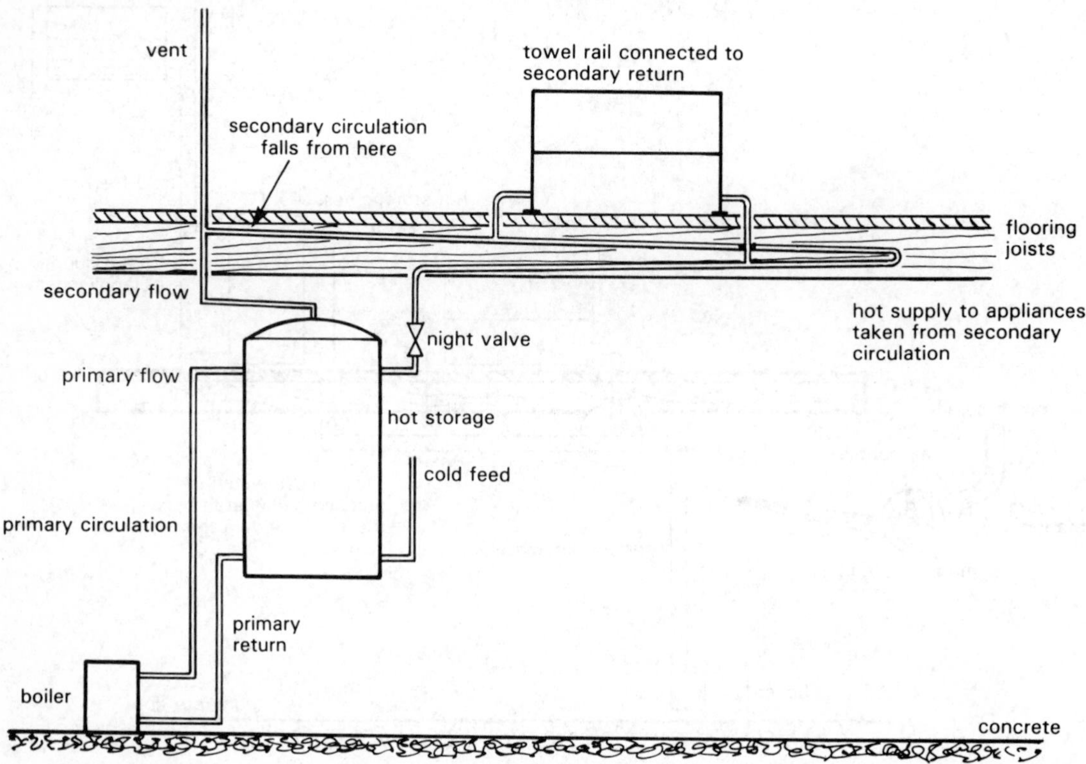

Figure 1.105 *Towel rail connected to secondary circuit*

the towel rail are equally applicable to the heating coil and the method of connection and air treatment will need the same treatment. Figure 1.107 shows one method of connecting the heating coil to a domestic hot water system where the coil forms a loop from a point in the vent pipe and reconnects back into the system in the top third of the hot water storage vessel. The only real disadvantage of this method of connection is that the heating coil will not become hot until the contents of the hot storage vessel becomes hot.

Faulty hot water supply

The main causes of a poor flow of hot water to appliances are generally the following:

1 Insufficient head,
2 Incorrect pipe sizing,
3 Incorrect connection and installation of pipes.

Insufficient head

This is the case for bungalows and flats where the head of water (cistern) is on the same level as the appliances, particularly when the bathroom may be situated some distance away from the hot storage and cistern. Figure 1.108 illustrates a typical bungalow layout, indicating the problems and how they may be overcome.

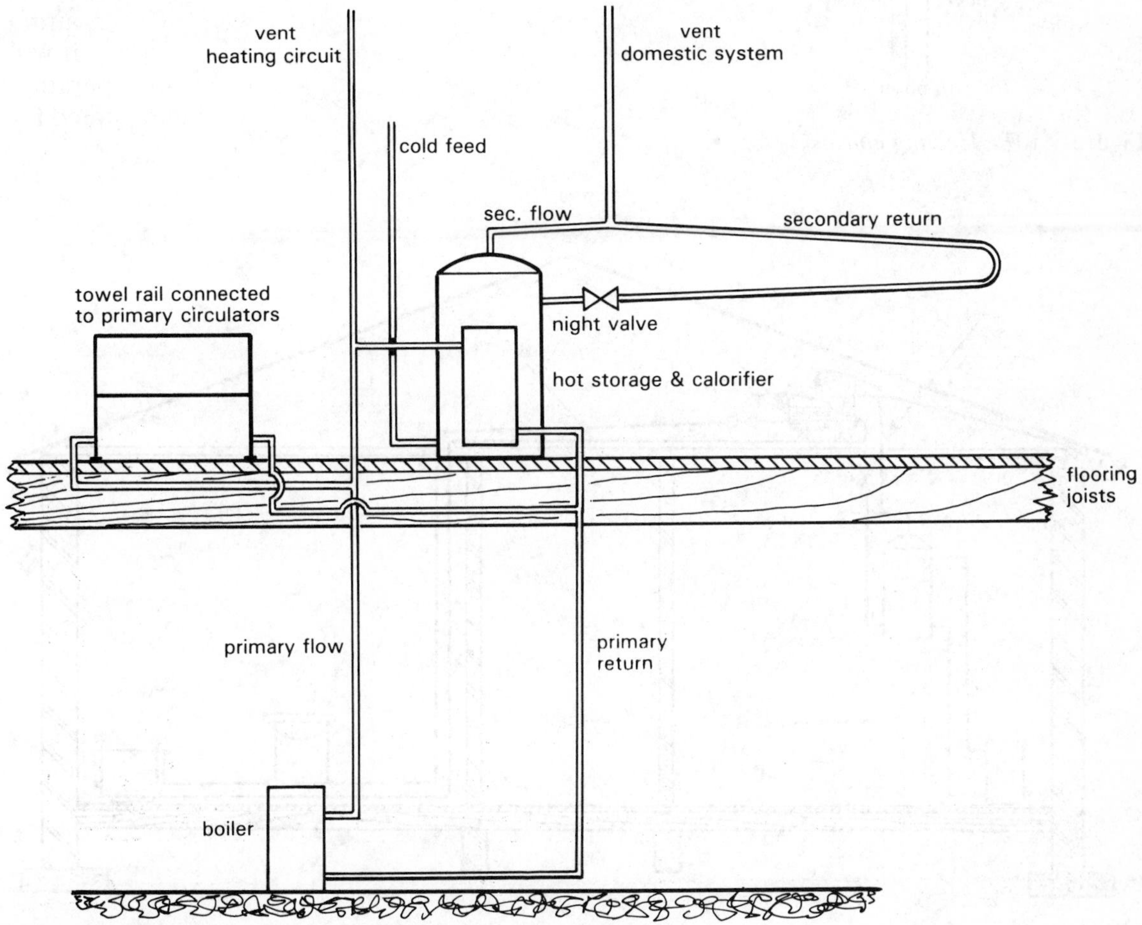

Figure 1.106 *Towel rail connected to primary circuit*

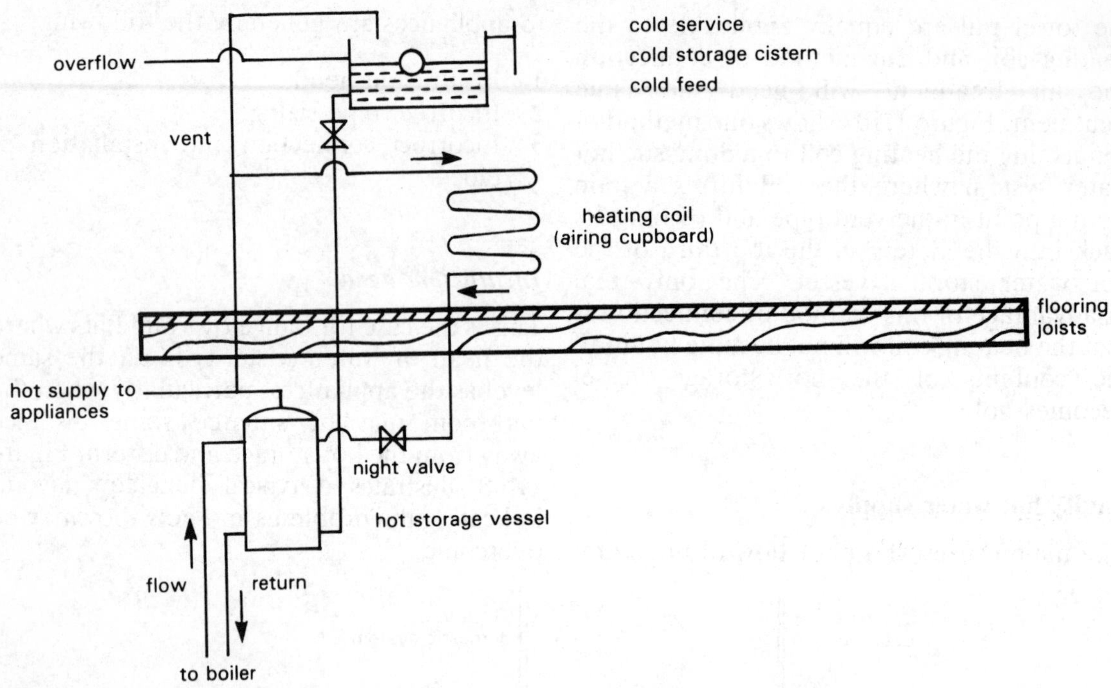

Figure 1.107 *Heating coil*

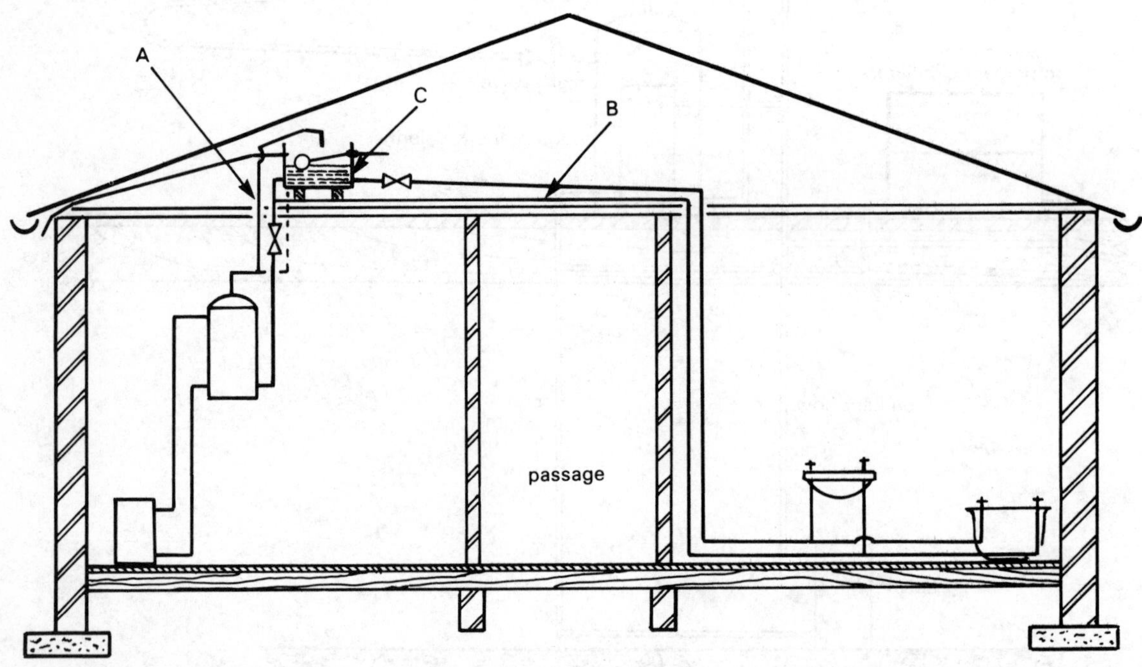

Figure 1.108 *Typical bungalow layout*

(A) Connection of hot water supply to appliances as shown: when the hot water is drawn the water level in the vent pipe will fall. This could result in air being drawn into the system causing spasmodic flow and a risk of complete blockage due to an air lock.

Methods of overcoming problem A

1 The connection to be made as near as possible to the top of the cylinder as shown by the dotted line in Figure 1.108, then taken up into roof space as previously,
2 Where possible the distribution pipes to be taken down below the floor (suitably insulated and to rise to the appliances).

(B) Pipes not laid to a fall

All pipework should be laid to falls whether it be to facilitate the removal of air, so preventing air locks, or the complete draining of a system for maintenance work or frost damage protection.

Methods of overcoming problem B

By careful design and fully supporting and clipping all horizontally placed pipes ensuring that they have a slight fall.

(C) Position of cistern

This position gives very little head pressure, particularly if fitted in a room.

Methods of overcoming problem C

Raise the level of the cistern as high as is practical.

Other factors to be considered:

1 The supply to the bath to be not less than 22 mm tube,
2 Use only fullway control valves,
3 The grouping of appliances (pipe runs should be as short as possible),
4 The cold feed to be one size larger in diameter than the hot distribution pipe, as water can only run out as fast as it can run in (no restrictions in pipework, elbows etc.).

Self-assessment questions

Heating

1 A chromium-plated towel rail having a panel radiator made of ferrous metal inset and connected to a secondary flow and return could cause:
(a) stagnation of the water
(b) discoloration of the water
(c) a reduced flow rate
(d) noise in the system

2 When connecting an oil-fired boiler to an existing brick flue it is necessary to:
(a) parge the inside of the flue with Portland cement
(b) parge the inside of the flue with fire cement
(c) install an aluminium flue liner
(d) install a stainless steel flue liner

3 The type of isolating valve fitted to each side of a domestic central heating pump must allow:
(a) sludge to be removed
(b) instantaneous shut off
(c) the system to be drained
(d) full flow of water

4 In a pumped system of heating the necessity for frequent venting of radiators may be due to:
(a) feed/expansion cistern fixed too low
(b) incorrect position of radiators
(c) air being drawn through the flue pipe
(d) the position of the safety valve

5 A pressed steel radiator should not be connected to a secondary flow and return because:
(a) there will be repeated air locking
(b) it will not heat up quickly
(c) electrolytic action will take place
(d) it will always be cold at the bottom

6 Which one of the following groups of features is most important to the householder in the way a heating system works:
(a) high velocity flow, maximum heat output, minimum installation cost
(b) low velocity flow, high temperature emitters, surface thermometers
(c) freedom from noise, temperature control, economy in fuel
(d) full pump control, low boiler temperature, restriction on venting

7 A bi-metallic thermostat operates by the expansion and contraction of two different metals, one of which has a low coefficient of expansion. It is:
(a) copper
(b) invar steel
(c) aluminium
(d) stainless steel

8 Heating temporary hard water above 60 °C causes:
(a) electrolysis
(b) erosion
(c) encrustation
(d) corrosion

9 The presence of encrustation in a boiler results in the:
(a) increase of circulation pressure
(b) increase of electrolytic action
(c) reduction of heat transfer rate
(d) decrease of plumbo-solvency

10 Hard waters contain:
(a) common salt
(b) calcium salts
(c) ferric salts
(d) saline salts

Self-assessment questions

Domestic hot water

1 On an indirect system of hot water supply a suitable position for the safety valve is on the:
(a) primary return
(b) secondary return
(c) secondary flow
(d) cold feed

2 An airlock is used to advantage in a:
(a) single feed indirect cylinder
(b) room sealed gas boiler
(c) balanced flue heater
(d) flue draught stabiliser

3 In a hot water supply installation the primary circulation is between:
(a) towel rail and cylinder
(b) boiler and cylinder
(c) cylinder and taps
(d) feed cistern and cylinder

4 The main reason for installing secondary circulation to a hot water supply system is to:
(a) achieve economy in pipework
(b) give a better flow rate to the appliance
(c) avoid using larger diameter pipework
(d) avoid long dead legs

5 The connection of the secondary return to a hot water cylinder should be:
(a) adjacent to the secondary flow
(b) level with the primary return
(c) opposite the cold feed pipe
(d) not more than a third from the top of the cylinder

6 In a temporary hard water area, which one of the following types of hot water system should be used:
(a) a tank system
(b) the direct system
(c) the indirect system
(d) a combination tank system

7 The recommended maximum temperature for a hot shower is:
(a) 24 °C
(b) 40 °C
(c) 60 °C
(d) 105 °C

8 The diameter of the cold feed pipe in a domestic hot water system should be:
(a) The same diameter as the supply to the feed cistern
(b) The same diameter as the bath supply
(c) Not less than the diameter of the service pipe
(d) Not less than the diameter of the hot water draw off pipe

9 In temporary hard water districts the concentration of scale in hot water systems occurs mainly in:
(a) the cold feed to the cylinder
(b) boilers and circulating pipes
(c) the hot water draw off
(d) the cold feed cistern

10 Heat is transferred through a metal by:
(a) convection
(b) radiation
(c) emission
(d) conduction

2 Sanitation and sanitary pipework systems

After reading this chapter you should be able to:

1. Demonstrate knowledge of the main provisions of the Building Regulations and British Standard Codes of Practice relating to sanitary pipework systems.
2. Identify different types of traps, understand and state the reasons for loss of seal in traps.
3. Describe preventative measures to eliminate loss of water seal in traps.
4. Recognise and identify systems of above-ground discharge pipework and sanitation for dwellings, small industrial, commercial and public buildings.
5. Name common materials used for pipework systems and understand jointing techniques, including connections between above-ground and below-ground discharge systems.
6. Make line diagrams to illustrate the layout of above-ground discharge pipework and sanitation systems including appliances and components.
7. Recognise and identify different types of urinal.
8. State the purpose of domestic waste disposal units and describe the operating cycle of these units.
9. State the purpose and functional requirements of above-ground discharge pipework and sanitation systems.
10. Describe and apply the correct procedures for testing and commissioning sanitary pipework installations.

Terminology

To enable the reader to understand sanitary pipework systems, some knowledge of the terms used is necessary.

The Building Regulations and BS 5572 (1978), give the following definitions relevant to the design and installation of soil and waste discharge systems:

Definitions

1. *Access cover* A removable cover on pipes and fittings providing access to the interior of pipework for the purposes of inspection, testing and cleansing (see Figure 2.1).
2. *Branch discharge pipe* A discharge pipe connecting sanitary appliances to a discharge stack (see Figure 2.1).
3. *Crown of trap* The topmost point of the inside of a trap outlet (see Figure 2.1).
4. *Discharge pipe* A pipe which conveys the discharges from sanitary appliances (see Figure 2.1).
5. *Soil appliances* Fittings which receive the waste products of the human body, including water closet, urinal and slop sink.
6. *Soil pipe* Means a pipe (not being a drain) which conveys soil water either alone or together only with waste water, or rainwater or both (this term is not

generally used now, the pipe being designated a discharge pipe).

7 *Stack* A main vertical discharge or ventilating pipe (see Figure 2.1).

8 *Trap* A fitting or part of an appliance to retain water or fluids so as to prevent the passage of foul air (see Figure 2.1).

9 *Ventilation pipe* Means a pipe (not being a drain) open to the external air at its highest point (see Figure 2.1), which ventilates a drainage system, either by connection to a drain, or to a soil pipe or waste pipe, and does not convey any soil water, waste water or rainwater.

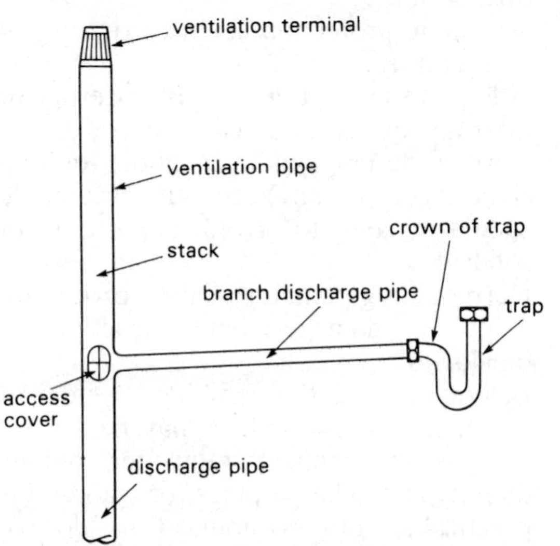

Figure 2.1 *Definition of sanitation terms*

10 *Waste appliance* A fitting which receives waste water, including wash basins, bath, shower tray, bidet, drinking fountain, sink.

11 *Waste pipe* Means a pipe (not being a drain, or overflow pipe) which conveys waste water, either alone or together only with rainwater.

Regulations relevant to sanitary discharge pipe systems

Regulations are necessary in the construction and building industry to ensure that a structure and its component parts are suitable and safe for the purpose for which they are designed. Building Regulations cover the main legislation relating to building drainage for above- and below-ground systems of discharge pipework. The following is a summary related to above-ground discharge pipework and sanitation systems.

(a) All discharge pipes must be of adequate size for their purpose and must not be smaller in diameter than that of the largest trap or branch discharge pipe connected to and discharging into it.

(b) Provision must be made, where necessary, to prevent the loss or destruction of trap seals.

(c) All pipes and fittings used for the discharge of soil or waste and the ventilation of above-ground discharge systems must be made of suitable materials, these having the required strength and durability for this purpose.

(d) All joints must be made in such a way as to avoid obstructions, leaks or corrosion.

(e) Bends must have an easy radius and should not have any change of cross-sectional area throughout their length.

(f) Pipes must be adequately secured to the building structure or fabric without restricting their movement due to thermal expansion or contraction.

(g) The discharge system must be capable of withstanding an air test when subjected to a minimum pressure equivalent to 38 mm head of water for 3 minutes minimum.

(h) Pipework and components must be accessible for repair and maintenance, and means of access must be provided for clearing blockages in the system.

(i) Every sanitary appliance must be fitted with a suitable trap as close as possible to its outlet. Each trap must have an adequate water seal and access for cleaning. This does not apply to appliances which have an integral trap as part of the appliance, e.g. water closets, or where appliances such as wash basins, sinks or baths are connected to form a range and each appliance discharges into an open half-round channel which discharges to its own trap – usually a gully – or where the range connects to a common waste pipe which is itself trapped.

(j) No discharge pipes on the exterior of a building may discharge into a hopper head or above the grating of an open drain inlet or gully.

(k) Discharge pipes carrying soil or waste water from a sanitary appliance must not be fitted on the exterior of a building except (i) on low-rise buildings of up to three storeys in height; or (ii) where the building was erected before the 1976 Building Regulations came into force, and the discharge system is being extended or altered; or (iii) on the lower three storeys of a high-rise building.

Codes of Practice related to sanitation and sanitary pipework systems

In addition to the Building Regulations, with which it is obligatory to comply, the other main document dealing with Sanitary Pipework is British Standards Code of Practice 5572: 1978. This code is based on the findings of research conducted mainly by the Building Research Centre over many years. Although the Code is only a recommendation it has become the basis for the design and installation of modern sanitary pipework systems and consequently its main recommendations are explained in this chapter.

Above ground discharge pipework and sanitation systems

General principles

Soil and waste pipe discharge systems should comprise the minimum of pipework necessary to carry away the foul and waste water from the building quickly and quietly and with freedom from nuisance or risk of injury to health. The system should satisfy the following requirements:

1 Efficient and speedy removal of excremental matter and urine plus other liquids and solids without leakage;
2 The prevention of ingress of foul air to any building whilst providing for their escape from the pipework into a 'safe' position;
3 The adequate and easy access to the interior of the pipe for the clearance of obstructions;
4 Adequate protection against extremes of temperature;
5 Adequate protection against external or internal corrosion attack;
6 Correct design and installation to limit siphonage (if any) to an acceptable standard, and to avoid deposition of solids;
7 Correct design and installation procedure to prevent damage from obstructions or blockages;
8 In areas where the combined system of drainage is permitted, it may be advantageous to connect rainwater outlets directly to discharge pipes, providing it is practicable and economical to do so; ventilation must be able to take place even if the rainwater outlet is obstructed or blocked;
9 Economy and good design are essentials: both are aided by compact grouping of sanitary appliances in both horizontal and vertical positions.

Domestic sanitation systems (historical development)

The function of a well-designed soil and waste discharge system in a building is to take away efficiently all waste from the sanitary fitments to the main drains, without allowing foul air to enter the building via the system of sanitation pipework.

Three basic systems have evolved over the years to fulfil these requirements. The order in which the systems were developed was:

1. The two-pipe system (see Figure 2.2),
2. The one-pipe system (see Figure 2.3),
3. The single-stack system (see Figure 2.4).

Each system is different, and consequently has its own respective merits and limitations, which must be observed to ensure correct functioning.

The two-pipe system

From the beginning of this century until the late 1930s most of the buildings above one storey in height utilised the two-pipe system of sanitation. In this system, the discharges from the ablutionary fitments, such as baths, bidets, wash basins, showers, sinks, etc., were kept separate from the discharges from soil appliances such as water closets, urinals, and slop sinks.

1. The waste stack received the discharge from the ablutionary fitments and conveyed this to ground level where it was delivered above the water seal in a trapped gully connected to the drainage system;
2. The soil stack received the discharge from soil appliances and delivered it direct to the underground drainage system.

The waste and soil waters did not combine until they reached the below-ground drainage system.

All pipework was fully vented to avoid the

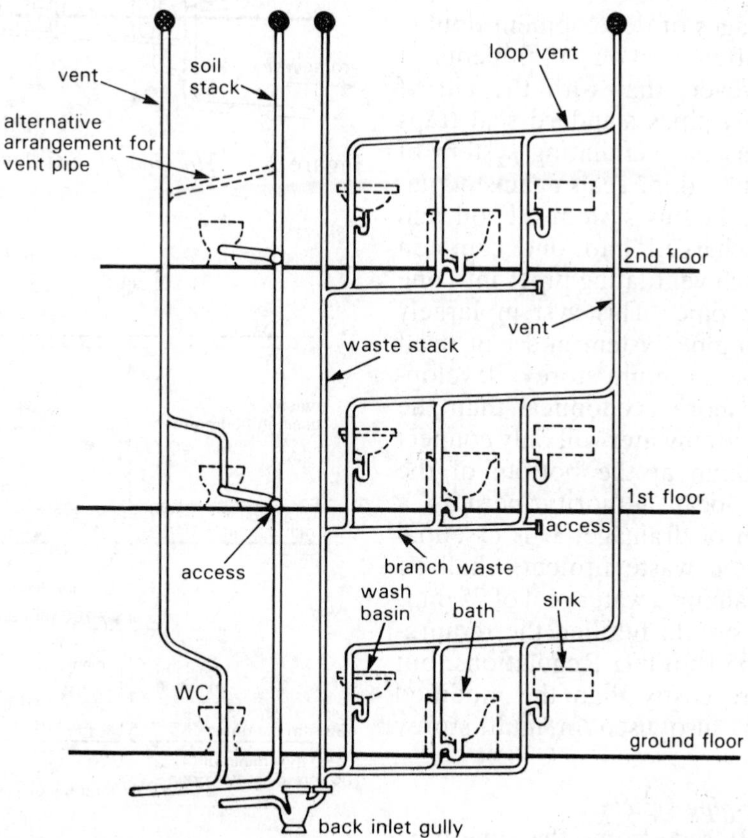

Figure 2.2 *Two-pipe or dual-pipe system*

unsealing of traps and the risk of foul gases entering the building. Because the waste system was fully ventilated, shallow seal traps were adequate on baths, sinks, wash basins, etc. The two-pipe system functioned efficiently, but due to the duplication of pipework and the excessive labour involved in installation, it was expensive and was gradually replaced by the one-pipe system.

The one-pipe system

Since the late 1930s this system of sanitation has grown in popularity, particularly in multi-storey buildings. Originally used for domestic installation, it had virtually become standard practice in the non-domestic sector (hospitals, schools, offices, etc.). The name of the system is misleading, as two stack pipes are required, one for the combination of waste and soil, and the other as the main ventilating stack.

In the early stages of development double seal traps were often used on soil fitments. It was found, however, that with the aid of ventilating (relief) pipes standard seal traps could be used, as the ventilating system of pipework safeguarded the seals against variation in pressure. In this system all soil and waste water discharged into one common pipe and all branch ventilating pipes into one main ventilating pipe. This system largely replaced the two-pipe system and lent itself very well to use in multi-storey developments. It is far more economical than the two-pipe system. Rainwater will only connect into the same drain at the bottom of the building if the local authority permits a combined system of drainage. It is essential that all traps from waste fitments shall be capable of maintaining a water seal of 75 mm.

The one-pipe system fulfilled the requirements of the 1965 Building Regulations, but still proved more costly than the modified one-pipe system, also used in multi-storey buildings.

The modified one-pipe system

This system differed from the one-pipe system only in the provision of branch

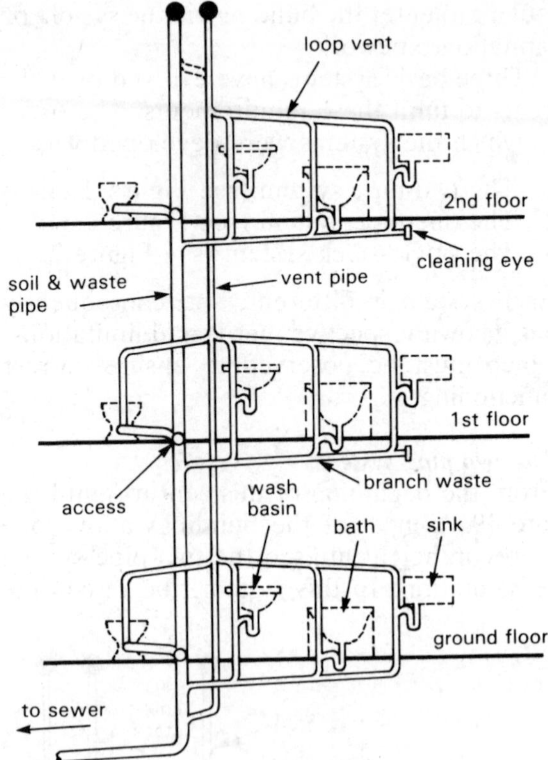

Figure 2.3 *The one-pipe system*

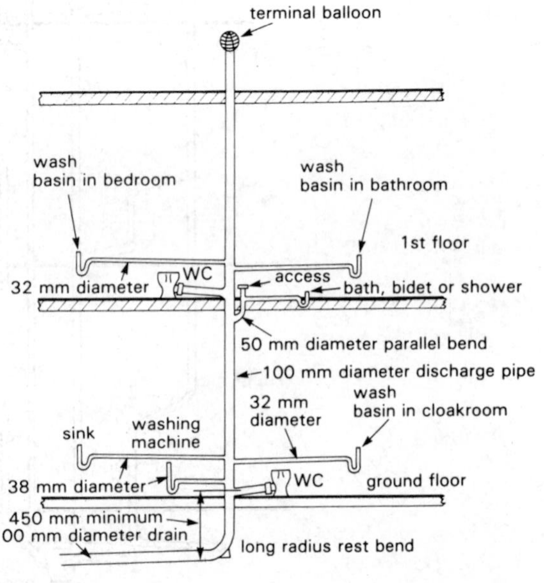

Figure 2.4 *Single stack system*

ventilating pipes. The modified system is extensively used in multi-storey buildings, and, where advantage can be taken of the design considerations of the single-stack system, enables certain branch ventilating pipes to be omitted.

The single-stack system

The single-stack system is a one-pipe system from which, subject to the observance of certain rules, all or most of the trap ventilating pipes are omitted.

The single-stack system was developed by the Building Research Establishment and was formerly identified as the simplified system. The research which commenced in the mid 1940s soon proved that the unsealing of traps did not occur as readily as had been presumed previously and that, under certain circumstances, trap ventilating pipes could be omitted from the one-pipe system. It also proved that when properly designed it becomes basically a one-pipe system without vent pipes, hence the name *single stack*.

Originally this system was developed for low-cost housing, but, having proved successful, it was further developed for multi-storey buildings. The design of the pipework is very important and success relies upon close grouping of single appliances (each with a separate branch) around the stack. Providing the design is correct, vent pipes can be omitted except venting via the main stack, which is continued upwards to a safe position. The single-stack system is a simple system particularly suitable for housing, and for multi-storey dwellings distinct savings are claimed over the one-pipe system. The single-stack system is the most common discharge pipe system used for domestic properties.

Modern above-ground discharge systems

BS 5572 (1978) describes and classifies discharge systems within three categories which are as follows:

Ventilated system (Figure 2.5) This system is used in situations where there are large numbers of sanitary appliances either in ranges or where they have to be widely dispersed and it is impracticable to provide discharge stacks in close proximity to each appliance.

Trap seals are safeguarded by extending the discharge and ventilating stacks to the atmosphere and providing individual ventilating pipes to each branch discharge pipe. To prevent back pressure at the base of the stack the vent stack is connected either into the vertical discharge stack near to the bend or into the horizontal drain.

Ventilated stack system (Figure 2.6) A ventilated stack system is used in situations where close grouping of sanitary appliances makes it practicable to provide branch discharge pipes without the need for individual ventilating pipes.

Trap seals are safeguarded by extending the stacks to the atmosphere and by cross-connecting the ventilating stack to the discharge stack on each or alternate floor levels, depending upon the discharge stack loading.

Single-stack system (Figure 2.4) A single-stack system is used in similar situations to those applicable to the ventilated stack system. Sanitary appliances must be connected within the limitations and recommendations for this system and the discharge stack must be large enough to limit pressure fluctuations without the need for a separate ventilating stack. A modified single-stack system (see Figure 2.7), providing ventilating pipework extended to the atmosphere or connected to a ventilating stack, can be used where the location of appliances on branch discharge pipes could cause loss of their trap seals. The ventilating stack need not be connected directly to the discharge stack and can be smaller in diameter than that required for a ventilated stack system.

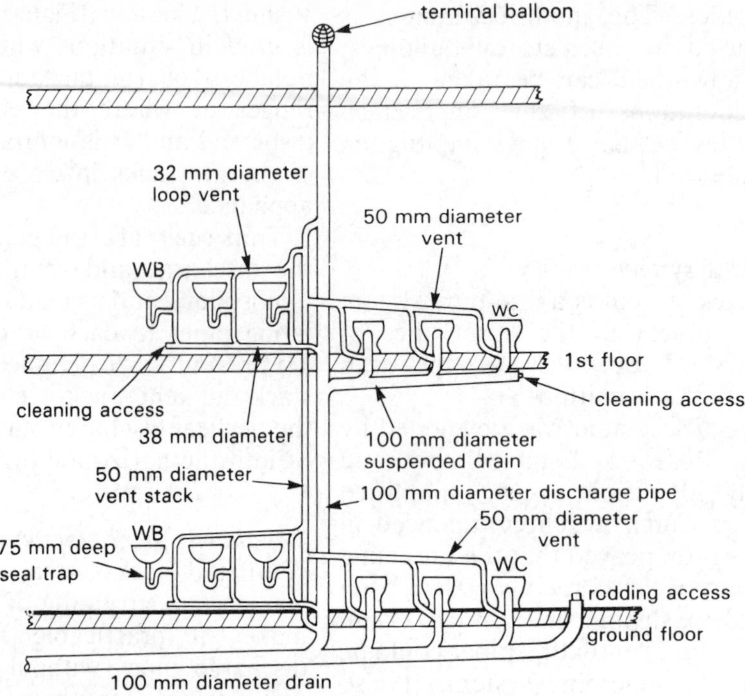

Figure 2.5 *Ventilated or fully vented one-pipe system*

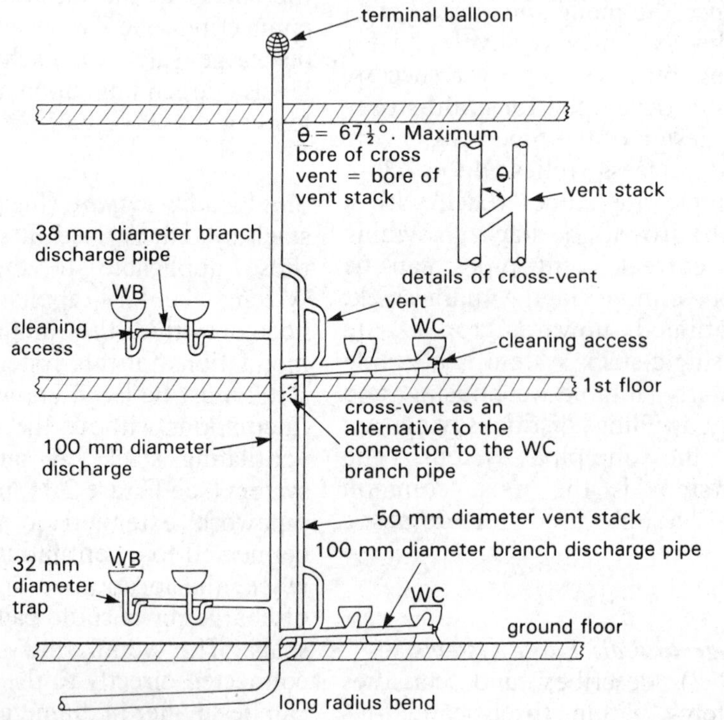

Figure 2.6 *Ventilated-stack system*

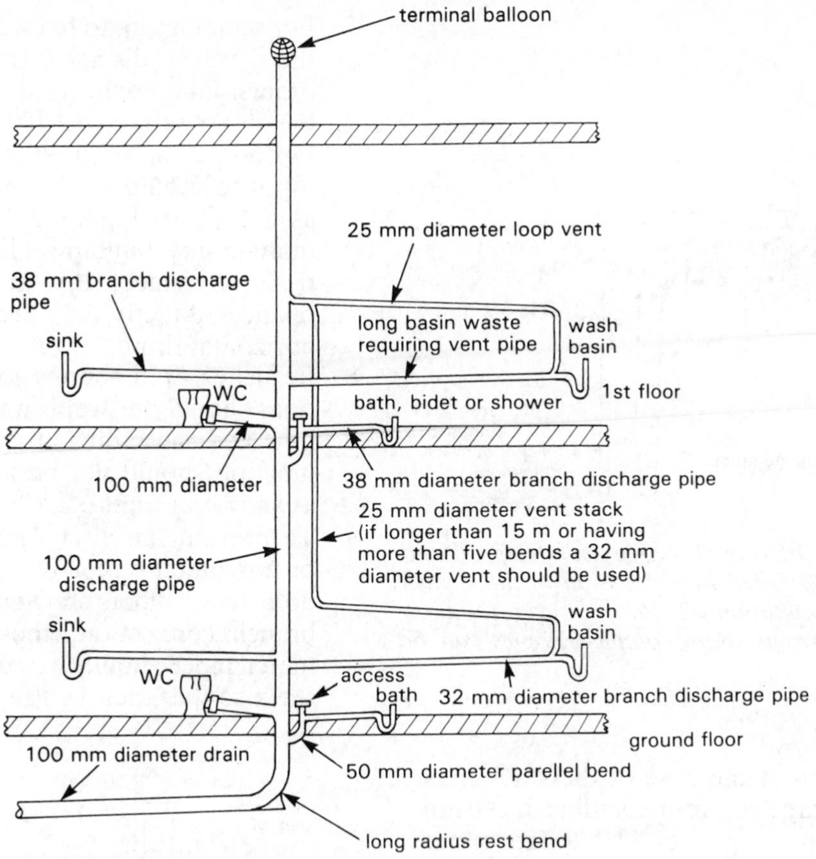

Figure 2.7 *Modified single-stack system*

In modern installations the single-stack system is used wherever possible. The absence of ventilating pipes makes it less costly to install and its compact characteristics allow for it to be located within internal plumbing and pipe ducts. For domestic buildings of up to five storeys having one or two standard groups of sanitary appliances on each floor, a 100 mm diameter main discharge stack is usually adequate. The main requirements for this system are as follows:

1 Sanitary appliances must be arranged close to the discharge stack so that the branch waste discharge pipes are as short as possible.
2 Sanitary appliances must be individually connected to the main stack.
3 The main stack should be at least 100 mm diameter, except for two-storey housing where 75 mm may be satisfactory, providing the diameter of the WC outlet does not exceed the diameter of the discharge stack.
4 For buildings more than five storeys high, ground-floor appliances should be connected separately to the drain.
5 The discharge pipe must be vertical below the highest sanitary appliance; bends or offsets in the 'wet' portion of the discharge pipe must be avoided. If an offset in the discharge stack is unavoidable it will be necessary to prevent pressure build-up in this section of the stack; this may be achieved by connecting cross vents to the discharge stack close to bend or bends as shown in Figure 2.8.

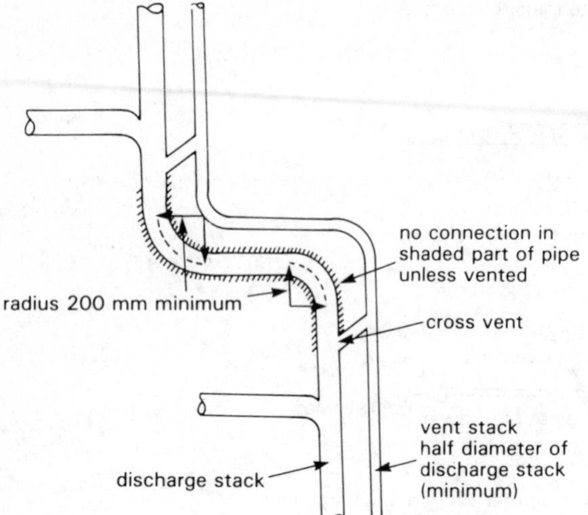

Figure 2.8 *Offset in discharge stack (Reproduced from BS 5572: 1978 by permission of the British Standards Institution, 2 Park Street, London, W1A 2BS, from whom complete copies can be obtained)*

6 The bend at the base of the stack must have a large centre-line radius of 200 mm (minimum) or two 135° large radius bends may be used as shown in Figure 2.9. This procedure is necessary to eliminate the possibility of back pressure at the base of the discharge stack.

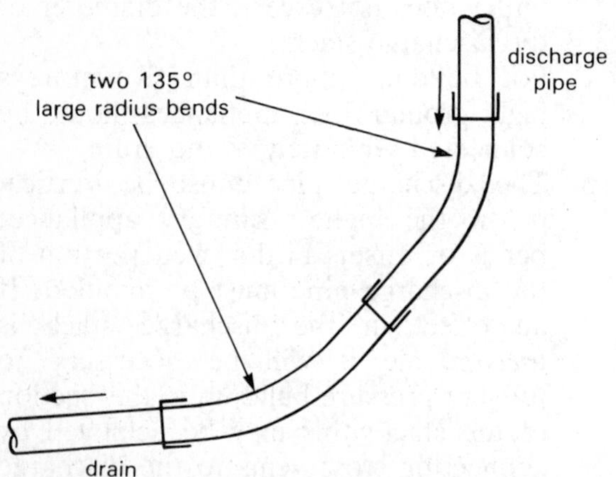

Figure 2.9 *Base of discharge pipe detail*

7 For buildings up to five storeys in height, the vertical distance from the lowest branch inlet connection to the invert of the drain should be 750 mm minimum. For houses up to three storeys high, this distance should not be less than 450 mm as shown in Figure 2.4. For large or multi-storey buildings, it is usual practice to connect the ground-floor appliances directly into the below-ground horizontal drain.

8 Branches or junctions for WC connections should be swept in the direction of flow and the radius at the invert of the junction should not be less than 50 mm as shown in Figure 2.10a.

9 To prevent the discharge from a P trap or horizontal outlet WC branch backing up a bath, bidet, shower or wash basin branch connection, these smaller diameter pipes should be connected to the stack so that their centre line meets that

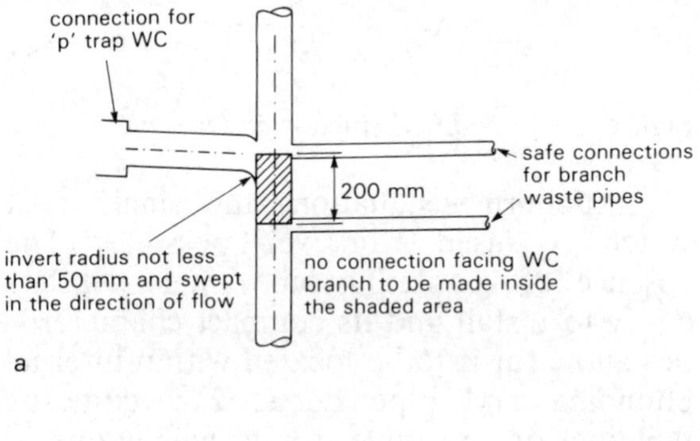

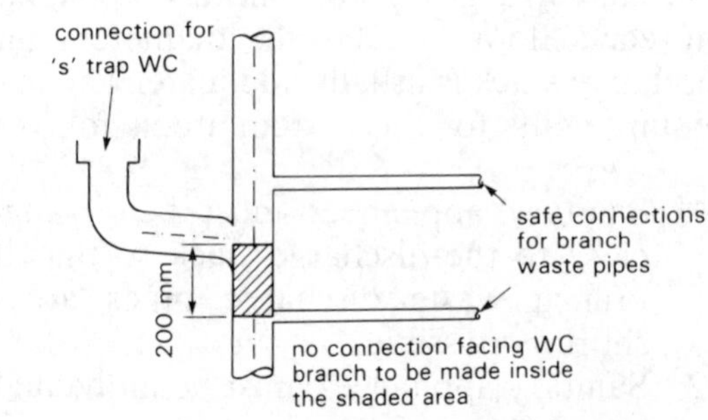

Figure 2.10 *Connections near WC branch*

of the WC branch or is above this level. Alternatively the centre line of these branch discharge pipes should be at least 200 mm below that of the WC branch connections (see Figure 2.10b).

10 To eliminate the possibility of crossflow between branch discharge pipe connections the minimum distances between the centre lines of the opposing connections should be as shown in Table 2.1; alternatively a collar boss fitting (see Figure 2.11) may be used to prevent backing up or crossflow in situations where several connections have to be made in close proximity to each other.

Table 2.1

Stack diameter	*Distance between opposing connections*
75 mm	90 mm
100 mm	110 mm
125 mm	210 mm
150 mm	250 mm

11 To reduce the possibility of self-siphonage and discharge noise all waste appliances should be fitted with P outlet-type traps.

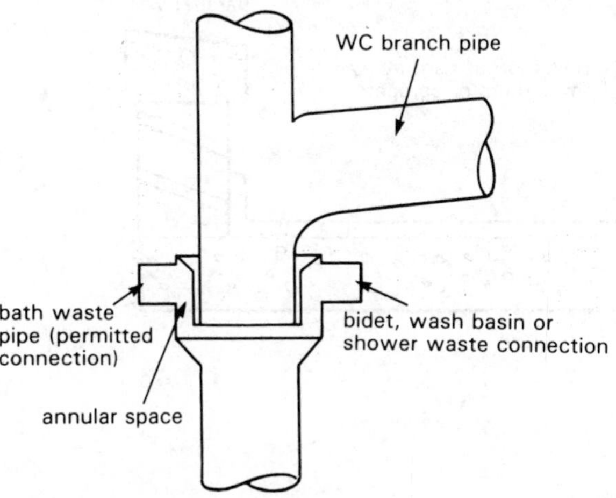

Figure 2.11 *Collar boss fitting*

12 Branch discharge pipes should not exceed the lengths as shown in Table 2.2.

Table 2.2

Appliance	*Maximum length*
Baths	3 m
Shower trays	3 m
Sinks	3 m
Bidets	1.7 m
Wash basins	1.7 m
Wash basins with 38 mm dia. waste pipes	3 m
WCs	6 m
Urinals with 38 mm dia. waste pipes	3 m
(All urinal wastes should be kept as short as possible)	

13 Branch discharge pipes should have falls as shown in Table 2.3.

Table 2.3

Appliance	*Fall/gradient*	
Baths	1°–5°	18–90 mm/m
Shower trays	1°–5°	18–90 mm/m
Sinks	1°–5°	18–90 mm/m
Bidets	1°–2½°	18–45 mm/m
Wash basins	1°–2½°	18–45 mm/m
WCs	1° minimum	
Urinals	1°–5°	18–90 mm/m

14 The discharge stack must be terminated at its highest point:

(a) With a domical cage or mesh balloon which does not restrict the flow of air through it.

(b) In a safe position so that foul air does not cause a nuisance or health hazard. This is usually achieved if the termination of the stack is not less than 900 mm above the top of any ventilating window, or other opening into the building that is within a horizontal distance of 3 m of the termination point as in Figure 2.12a.

(c) The termination position should be away from the exposed edges of parapet walls or the corners of buildings. This is to prevent possible loss of trap seal due to wavering out and the nuisance of noise in the discharge pipework when windy conditions prevail (see Figure 2.12b).

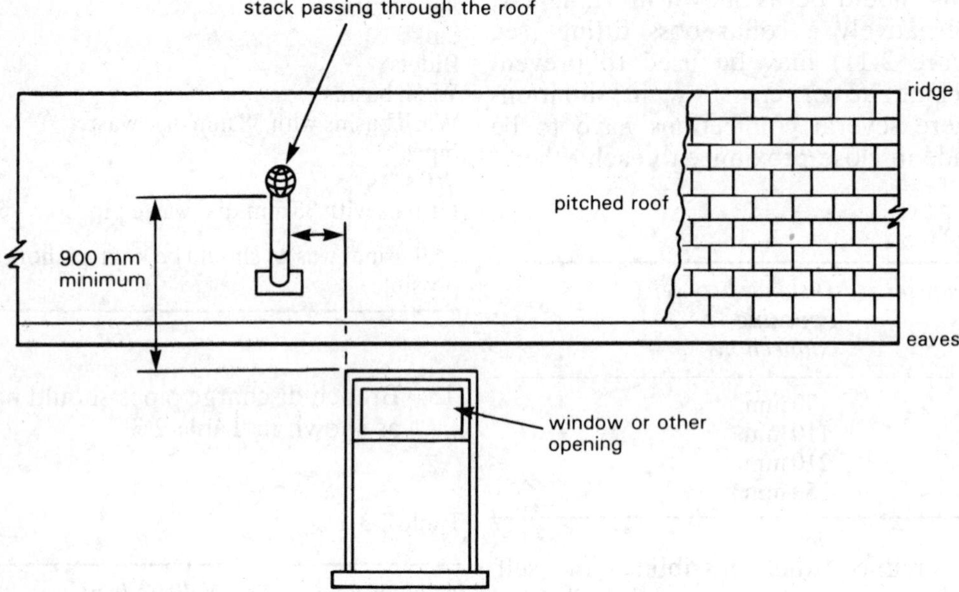

(a) pitched roof. (Requirement if the vent is within 3m of the opening into the building)

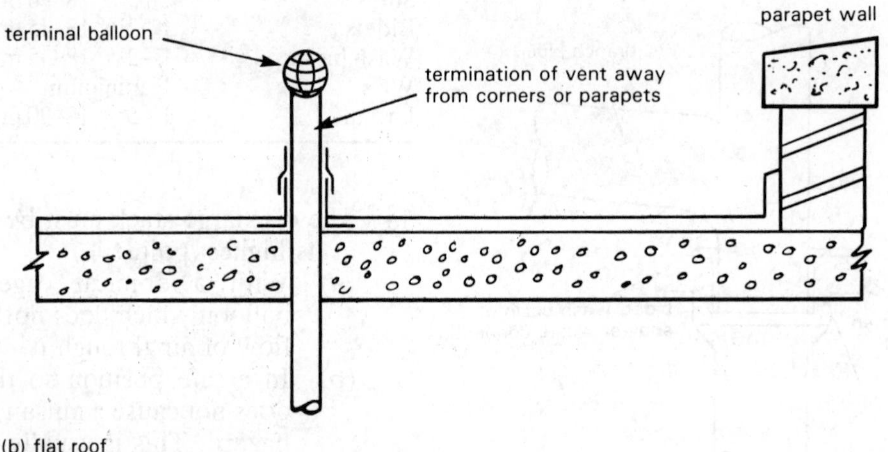

(b) flat roof

Figure 2.12 *Termination of ventilating stacks*

Loss of water seal in traps

The loss of a trap seal will usually result in foul and objectionable gases and odours entering a building from the system of discharge pipework. These gases will at least be a nuisance and in extreme cases create a health hazard to the occupants of the building. It is for these reasons that much importance is placed upon retention of the water seal in all traps.

Trap seals may be lost in several ways. Some reasons may be attributed to bad system design or installation techniques, others are more natural or the result of particular circumstances. It is important that a plumber understands the particular circumstances which may cause the loss of water seal, the common circumstances are:

1 Self-siphonage,
2 Induced siphonage,
3 Compression or back pressure,
4 Capillary action,
5 Evaporation,
6 Leakage,
7 Momentum,
8 Wavering out.

Self-siphonage

This is caused by a moving plug or charge of water running out of a steep-sided sanitary appliance such as a wash basin. The plug of water moves through the trap, pushing the air on the outlet side of the trap in front of it, thereby creating a partial vacuum in the branch discharge pipe which causes siphonage to occur and loss of the trap seal. The technical term for the partial vacuum or negative pressure zone is 'hydraulic jump' as illustrated in Figure 2.13. Self-siphonage of a trap seal is usually indicated by an excessive amount of noise as the sanitary appliance discharges its final quantity of waste water. Self-siphonage is usually prevented by one or more of the following:

1 By fitting a P trap to the sanitary appliance (thus avoiding vertical branch discharge pipes),
2 Ensuring that the branch discharge pipe length and slope do not exceed those recommended in BS 5572: 1978,
3 By fitting a ventilating or anti-siphon pipe adjacent to the trap outlet,
4 By fitting a larger diameter branch discharge pipe to the trap outlet,
5 By correct design of sanitary appliance,
6 By fitting a resealing or anti-siphon trap to the sanitary appliance.

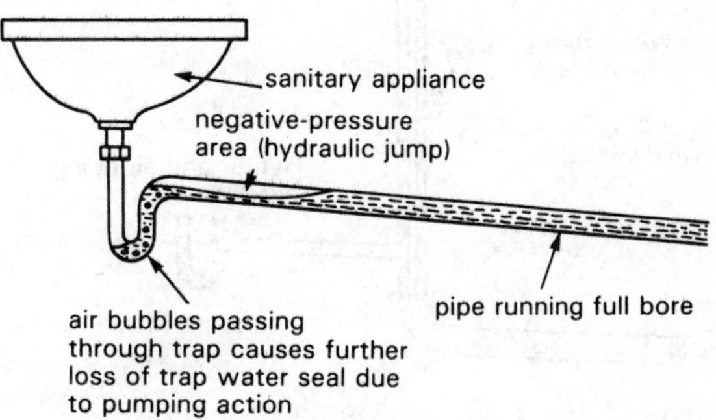

Figure 2.13 *Self-siphonage*

Induced siphonage

This is caused by the discharge of waste water from one sanitary appliance, pulling or siphoning the seal of a trap of another appliance connected to the same branch discharge pipe as shown in Figure 2.14. This form of siphonage is most common in

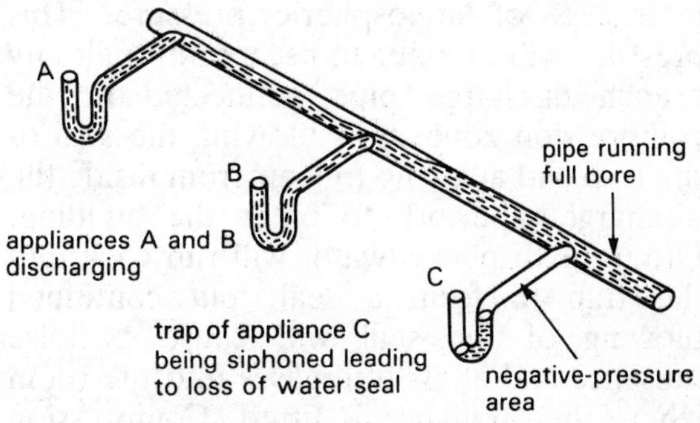

Figure 2.14 *Induced siphonage*

buildings where ranges of appliances are fitted or where it is necessary to connect several appliances to a common branch discharge pipe. The main causes of induced siphonage are poor system design, inadequate pipe sizes or bad installation techniques.

Compression or back pressure

Compression of air at or near the base of a discharge pipe may occur as shown in Figure 2.15. As water flows down a vertical stack it

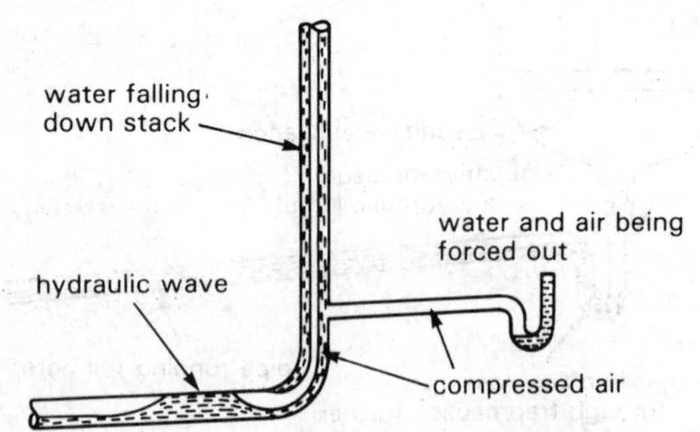

Figure 2.15 *Compression or back-pressure*

takes some air with it. As this water changes direction at the base of the stack (from the outside of the vertical pipe to the invert of the near-horizontal drain pipe) the pipe bore becomes momentarily full of water, so preventing the free flow of air up and down the discharge stack. Under these conditions a hydraulic wave is formed and air at the base of the stack is compressed by the water falling from above, thus creating air pressure in excess of atmospheric pressure. This pressure will attempt to escape through any branch discharge pipe connected in the compression zone, thus blowing the seal of the trap and allowing foul air from inside the discharge pipework to enter the building. Often the displaced water will run back into the trap to form a seal, but continued blowing of the seal will cause a noise nuisance and allow foul gases into the room where the appliance is fitted. Compression and back pressure are prevented by not fitting small or sharp radius bends or having branch discharge pipes connected near to the base of a discharge stack. Correct practice is to fit large radius bends and to ensure inlet connections are not made close to the base of the stack.

Capillary action

Loss of seal by capillarity occurs when a piece of porous material such as threads or string from a mop or dishcloth are deposited into the water seal and over the outlet invert of a trap, as shown in Figure 2.16. This will most commonly happen to appliances such as kitchen or cleaners' sinks fitted with S pattern traps. Capillarity is prevented by regular cleaning of the inside of the trap and branch discharge pipe, or by laying a loose mesh strainer into the waste fitting of the appliance to catch and retain loose porous strands which may be contained in certain types of waste water.

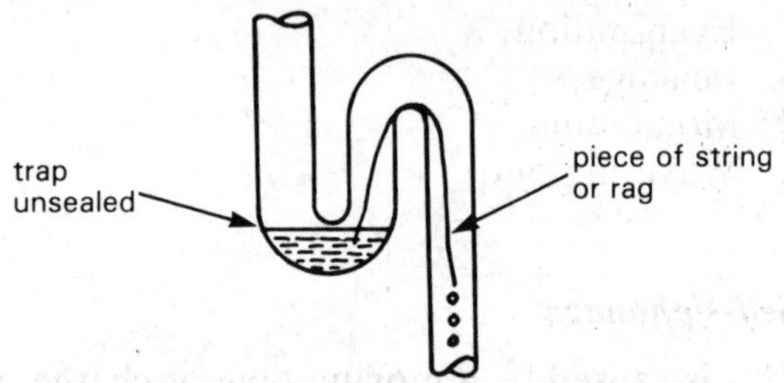

Figure 2.16 *Capillary attraction*

Evaporation

This occurs usually in traps connected to sanitary appliances which are not used regularly or where the ambient temperatures are relatively high. Evaporation is more usual in the summer months when temperatures are higher or when buildings are left empty or unattended due to holiday periods.

Leakage

Traps occasionally loose their seals due to a leak on the fitting below water-seal level. The leak may be due to a loose or badly jointed access bowl or cleaning eye, or, in the case of soft materials, caused by impact resulting in fracture and damage to the trap body.

Momentum
The most usual cause of loss of trap seal by momentum is when a quantity of water is quickly discharged into a gully or washdown WC pan and the discharging water carries away the water forming the trap seal.

Wavering out
The effect of a high-velocity fluctuating wind passing over the top of an exposed discharge stack ventilating pipe (see Figure 2.17) will create varying air pressures and draughts within the pipework system which may cause trap seals to fluctuate or waver, resulting in loss of water from the seal. Fluctuating

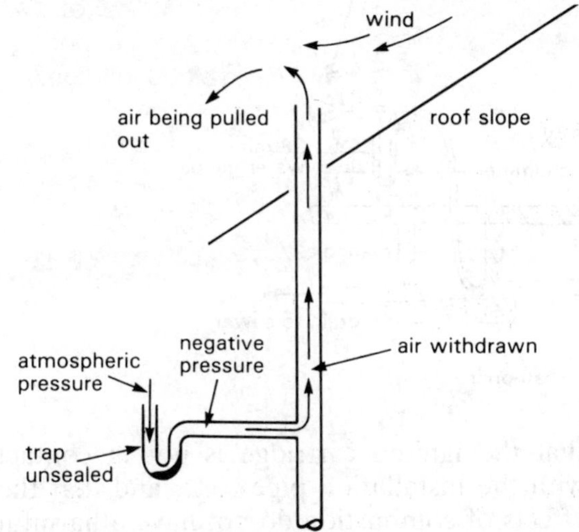

Figure 2.17 *Wavering out*

pressures caused by wind are best prevented by locating the vent terminal position away from exposed locations and ensuring that a terminal grating is fitted to all ventilating pipework.

Inspection and testing

Inspections and testing should be carried out during the installation of a discharge system to ensure that the pipework is properly secured and clear of obstructing debris, and that all work which is to be concealed is sound and free from defects before it is finally concealed. On completion, the discharge system installation should be visually inspected for damage and then tested for soundness and performance.

As mentioned earlier in this chapter, the Building Regulations require that a completed discharge system should be capable of withstanding an air (pneumatic) test for a minimum period of 3 minutes at a pressure equivalent to a head of not less than 38 mm of water. BS 5572: 1978 states the following procedures:

Air test (Figure 2.18) Normally this test should be carried out as one operation, although on large installations it may have to be done in stages or sections or as the work proceeds. The water seals of the traps of all connected appliances should be fully charged and test plugs or inflatable bags inserted into the open ends of the pipework to be tested. To ensure that there is a complete air seal at plugs or bags at the base of the stack, a small quantity of water can be emptied into the system. Plugs located inside vertical ventilating pipework can be sealed by a small quantity of water poured around the outside of the plug. A flexible tube should be fitted to one of the test plugs as shown and be complete with a valved T piece, manometer and air pump. An air test is very thorough and searching and will indicate any leak in the system. It is therefore necessary to ensure that all the testing equipment is well sealed and sound before testing of the pipework commences. Air is then pumped into the system until the required test pressure is recorded on the manometer. The valve adjacent to the air pump should then be closed to isolate the air pump while the test is in progress; the pressure within the installation should remain constant for a period of not less than 3 minutes. If it is not convenient to connect the flexible tube to one of the test plugs, the tube can be passed through the

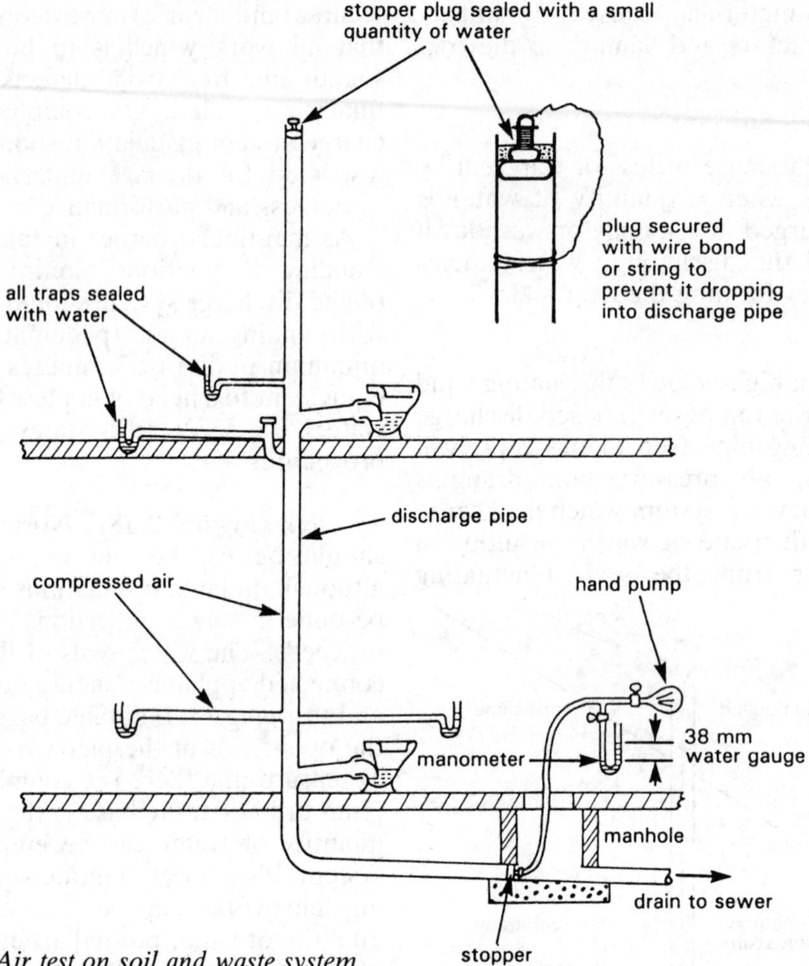

Figure 2.18 *Air test on soil and waste system*

water seal of a sanitary appliance and the test procedure applied from this position.

Leak detection Any leak in the system will be indicated by a drop in pressure reading at the manometer. Detection is best carried out with the aid of a soap solution which is brushed on to the joints and pipes, the leak position being indicated by the formation of air bubbles in the soap solution liquid. Smoke may be used as a trace to detect a leak. Either a smoke-producing machine or a smoke-generating cartridge can be used. The leak is detected visually as the smoke escapes from the system.

Smoke cartridges should be used with caution, and care must be taken to ensure that the ignited cartridge is not in contact with the installation pipework, and that the effects of combustion do not have a harmful effect upon the material used for the discharge pipe system. Testing or tracing of plastics pipework using smoke is not recommended and should be avoided due to naptha having a detrimental effect, particularly on ABS, UPVC and MUPVC. Rubber jointing components can also be adversely affected.

Water test There is no justification for a water test to be applied to a complete above-ground discharge system as the part of the system mainly at risk is that below the lowest sanitary appliance connection. Some authorities may request that this lower part of the system is water-tested and this is carried out by inserting a test plug or sealing bag into the lowest point of the pipework system and filling the pipe with water up to the flood level of the lowest sanitary appliance, provided that the static head imposed does not exceed 6 m. The water level is observed during the test period and any leak will be indicated by a drop or lowering of the water level; the leak itself is detected by a visual inspection of the installation.

Performance test

The main purpose of these tests is to check the stability and retention of trap seals under peak working conditions and that all appliances, whether discharged singly or in groups, drain speedily, quietly and completely. Immediately after each test a minimum of 25 mm of water seal should be retained in every trap. Each test should be repeated at least three times, with the trap or traps being fully recharged before each subsequent test. The maximum loss of water seal in any one test, measured by a small diameter transparent tube, or dry dip-stick, should be taken as the significant result.

Tests for self-siphonage and induced siphonage in branch discharge pipes

To test for the effect of self-siphonage the sanitary appliance should be filled to overflow level and then discharged by removing the waste outlet plug; WC pans should be flushed. The seal remaining in the trap should be measured and recorded when the discharge of water has finished. Ranges or multiple appliances, connected to a common discharge pipe, should also be tested for induced siphonage in a similar way, and the seals remaining in all the traps measured at the end of the discharge period.

Table 2.4 has been produced from BS 5572: 1978 and shows the number of appliances that should be discharged simultaneously during performance tests in different types of building.

Table 2.4

Type of use	*Number of appliances of each kind on the stack*	*Number of appliances to be discharged simultaneously*		
		9 litres WC	*Wash basin*	*Kitchen sink*
Domestic	1 to 9	1	1	1
	10 to 24	1	1	2
	25 to 35	1	2	3
	36 to 50	2	2	3
	51 to 65	2	2	4
Commercial or public	1 to 9	1	1	
	10 to 18	1	2	
	19 to 26	2	2	
	27 to 52	2	3	
	53 to 78	3	4	
	79 to 100	3	5	
Congested	1 to 4	1	1	
	5 to 9	1	2	
	10 to 13	2	2	
	14 to 26	2	3	
	27 to 39	3	4	
	40 to 50	3	5	
	51 to 55	4	5	
	56 to 70	4	6	
	71 to 78	4	7	
	79 to 90	5	7	
	90 to 100	5	8	

NOTE. These figures are based on a criterion of satisfactory service of 99%. In practice, for systems serving mixed appliances, this slightly overestimates the probable hydraulic loading. The flow load from urinals, spray tap basins and showers is usually small in most mixed systems, hence these appliances need not normally be discharged.

Waste disposal units

These mechanical units are designed to macerate and dispose of organic food such as vegetable or fruit peelings. The units must not be used for the disposal of wooden objects, plastics, string, rag, glass or metal. The disposal unit connects to a sink outlet and is housed beneath the sink. Most units require a larger diameter sink waste outlet hole than the standard opening, and it is normal practice to connect waste disposal units to sinks specially manufactured to accommodate these units.

The operating cycle of the unit is as follows:

1 Cold water is run into the sink and the unit is switched on;
2 Waste food is pushed through the rubber splash guard and is washed down by the flowing water on to the cutter rotor;
3 The waste food is thrown by centrifugal force (the tendency of a rotating body to throw liquids and solids away from the force centre) on to the cutting blades, thereby cutting and shredding the waste substances into very small particles.
4 The small particles are washed by the flowing water through the cutting ring into the sloping discharge chamber and pass out of the unit and into the waste pipe via the tubular trap connected to the outlet connection.

The outlet invert of the connected trap should be lower than the inlet so that the water or liquified waste is not trapped inside the discharge chamber. Tubular traps should always be used for waste disposal units. Bottle traps are not recommended as they tend to clog and block, due to the nature of the waste discharged from the unit. The slope of the discharge pipe must be enough to ensure good velocity of waste flow, thereby preventing deposition of solid particles in the pipe. The discharge pipe should be as short and straight as possible and any bends or changes of direction should be large-radius with cleaning access available. Disposal units connected to ground-floor sinks should

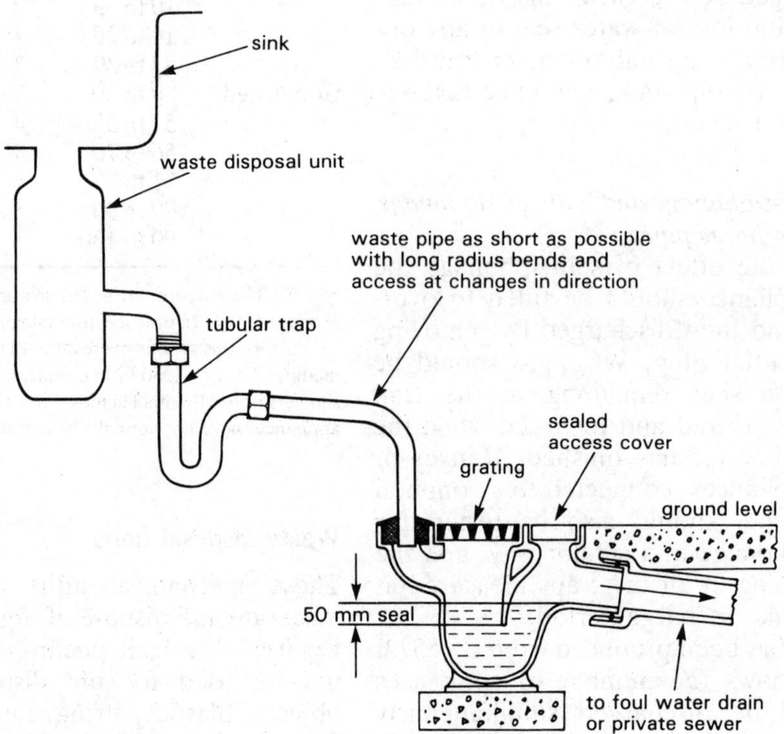

Figure 2.19 *Waste disposal unit discharging direct to below ground discharge pipe system*

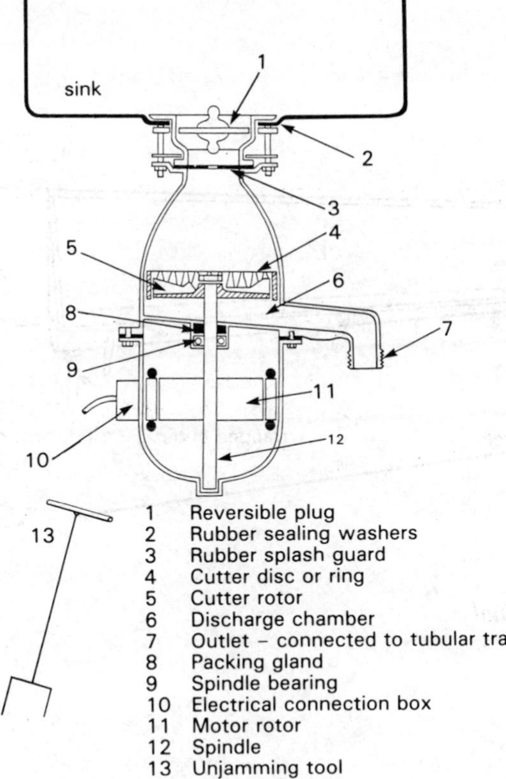

Figure 2.20 *Domestic waste disposal unit*

discharge to the drainage system via a back inlet gully as shown in Figure 2.19. Units located on upper floors in a building will connect to the discharge system. Figure 2.20 shows a typical domestic waste disposal unit. Modern units are provided with a thermal overload cut-out switch in case of jamming, and a de-jamming tool is provided should the unit lock. The unit is supplied with electricity via a fused control switch, usually incorporating a pilot light which illuminates when power is on and the motor is running.

Urinals

Urinals are sanitary appliances for the reception and flushing away of urine.

Bowl urinal (Figure 2.21) This is a wall-hung receptacle shaped like a bowl, frequently with an extended lip. It should be fitted with an individual spreader for flushing water which should wash the whole internal surface of the bowl. These appliances can be fitted on their own or in a range (see Figure 2.22).

Pedestal urinal A pedestal urinal is a bowl-type urinal supported on a pedestal.

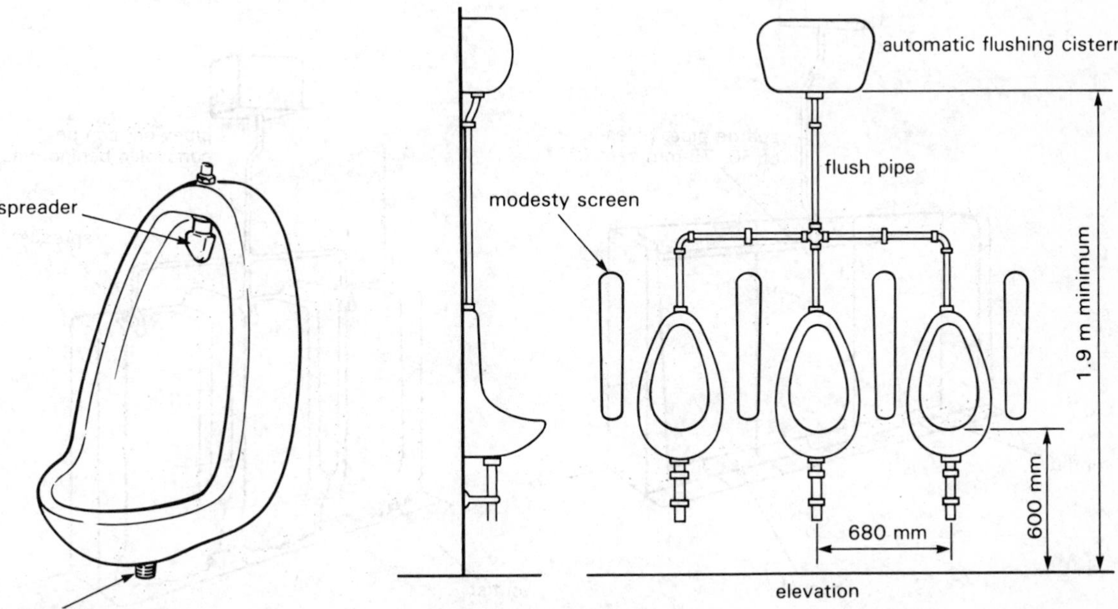

Figure 2.21 *Bowl urinal*

Figure 2.22 *A range of bowl urinals with fixing dimensions for adult use*

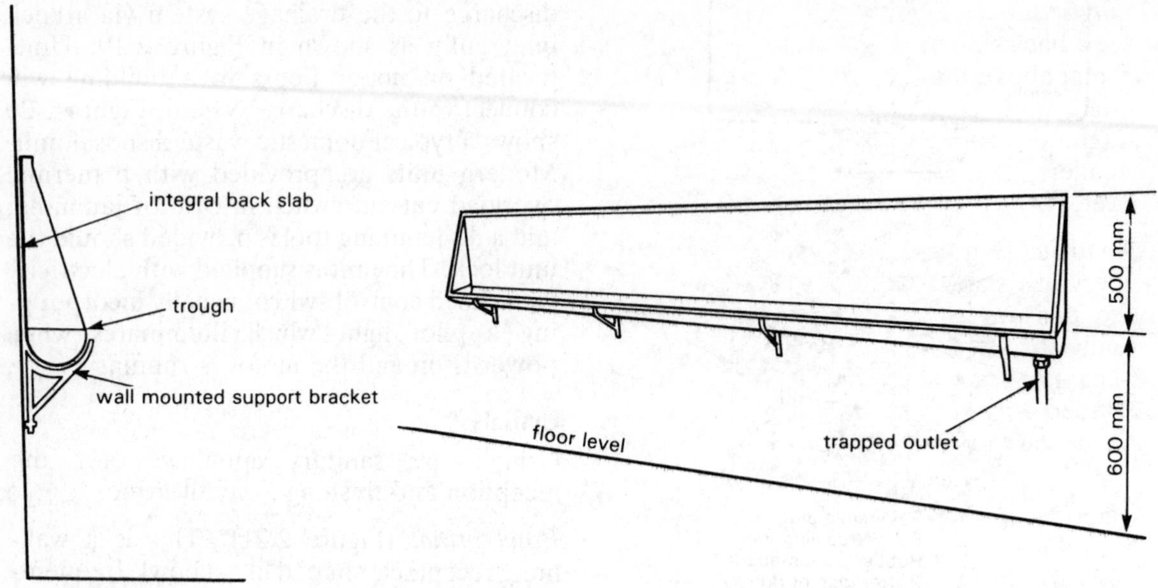

Figure 2.23 *Wall-mounted stainless steel trough urinal*

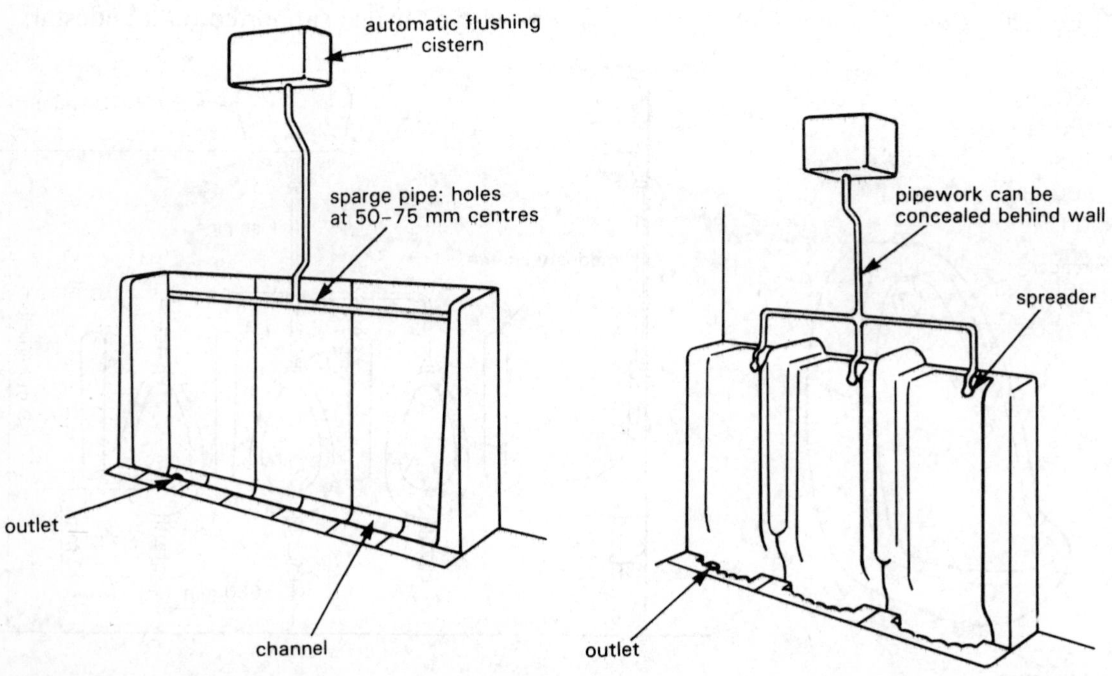

Figure 2.24 *Slab type urinal with automatic flushing cistern, sparge pipe and channel*

Figure 2.25 *Stall type urinal*

Trough urinal (Figure 2.23) These should have a back slab or edge extending to at least 450 mm above the level of the front lip of the trough and preferably integral with the trough. Flushing is by sparge pipe or spreaders, and discharge is to a trapped outlet.

Slab urinal (Figure 2.24) This consists of an impervious slab fixed against a wall. It may have division slabs of the same material, usually for only part of its height. The slab discharges to a channel which has an outlet provided with a trap. Flushing is by sparge pipe or individual spreaders.

Stall urinal (Figure 2.25) This consists of a back slab curved in plan with integral side divisions. These divisions are generally supplemented by separate extension rolls or wings to cover the joints when stalls are fixed in ranges. Stalls are complete with integral channels. A spreader should be fitted to each stall for flushing. The discharge is via a trapped outlet.

Bowl, trough and slab urinals are made from glazed ceramic materials, stainless steel or suitable plastics. Stall urinals are made from glazed fireclay or stainless steel. Outlet grids should be made of gun-metal, brass, glazed ceramic or suitable plastics, all preferably domed, and either hinged or capable of being fixed (Figure 2.26). Sparge pipes and spreaders should be of corrosion-resisting metal or suitable plastics.

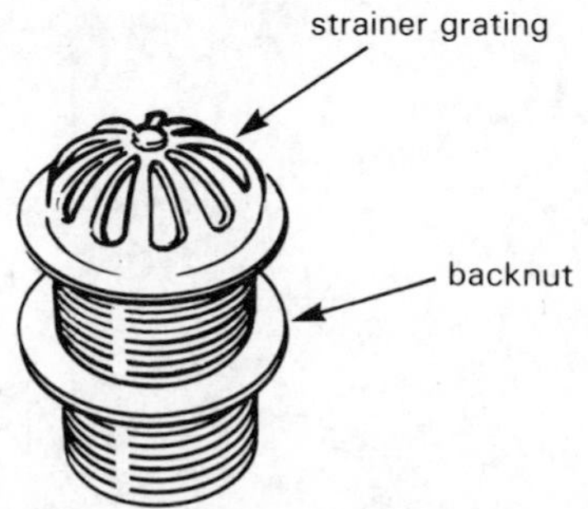

Figure 2.26 *Domed and hinged urinal outlet fitting*

Selection

The slab type is simple and usually less difficult to clean than the others.

The stall type is heavy, takes up more space and is more difficult to clean. It affords greater privacy than other types and is better able to withstand rough usage.

Bowl urinals are less restrictive to planning and are more suitable for use where floor/wall movement may occur.

In hard water districts, the waste pipes are more liable to blockage.

The combination of hard water and urine results in rapid deposition of encrustation which should be removed by frequent and thorough cleaning. Acid-based cleaning powders and fluids should be used with caution to avoid damage to the appliance or injury to the cleaning operatives.

Urinal flushing

All urinals should be fitted with a means of flushing. Where a single unit is installed a hand-operated flushing cistern may be used.

Where more than one urinal is fitted, flushing is achieved by the use of an automatic flushing cistern. This is designed to discharge its contents, by siphonage, at intervals determined by the rate at which water is fed into the cistern (see Figure 2.27).

The cistern should hold at least 4.5 litres of water per bowl, stall or 0.6 m length of slab. All flushing pipework and spreaders should be manufactured from non-ferrous materials. The cistern should fill at such a rate as will produce a flush of water at 20 minute intervals or depending on frequency of use. The supply of water into the cistern should be controlled by a valve approved by the Regional Water Authority. The most common valves are dribble or pet cocks, or calibrated disc valves (see Figure 2.28).

Urinals and waste of water

Since 1981 all supplies to urinals in new buildings have required some form of electrically or hydraulically operated control valve

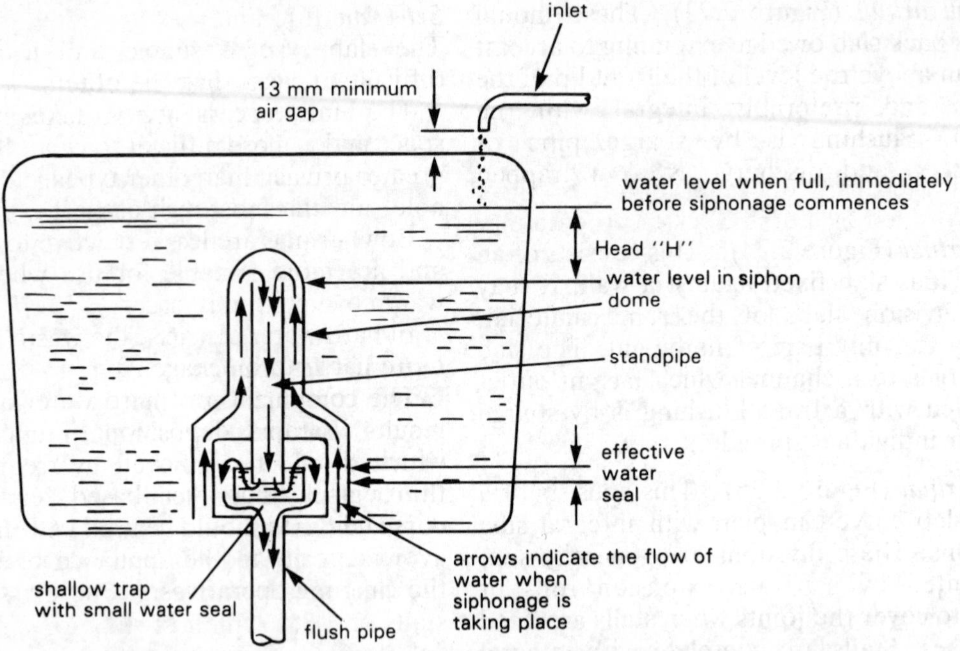

Operation – As the water level rises pressure—shown as head 'H'—increases until it overcomes the water seal in the shallow trap, causing it to overflow into the flush pipe. This causes a lowering of the air pressure in the standpipe and the dome and the water contained in the cistern is forced under the dome and into the flush pipe by atmospheric pressure. The action continues until the cistern is emptied and air enters the base of the dome. The shallow trap is resealed during the process of the flush

Figure 2.27 *Automatic flushing cistern*

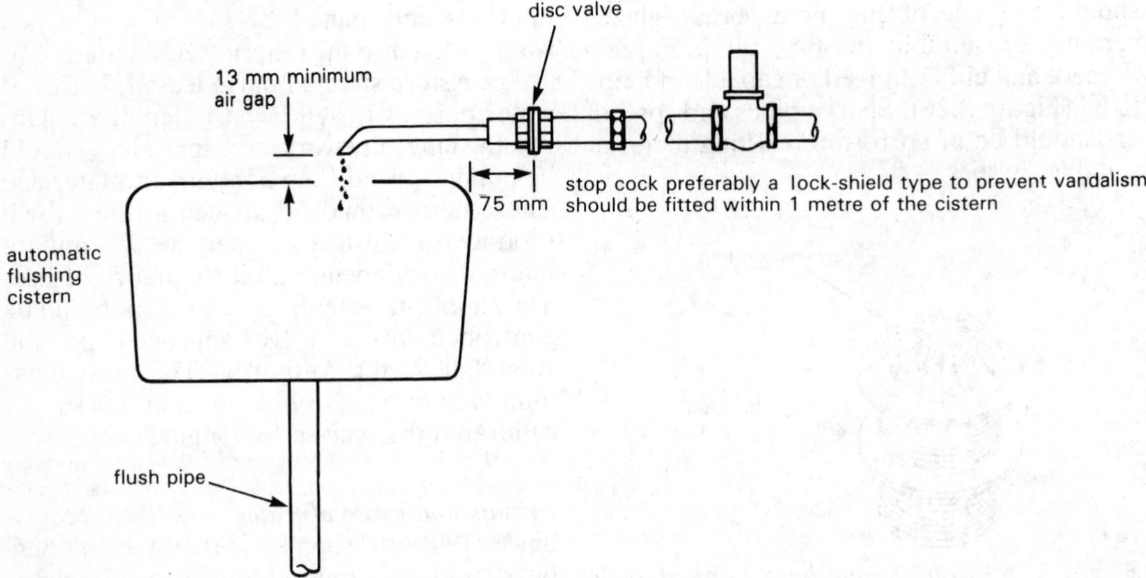

Figure 2.28 *Disc valve*

to shut off the supply of cold water to urinal flushing cisterns where the sanitary accommodation or the building is not in use (By-law 52A section 2).

The usual way of achieving this control is:

1 An electrically operated motorised valve controlled by a time clock or programmer (the clock closing the valve when the building is vacant.)
2 A valve or flushing cycle operated and actuated by the breaking of a beam or ray of light (photo-electric).
3 A hydraulically operated flush control valve. The 'cistermiser' (see Figure 2.29) is a valve which operates as follows:

   When adjacent sanitary appliances (wash basins, toilets) are in operation (i.e. the building is in use) the cistermiser opens and fills the automatic flushing cistern at a pre-set rate. When these appliances cease to be used, and the flow and pressure in the cold water supply pipe becomes static and constant, the cistermiser closes and remains shut until the pressure in the cold water supply drops (i.e. when wash basins and toilets are being used). This indicates that the building is back in use and the normal automatic flushing cycle commences.

Urinal outlet details

Urinal compartments should have an impervious floor, draining into the urinal channel (if fitted) or into a separate gully let into the floor. This is to facilitate cleaning. All compartments should be well ventilated.

With stall and slab ranges one outlet (75 mm) should not serve more than seven stalls or 4.3 m length of slabbing and should be at or near the middle of the range if there are more than three stalls or 1.8 m length of

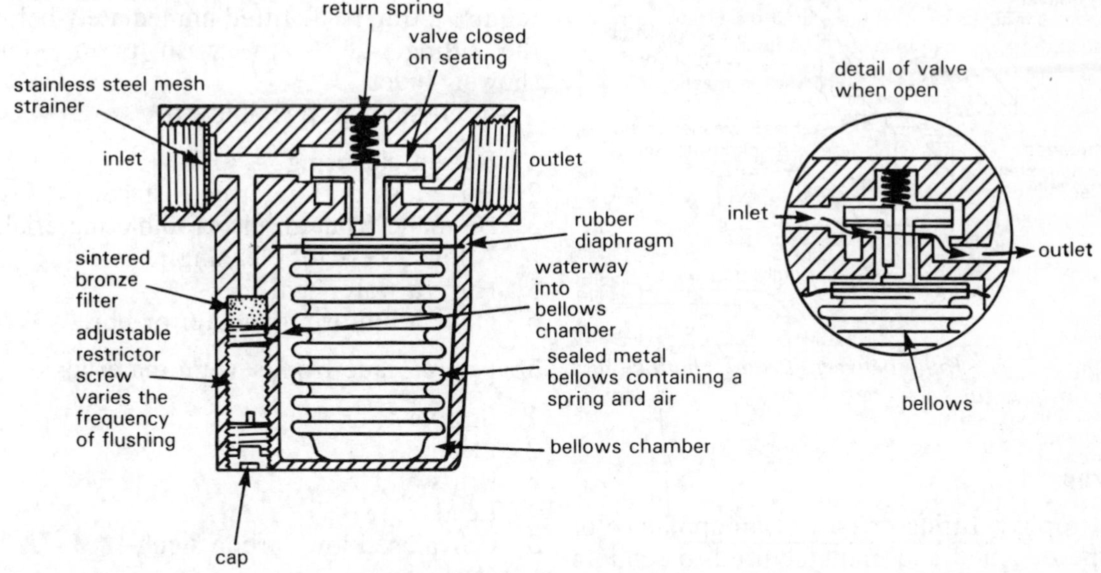

Operation – When draw-off points are closed the pressure of the water holds the valve down on to its seating. At the same time water passes through the sintered bronze filter and in to the bellows chamber where its pressure compresses the bellows. The water pressure on both sides of the diaphragm is now equal. The valve is held in the closed position by the return spring and the water pressure. If a tap is opened and the water pressure is reduced above the diaphragm the bellows can now exert an upward pressure lifting the valve off its seating, as shown, and water will flow into the flushing cistern

Figure 2.29 *Hydraulically operated valve (cistermiser)*

slab. The chart below indicates the recommended outlet sizes.

1 or 2 stall	0.6–1.2 m slab	50 mm outlet
3 or 4 stall	1.8–2.4 m slab	65 mm outlet
5 to 7 stall	3.0–4.3 m slab	75 mm outlet

Bowl or trough urinals should be provided with an outlet fitting of not less than 32 mm diameter. Outlet fittings for all types of urinal should incorporate either an inbuilt strainer or a hinged dome (Figure 2.30) to prevent possible blockage of the outlet, trap or waste pipe.

Tubular traps are best for all types of urinal as they offer less restriction to flow of water than other types of trap.

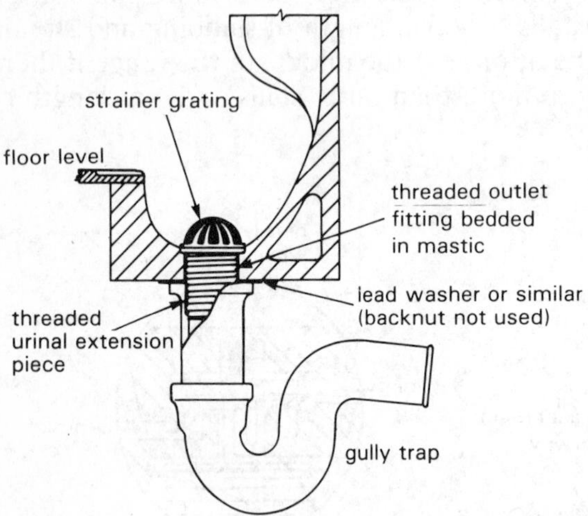

Figure 2.30 *Joint between urinal channel and cast-iron drain*

Traps

A trap is a fitting or part of an appliance or pipe designed and manufactured to retain a quantity of water which forms a seal to prevent the flow of gases or foul air from the discharge pipe into the building. The 'water seal' is that depth and quantity of water which would have to be removed from a fully charged trap before gases or air could pass through.

For identification purposes traps are usually classified in three distinct types:

1 Common traps (see Figure 2.31),
2 Resealing traps (see Figure 2.32),
3 Anti-siphon traps (see Figure 2.33).

A trap should be fitted to every sanitary appliance and omitted only in special cases.

Principles of operation. Figure 2.33a shows the trap under normal operating conditions with full water seal. Figure 2.33b when subjected to severe siphonage conditions the automatic hydraulic action allows air through the by-pass tube without any major loss of water. Figure 2.33c when normal conditions return the remaining water falls back to re-seal the trap.

Some sanitary appliances are manufactured with the trap as part of the appliance, i.e. WC pan. This arrangement is called an *integral trap*. The majority of sanitary appliances do not have integral traps and these require a trap to be fitted immediately below the fitting, i.e. bath, wash basin, sink, shower, bidet.

An efficient trap should:

1 Be self-cleansing,
2 Have a smooth internal surface,
3 Be made from an incorrodible material,
4 Allow access for cleaning,
5 Have a water seal,
6 Have a uniform diameter or bore.

Traps are made from several materials:

1 Copper,
2 Plastics,
3 Brass,
4 Lead,
5 Galvanised low carbon steel,
6 Cast iron,
7 Aluminium,
8 Glass.

The type of trap required for a particular location depends on several factors:

1 The type of sanitary appliance,

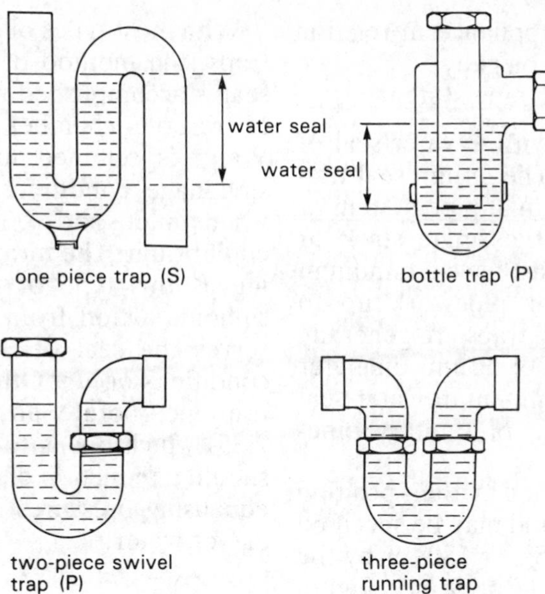

Figure 2.31 *Common traps*

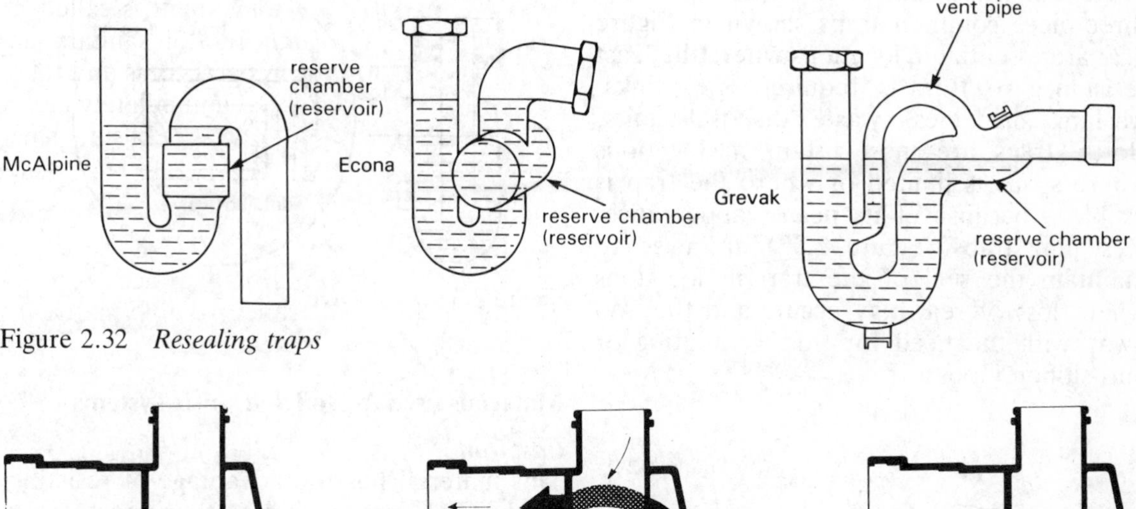

Figure 2.32 *Resealing traps*

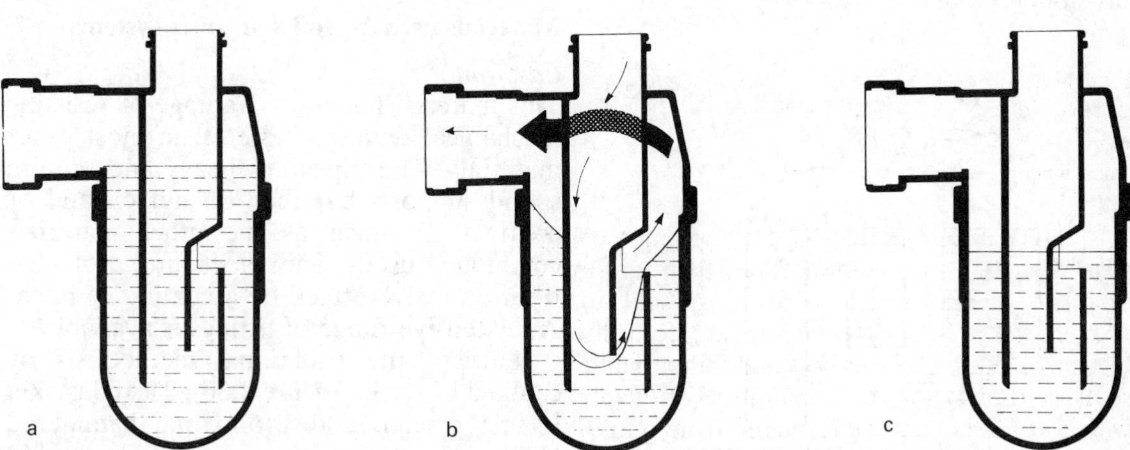

Figure 2.33 *Anti-syphon bottle trap*

2 The position of the appliance in relation to the discharge pipe or gully,
3 The type of discharge pipe system.

Traps may be obtained with a water seal of either 38 mm or 75 mm. The 38 mm seal trap is for use on the two- or dual-pipe system of sanitary pipework. For the single stack or one-pipe system, traps must have a minimum water seal of 75 mm for pipework up to 50 mm in diameter. These traps are generally called 'deep seal'. Traps of 50 mm diameter and above must have a minimum water seal of 50 mm for any system of sanitary pipework.

Trap outlets are identified by their position relative to the trap inlet and may be specified in degrees, i.e. 92½°, 135°, or 180° or by the corresponding letters P, Q or S (see Figure 2.34).

Tubular traps such as the one-, two- and three-piece common traps shown in Figure 2.31 are essential in locations where the least resistance to flow is required, i.e. sinks, washing machines, waste disposal units. Bottle traps are most useful in locations where space is limited or where the trap is visible – because of its neater appearance. Resealing traps (Figure 2.32) are used to maintain the seal of the trap in locations where loss of seal may occur, and they do away with the need for trap ventilating or anti-siphon pipes.

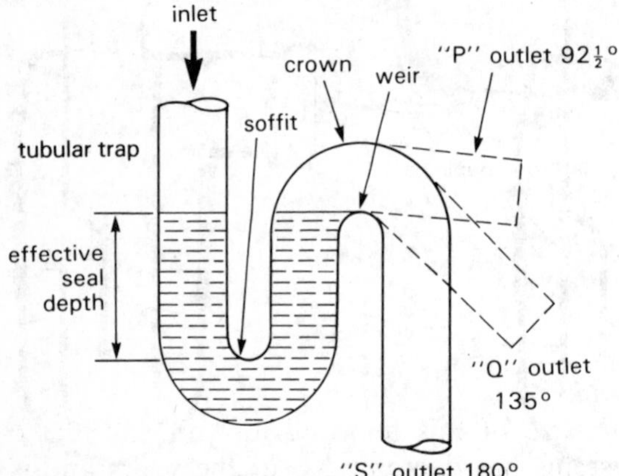

Figure 2.34 *Trap outlets and definitions*

With most types of anti-siphon or resealing traps, the method of maintaining the water seal is accomplished by means of a reservoir or reserve chamber integral to the trap. Water is retained in this chamber during siphonage conditions and reseals the trap when inlet and outlet pressures are in equilibrium. The term anti-siphon is misleading as this type of trap does not prevent siphonic action from taking place, but preserves the seal of the trap when siphonic conditions occur. Other types of anti-siphon trap incorporate an air valve (see Figure 2.35) which opens to allow air into the trap should a reduction of air pressure occur, thus equalising pressures and preventing siphonage of water.

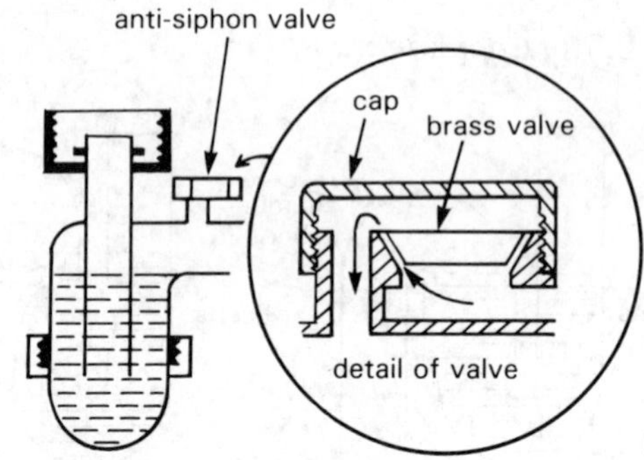

Figure 2.35 *Air valve*

Materials used for soil and waste systems

Cast iron

This material has the advantage of resisting mechanical damage better than most other materials. The pipes are heavy and require sound support, but they do not expand or contract as much as the other materials commonly used. The pipes are protected from corrosive attack by a coating of pitch. An extensive range of fittings is available.

Jointing, the traditional method, is by caulked lead (see Figure 2.36). Tarred gaskin is first placed in the joint and hammered firmly in with a yarning chisel. The gaskin helps to centralise the spigot in the socket

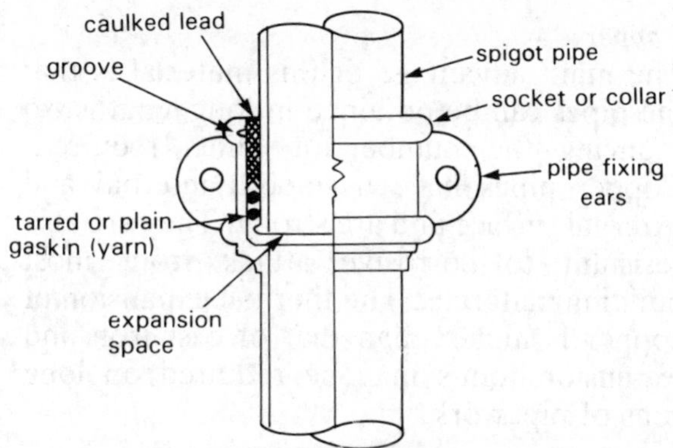

Figure 2.36 *Caulked lead joint*

and prevents molten lead from entering into the bore of the pipe. Molten lead is then poured in to fill the remainder of the joint (approx. ⅔ lead to ⅓ gaskin). When the lead has cooled, it must be caulked with special chisels (see Figure 2.37) to compress the poured lead. A semicircular groove inside

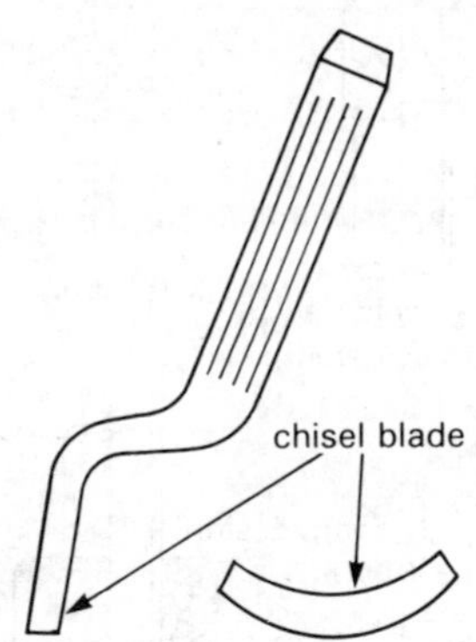

Figure 2.37 *Caulking chisel*

the collar of the pipe or fitting provides a key for the lead jointing, preventing it from slipping out over a period of time due to expansion and contraction.

Note An alternative type of caulked joint may be made with a cold caulking compound. This joint consists basically of cement and asbestos fibre which is caulked into the joint with a damp caulking chisel – the moisture causing a chemical action with the cement, so making the joint material 'set'. (The manufacturers' instructions should be adhered to when making this type of joint.)

Flexible joints may also be used: these are quicker to make and permit easier thermal movement than caulked joints. The joint is also easier to disconnect should this be necessary.

Figure 2.38 shows a 'Rollring' type joint manufactured by Burn Bros (London) Ltd.

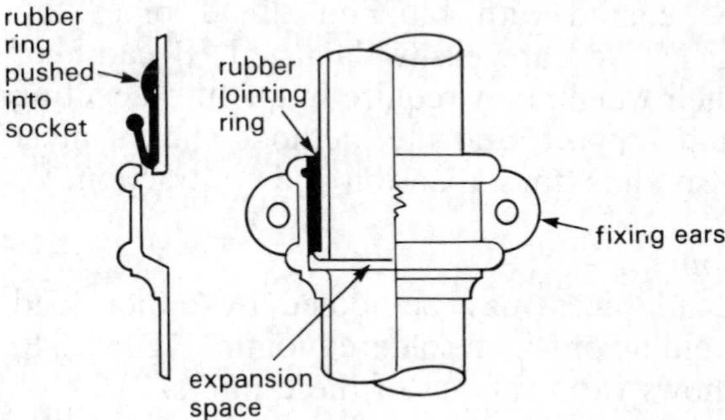

Figure 2.38 *Synthetic rubber ring joint*

Figure 2.39 shows a 'Timesaver' joint, used for jointing spigot ends, manufactured by Gynwed Ltd.

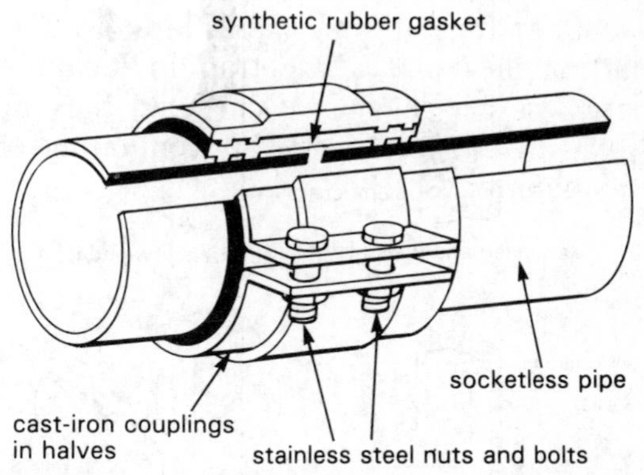

Figure 2.39 *Flexible (timesaver) joint for cast-iron pipe*

Lead

The use of this material for soil and waste systems has decreased over the years and is now generally only used for short branch pipes or connections to the main stack. The material is adaptable, especially where space is limited. It has a smooth internal and external surface and is resistant to acid attack, although materials such as lime, plaster and Portland cement may cause corrosion and therefore lead must be protected when it comes into contact with these. The normal method of protection is by wrapping with waterproof paper or petroleum jelly tape. Alternatively the pipe can be coated with bitumen. Lead pipes are heavy and are easily damaged. Because of their weight they require frequent bracketing and support, and the methods chosen must also allow for expansion and contraction.

Jointing

Lead pipes may be joined by either lead welding or wiped soldered joints. Figure 2.40 shows various types of these joints.

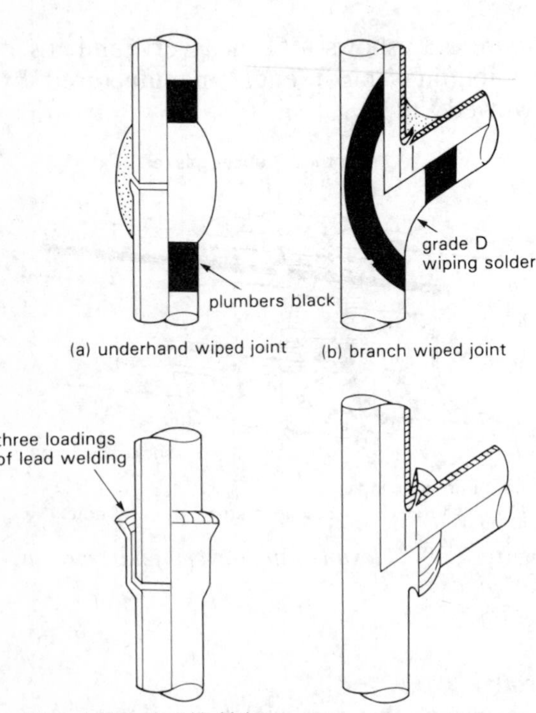

Figure 2.40 *Joints on lead pipes*

Copper

The main advantage of this material is that the pipes can be obtained in long lengths, so reducing the number of joints required. Copper pipes have a smooth internal and external surface and are strong. They are also resistant to corrosive attack from most building materials. The thermal expansion of copper is higher than that of cast iron and expansion joints may be required on long runs of pipework.

Jointing

Copper pipes may be jointed by several techniques, including hard and soft-soldered, capillary and non-manipulative fittings. Figure 2.41 shows a selection of these methods of jointing.

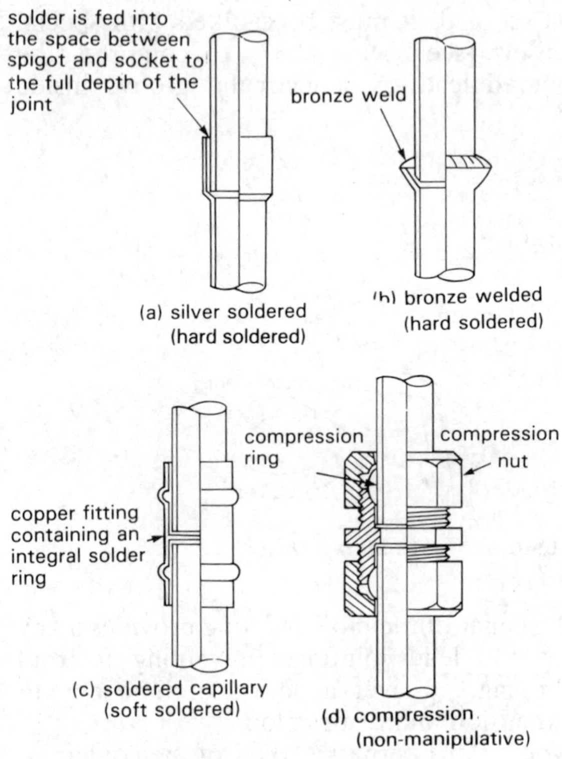

Figure 2.41 *Joints on copper pipes*

Galvanised steel

This material is lighter than cast iron and less liable to mechanical damage. As with copper, the pipes are obtainable in long lengths which makes jointing economical. The pipe must be protected from compounds or other metals which corrode zinc or mild steel. Although the pipes are rigid and strong, adequate support and bracketing must be provided.

Jointing
Small diameter steel pipes have screwed joints (see Figure 2.42). Large diameter pipes have screwed joints or spigot and socket caulked joints.

Synthetic plastics

The advantage of these materials is their light weight and the simple methods used in the jointing process. The term 'plastics' can be applied to a very wide range of materials. It is important for a plumber to be able to identify the type of plastic which is being used as only a few of these materials are suitable for sanitary pipework systems.

Plastic materials used in plumbing have many common physical characteristics and these are described in Book 1, chap. 2, pp. 46–9.

The most popular plastic materials used for sanitary pipework systems and components are:

1 Polyvinyl chloride (PVC) to BS 4514,
2 Acrylinitrite butadiene styrene (ABS),
3 Polythene to BS 3284,
4 Polypropylene.

Polyvinyl chloride This is used for discharge system pipework and fittings, and is capable of receiving discharge water at high temperatures without softening. Unlike other plastic materials, it will not burn easily and, if set alight, usually self-extinguishes.

Acrylinitrite butadiene styrene (ABS) This is another material suitable for discharge system pipework and is usually employed for small diameter branch discharge and waste pipes. It has the advantage over PVC in that it can withstand higher water temperatures for a longer period of time than PVC. Some manufacturers do produce complete ABS

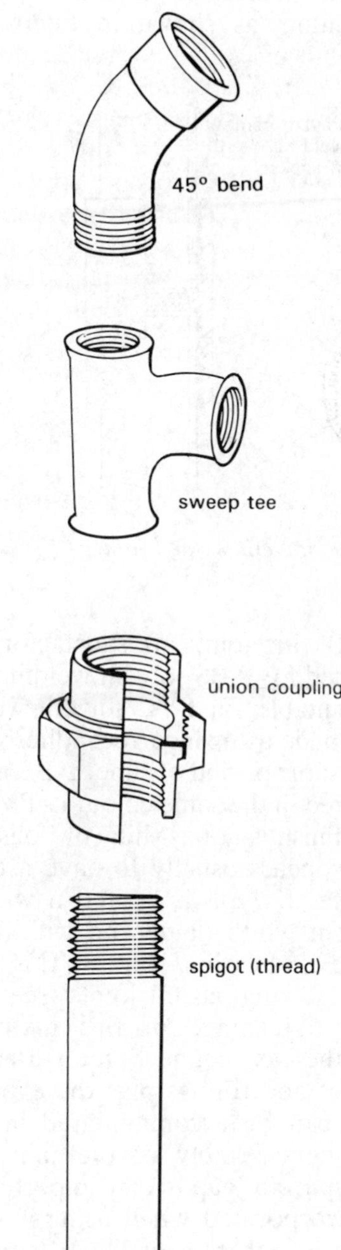

Figure 2.42 *Pipe and screwed joints manufactured from galvanised low carbon steel*

waste systems, but as it is a more expensive material than PVC its use is restricted to smaller diameter pipework.

The usual method of jointing ABS is by solvent welding as shown in Figure 2.43,

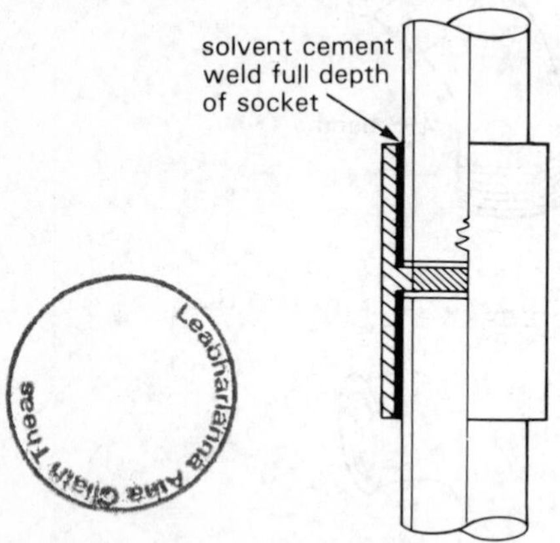

Figure 2.43 *Solvent welded joint*

although 'O' ring joints are satisfactory. The adhesive used for ABS solvent welding is not generally suitable for PVC solvent welding, and joints made using incorrect adhesive may fail after a short period of time. ABS is often manufactured in the same colour as PVC, but its surface finish is not as shiny or polished as PVC and appears usually to have a dull or matt surface. If ABS is ignited it will burn with a bright white flame and will give off small particles of carbon soot. PVC is usually jointed by 'O' ring pushfit joints (see Figure 2.44). The advantage of this method of jointing is the easy connection or disconnection of pipes and fittings plus the expansion gap which can be accommodated in every fitting during assembly of the joint. This expansion space or gap is very important and must be incorporated when using materials which have a high rate of linear expansion such as PVC and ABS.

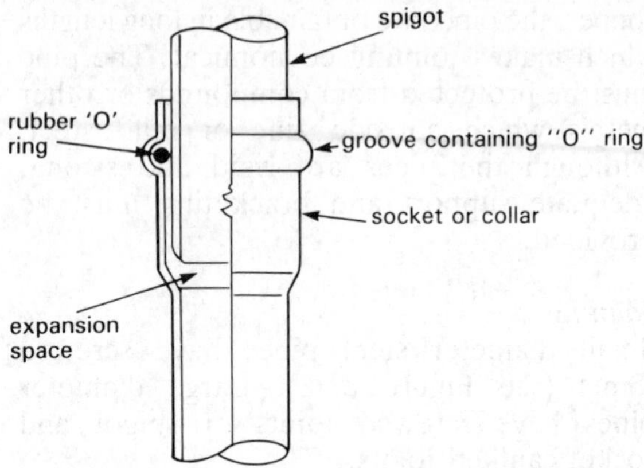

Figure 2.44 *'O'-ring pushfit joint*

Polythene These pipes are not so rigid as other thermoplastics used for sanitation discharge systems, and their use is usually limited to specialist work such as laboratory wastes because of their high resistance to chemical attack. Polythene pipes up to 150 mm diameter are manufactured to BS 3284 and are classified as 'high' or 'low' density. High-density polythene is more rigid and can withstand slightly higher temperatures than the low-density material. Pipes are usually jointed by 'O' ring pushfit-type joints or non-manipulative-type nut and rubber cone fittings.

Polypropylene This is a tough, rigid lightweight plastic in the same family group as polythene. Its main characteristics are its surface hardness and its ability to withstand high temperatures. Components manufactured from polypropylene are able to withstand boiling water for short periods of time without damage or deformation, which makes the material superior to PVC, ABS or polythene when high temperatures are a consideration. Manufacturers produce waste fittings and traps from this material, the methods of jointing being similar to those used for polythene. Solvent welded jointing is not suitable.

Self-assessment questions

1 The single-stack system of sanitary pipework can be used most conveniently where:
(a) the building is single-storey
(b) the sanitary appliances are grouped closely together
(c) there is a short run between the drains and the sewer
(d) an intercepting trap has been used

2 After a discharge pipe system has been installed in a partly completed building it is advisable to:
(a) leave all open ends and vents clear
(b) seal all inlets with cement mortar
(c) insert expanding drain plugs on the vents only
(d) fit temporary plugs or seals to all open ends

3 Where is compression most likely to occur in a six-storey discharge stack?
(a) At the ground floor waste connection?
(b) At the first offset above the wet-portion of the stack?
(c) At the foot of the stack?
(d) Where the branch from a WC enters the stack?

4 The minimum depth of water seal for a wash basin trap on a single stack system is:
(a) 23 mm
(b) 50 mm
(c) 75 mm
(d) 100 mm

5 Water discharged from a sanitary appliance and passing a branch discharge pipe lower down the stack may cause loss of the trap seal by:
(a) induced siphonage
(b) self-siphonage
(c) evaporation
(d) capillarity

6 The test applied to above-ground discharge pipework to ensure that adequate water seals are retained during working conditions is the:
(a) air test
(b) hydraulic test
(c) soundness test
(d) performance test

7 Ventilating pipes are connected to traps to prevent:
(a) loss of water seal
(b) capillarity
(c) evaporation
(d) corrosion

8 The Building Regulations require discharge pipe systems under test to withstand a water gauge pressure of:
(a) 28 mm
(b) 33 mm
(c) 38 mm
(d) 45 mm

9 The correct method of commissioning a domestic sink waste disposal unit is to:
(a) examine the waste outflow
(b) time the disposal of solid objects
(c) test until the motor has heated
(d) subject the machine to normal working conditions

10 An important consideration when fixing plastics discharge pipes is:
(a) to provide the required allowance for thermal movement
(b) not to use metal fixings which will corrode the discharge pipe
(c) to have as few joints as possible
(d) not to damage the protective coating

3 Working processes

After reading this chapter you should be able to:

1. Understand the principle of flow through pipes and its resistance.
2. Demonstrate knowledge of the various methods of bending pipes commonly used in plumbing.
3. Identify and name the usual bending machines and their component parts.
4. Appreciate the movement of metal during the bending process.
5. Understand the extent to which pipes should be bent.
6. Understand bending with the aid of a jig.
7. Demonstrate knowledge of plastic pipes, their bending capabilities and plastic memory.
8. Understand the bending techniques of (a) compression (b) draw (c) push.
9. Recognise and identify faults in bends.
10. Demonstrate knowledge of the GF dimensioning system.

Bending of pipes

General

Ideally for maximum flow through a pipe, the bore should be smooth, of uniform diameter and with no joints or bends. In practice, this of course is impossible. Therefore every effort must be made to create conditions as close as possible to perfection. As stated, changes in direction are unavoidable, and the use of purpose-made bends or elbows are generally fitted to accommodate for this. The bends used wherever possible should be purpose-made and be of large radius to minimise the frictional resistance to the flow through the system. In some instances it may be necessary for one pipe to pass over another pipe or obstruction, necessitating the use of offsets (double bends) or passovers. These could be made up from purpose-made fittings or the pipe can be bent to give a very effective result.

The methods of bending vary according both to the material from which the pipe is manufactured and to the size of pipe and the thickness of the pipe wall. The method of setting out bends to given radii, illustrations of bending springs and bending machines are fully illustrated in Book 1.

When pipes are subjected to the process of bending, particularly with small radius bends, tremendous stresses are set up in the material. The stresses will be either compression or tension, depending upon where the stress reading is taken.

Figure 3.1 illustrates the possible effect of bending a pipe, showing the thinning of the material at the heel and the thickening at the throat. In the case of small-diameter pipes, if these are of a heavy-gauge material, little or no adverse effect should be experienced with

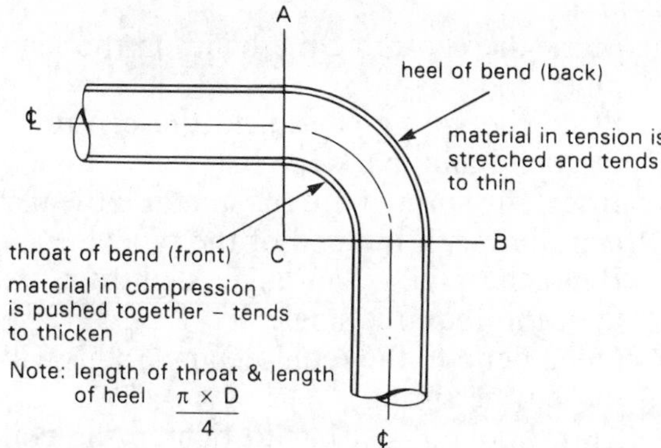

Figure 3.1 *Definitions*

bends of normal radii. As a guide this could be taken as four times the diameter of the pipe (4 × dia.).

The generally accepted methods of bending are classified as:

1 Loaded
2 Mechanical.

Loaded This could be by using a steel or rubber insert or loose fill material. It is also possible to a limited extent to use air pressure on certain pipes. For large diameter thin-walled lead pipes the manual use of tools and bobbins is required.

Mechanical The use of various types of bending machine is perhaps the most commonly accepted method. They are either manual (for the smaller diameter pipes) or hydraulic (for the large diameter pipes).

Note It must be remembered that it is not always the correct procedure to bend the pipe. As in the case of thin-walled copper tube (Table Z of BS 2871) and for certain plastic pipes, the change of direction must be performed by the use of purpose-made bends or elbows.

The materials most commonly used in the industry are:

1 Lead
2 Copper
3 Steel
4 Plastics.

Lead pipe bending

Thick-walled pipes These are generally small-diameter pipes which have been used for cold water service and distribution pipes. This type of lead pipe is now almost obsolete. The bending of thick-walled pipe is very simple and requires no special technique – only common sense and strength. The application of a little heat would assist greatly.

Simply straighten the pipe from the coil; place your knee or hand on the pipe to form the throat of the bend; and lift the free end of the pipe to give you the required bend (recommended radius 3 × outside diameter of pipe).

Thin-walled pipes These are used for sanitation work and because of the thinness of the lead the walls of the pipe must be supported during bending, or kinking of the pipe will occur. Before commencing to bend thin-walled pipes it is advisable and often necessary to check that the pipe is of uniform size and diameter. This is achieved by the use of a hardwood mandrel of the correct size, which is smeared with tallow and driven through the length of pipe to be bent (see Figure 3.4).

Spring bending (see note, in Book 1)

The use of a bending spring provides a quick and efficient way of bending lead pipes up to 50 mm in diameter. It is a time-saving method and can be used in bench work or *in situ* work where a bend is required near the end of a pipe. The spring is inserted into the pipe and centred to the approximate position of the required bend. The pipe is then pulled against the knee which is protected by a soft pad in a manner that permits maximum leverage. The bend is pulled to an easy radius which should not be less than four times the diameter of the pipe (4 × dia.). When bending pipes there is a tendency to thicken the material at the throat of the bend and to thin it at the heel. For this reason the pulling of bends with too tight a radius must be avoided. The completed bend should be slightly over-pulled, then eased to free the spring for its removal. The end of the spring is just protruding from the pipe and is extracted by turning the spring in a clockwise direction to tighten the coil of the spring. This reduces it slightly in diameter, enabling easy withdrawal of the spring.

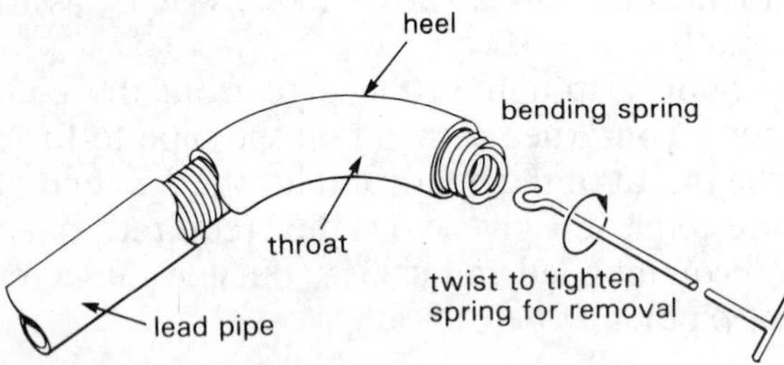

Figure 3.2 *Spring bending of lead pipe*

Where the bend is required further from the end of the pipe, i.e. greater than the length of the spring, the task is carried out in a similar manner. The only problem then is to ensure that the spring is located in the correct position prior to the bending and the extraction of the spring after bending. Figure 3.3 shows the method of forming the bend some distance from the end of the pipe.

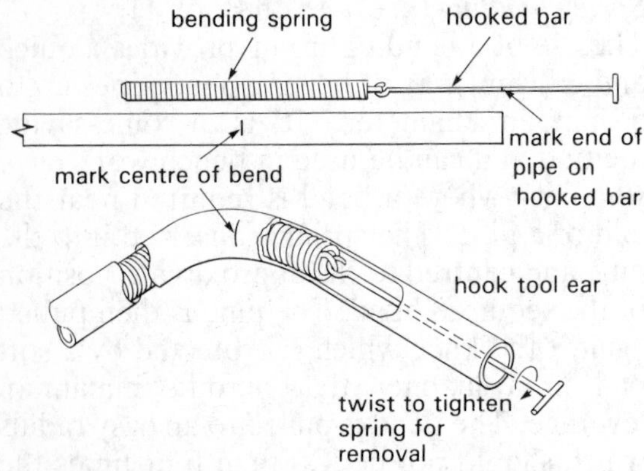

Figure 3.3 *Spring bending of lead pipe*

Method

1 Mark the centre of the bend on the lead pipe;
2 Place the spring central to the centre of the bend and mark the bar;
3 Insert the spring until the mark on the bar is in line with the end of the pipe;
4 Bend the pipe to an angle slightly more than the required angle;
5 Open bend to the required angle: this will loosen the spring;
6 Turn hooked tool bar to tighten the coil of the spring further, and withdraw.

Note When the pipe is initially bent, the sides of the pipe will nip on the spring, preventing its easy removal. By slightly opening the bend the side nip is released. This, with the greasing of tallow before use, will aid its removal. The bending spring method of supporting the walls of the pipe is the one recognised and used in the United Kingdom. In some countries it is quite an accepted practice to use dry sharp sand as the supporting material – which is the method we use for steel and copper tube bending and which is fully explained later in this chapter.

Bending using wooden bobbins

Method

1 Straighten the pipe and true the bore by driving a wooden mandrel of the appropriate size through the pipe (see Figure 3.4).
2 Set out the required bend as detailed in Book 1. Mark the beginning and the end of the bend.
3 Dress the lead from the inside (throat) of the bend to the back (heel) of the bend, making it slightly oval in cross-section (see Figures 3.5 and 3.6).

   Note The reason for making the pipe oval is it allows for a greater pulling of the bend before the pipe begins to kink.

4 Place your knee on the throat of the bend. Pull the pipe evenly, apply support on the throat of the bend. Note the change in the cross-section of the pipe: the flattened sides will rise and should be dressed towards the back (heel) of the bend. This prevents thinning of the bend and maintains a controlled shape.
5 The actions of 3 and 4 are repeated until the required angle and radius of bend is obtained (the angle should be slightly more than the required finished angle).
6 The wooden bobbins are now driven through the pipe, restoring the bend to the same uniform diameter as before. Great care must be taken in this action to prevent the bobbins from fracturing the back of the bend. Keep the heel of the bend cool and the throat warm. With care and skill this action need only be carried out once. The vee notch (Figure 3.8) enables the drive stick to fit snugly onto the bobbin without folding the cord.
7 The finished bend should be free of tool marks. These can be removed with the aid of a wooden dresser and a lead flapper.

Note The bending of large-diameter pipes will be dealt with at Advanced Craft level studies.

The two methods of lead pipe bending described, although completely different, have the same result. The bend formed with the aid of the bending spring is a much easier and quicker method than that of using bobbins.

Disadvantages of spring bending

1 It is not possible to make tight radius bends;
2 There is no control over working the lead to the heel of the bend;
3 There is more tendency to thin the lead at the heel of the bend.

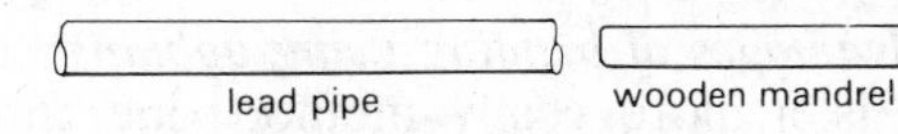

Figure 3.4 *Straightening the pipe*

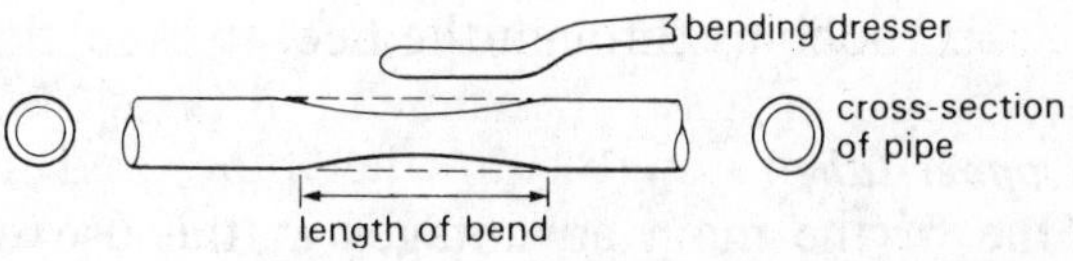

Figure 3.5 *Dressing the bend*

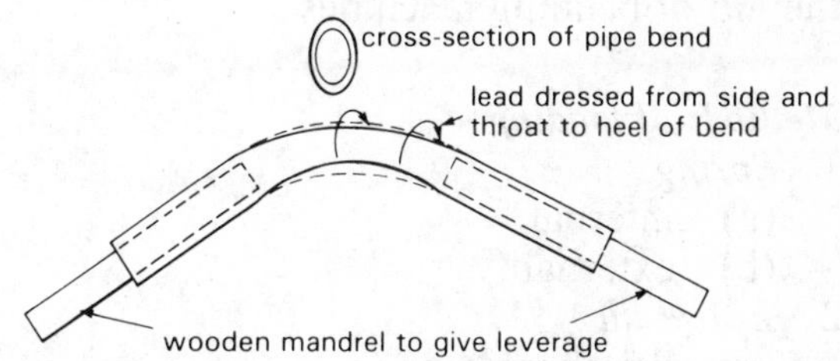

Figure 3.6 *Bending the dressed pipe*

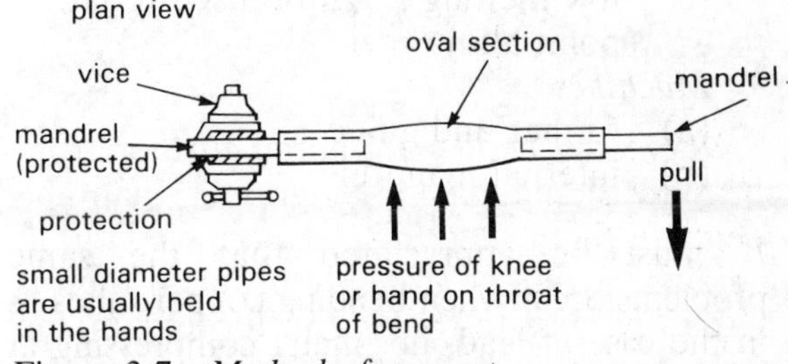

Figure 3.7 *Method of support*

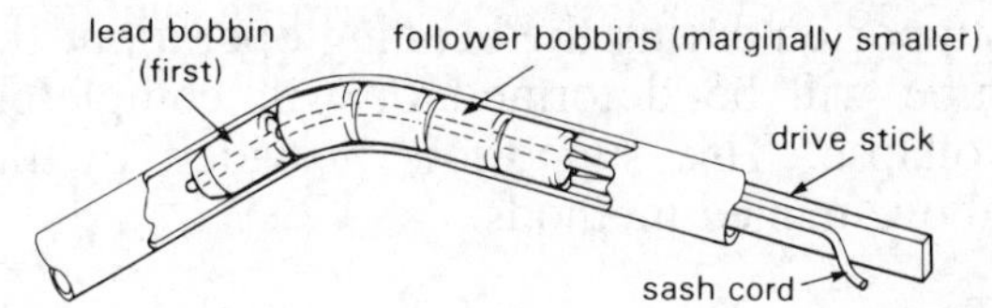

Figure 3.8 *Trueing the bend*

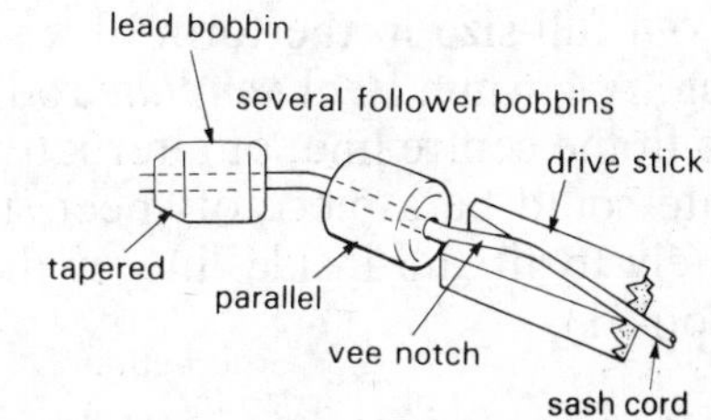

Figure 3.9 *Bobbins and drive stick*

Disadvantages of bending using bobbins

1 It is a more costly method due to the many operations involved;
2 More skill and know-how involved;
3 It is very easy to spoil the bend by driving the bobbins through the heel.

Copper tube

One of the main advantages in the use of copper tubes is the ease with which they can be bent, either by the loading method or with the aid of bending machines.

Methods of loading

1 *Springs*
   (a) internal
   (b) external
2 *Loose fill*
   (a) dry sharp sand
   (b) resin } little used today
   (c) low melting point alloy } little used today
3 *Machines*
   (a) former and guide
   (b) internal mandrel

It must be appreciated that the same problems occur when bending copper tubes as in the case of lead, i.e. metal compressing at the throat and stretching and thinning at the heel. Therefore, unless the wall of the pipe is supported during the bending operation, the tube will be deformed or will completely collapse. This support is provided by the above named methods.

Template Where bends have to be made to a given radius and accuracy and shape is important, the required bend/bends should be set out full-size in the form of a working drawing. A 4–6 mm steel *template* rod is then bent to fit the centre line, or alternatively the template could be a piece of sheet steel cut accurately to fit the inside line of the bend (see Book 1).

The British Standard for copper tubes is detailed in BS 2871: 1971. Part 1 states the metric dimensions. These tubes are suitable for connection by capillary or compression fittings (metric sizes) or other appropriate methods such as silver soldering (brazing) and bronze welding.

The tubes are standardised on the basis of the outside diameters. The dimensions and wall thicknesses of four varieties of tube are given in Tables X, Y and Z of BS 2871. within these ranges of tube there are three different tempers:

1 H = hard (as drawn),
2 ½H = half hard,
3 O = annealed.

BS 2871: Part 1 Tubes for water, gas and sanitation

Table X half hard light gauge copper tube – ½H

Table Y half hard and annealed copper tube – ½HO

Table Z hard drawn thin walled copper tube – H

Bending springs These are available in both internal and external types. Spring bending is by far the most commonly used method and is perhaps the easiest. The main advantage of spring bending is that the bend can with care be moved slightly should it be wrongly positioned.

Copper tube can be bent using internal springs up to 42 mm diameter, although at this size it requires strength, a fairly large radius and annealing after each pull or throw.

External springs These are used only on the smaller diameter size tubes up to 22 mm maximum. Their main advantage is that it is an easy operation to place and remove the spring in cases where the bend is required mid-way along the pipe.

Note Small diameter copper tubes can be bent satisfactorily without being annealed.

Internal spring bending of copper tube Care must be taken to ensure that you have selected the correct type of spring and to check whether the tightening effect is in a clockwise or anti-clockwise direction to assist in the removal of the spring after bending.

The bending of light gauge copper pipe up to 28 mm diameter can be performed fairly easily with the aid of bending springs (particularly 15 and 22 mm tube). The bends should be of an easy radius from 3 × dia. for the smaller size pipes up to 6 × dia. for the sizes up to 54 mm.

Before commencing to bend the pipe, first ensure that the bending spring passes easily down the pipe until the centre of the spring is positioned at the centre point of the bend. If satisfactory withdraw the spring.

Method

1 Set out the bend and make a template to fit the centre line of the bend;
2 Ensure uniformity of pipe and mark length of bend;
3 Anneal pipe (heat until red hot) from the *beginning* to the end of the bend. It is important not to heat either short of or outside the marks of the bend as this will affect the finished length of the bend making it either short or too long;
4 Insert the spring and pull the bend round the knee until it fits the template;
5 Bend slightly more than the required angle, then open the bend (this releases the pipe grip on the spring);
6 Turn the spring to tighten the coil and withdraw.

It is advisable to lubricate the spring with a smear of tallow or thin oil. The pipe can be bent while still hot or alternatively the pipe may be cooled, then bent. The softening is done by the heating until red hot and is equally annealed irrespective of the cooling.

Loose fill loading Sharp sand is by far the easiest and safest loose fill loading material for pipe bending (provided the sand is *dry*) and can be satisfactorily used for all the types of pipes mentioned and for all sizes (see Figure 3.10). It is possible to bend the pipes to much smaller radii and to more complex shapes. The removal of the sand is a simple operation of removing the bungs and tapping the sides of the bend, the sand being returned to a receptacle for further use.

Method

1 Set out the bend and make a template to fit the throat of the bend;
2 Seal one end of the pipe, then fill with dry sharp sand;
3 Compact the sand by tapping the side of the pipe, and gently bumping the sealed end of the pipe on the floor (this compacting is perhaps the most important point);
4 Seal the open end with a wooden bung;
5 Mark the beginning and end of the bend, heat until red hot the portion between the marks, allowing the heat to soak through the pipe into the sand (annealing process);
6 The bend can be formed while the pipe is hot or it can be allowed to cool (safer to handle);

   Note It is advisable to start the bending from the sealed end; the pipe may be held in a vice by means of purpose-made vice protectors.

7 When the bends are completed, remove the bungs and empty the sand back into the receptacle for re-use.

Note Care must be exercised in handling the completed work as the annealed work remains soft and can only be hardened by work hardening, which is explained under that heading.

Loading method It is advisable to have purpose-made hardwood plugs (bungs) with a fairly long taper to ensure maximum surface contact with the inside of the pipe. Figure 3.11 illustrates methods of sealing the pipe ends before the dry sharp sand is compacted by gently bumping the pipe on the floor and at the same time tapping the side of

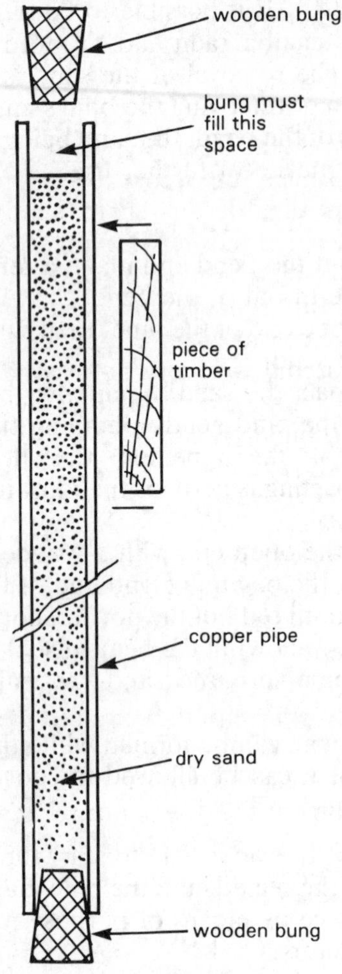

Figure 3.10 *Loose fill loading*

the pipe with a piece of wood. When the sand will not compress any more, remove sufficient sand to leave a void of approximately 30–40 mm. Drive in the wooden bung to seal the pipe, ensuring that there is no cavity under the bung. An alternative method is to use sand compressors as shown in Figure 3.12.

Vice holding method When holding a pipe in a vice some form of protection is essential or the pipe will become either marked or even damaged. Purpose-made clamps of cast lead are made which accurately fit the various sizes of pipe (see Figure 3.13). An alternative method is to make them from wood, but a much better method is to cut an old bending machine guide (slide) into pieces as shown in Figure 3.13.

Forming bend to a template

As stated earlier a template could be made from a 4–6 mm diameter wire or it could be made from a piece of sheet metal. The type shown in Figure 3.14 is perhaps the easiest to make. It also gives the most accurate check on the finished bend. The required bend is set out on the sheet material which is then cut out accurately to fit the internal size of the bend. The pipe is then pulled to form the bend, the template being held in position as shown.

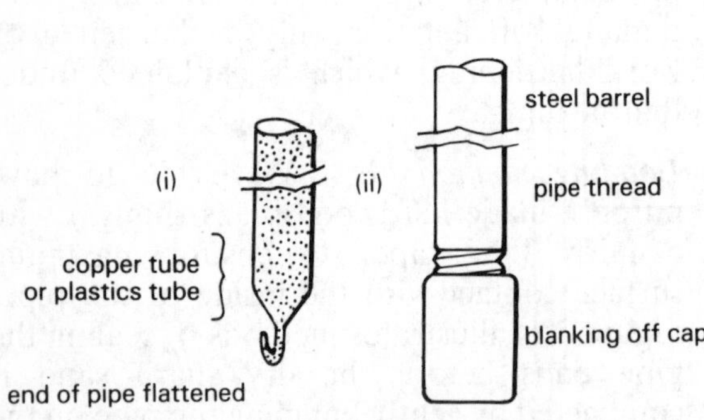

Figure 3.11 *Sealing on pipe*

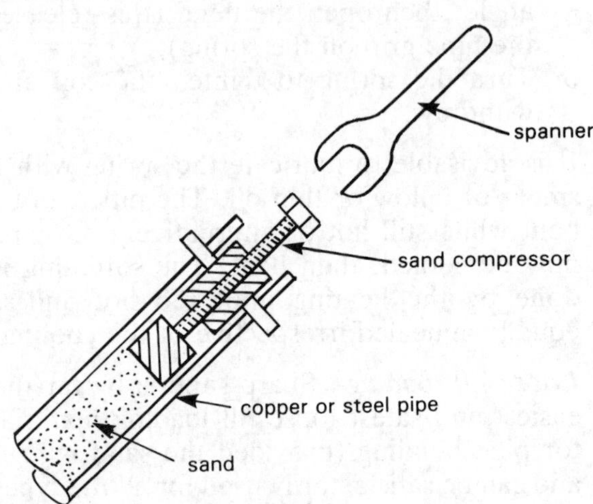

Figure 3.12 *Sand compressor*

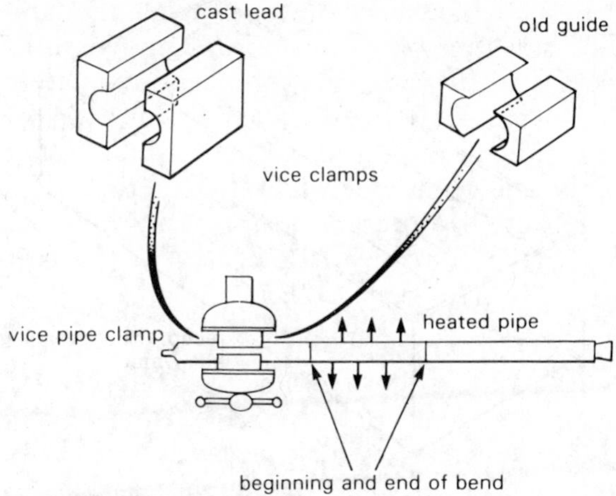

Figure 3.13 *Method of holding pipe while bending*

Plastic pipe bending

There are many different types of plastics on the market, some of which are not suitable for bending. Care must be exercised and consultation made with the manufacturers' literature if there is any doubt concerning the bending of a specific type of plastic. Where a particular plastic pipe is used which should not be bent the manufacturers have provided a comprehensive range of purpose-made fittings to enable all or most problems to be overcome.

There are several methods of heating and bending polythene and polyvinyl-chloride pipes. Perhaps the most commonly used method is that of dry sharp sand loading which is carried out in the manner described as for the loaded bending of steel and copper. The main difference is that plastic pipe does not change colour when heated and, because plastic material is a very poor conductor of heat, it is very easy to char or destroy the material completely.

Plastic memory

This is a term given to plastic material and can best be explained by using a straight length of plastic pipe as the subject. The plastic pipe when manufactured is in a straight length. It is then heated, softened and can be bent into the required position; it is held in this position until cold, whereupon it will remain in the new position so long as it remains below the temperature at which it was worked. Should the pipe be reheated the bend will tend to open and the pipe return to

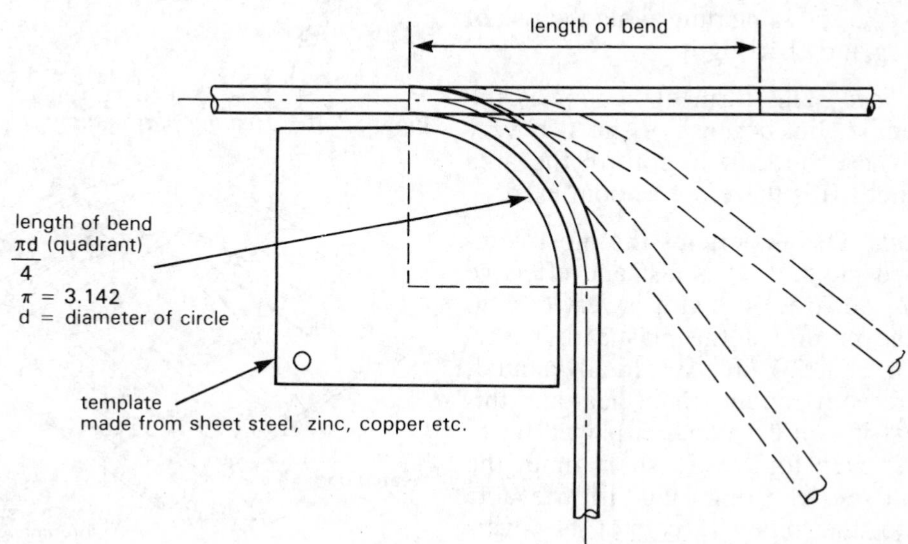

Figure 3.14 *Forming bend to template*

the original straight position. This is known as its memory, hence the term 'plastic memory'.

Accepted methods of heating plastic pipes prior to bending:

1 Immersing in boiling water (see Figure 3.15),
2 Heated air (see Figure 3.16),
3 Radiated heat,
4 Direct heat (see Figure 3.17).

Immersion in boiling water Although satisfactory from the softening point of view, it has its limitations in practical use: the availability of boiling water and the difficulty of manoeuvring, filling and emptying a large container of water being the main problems. It is also very difficult to cater for long lengths of pipe so that this method tends to be restricted to bending short lengths.

Heated air This is one of the better methods of heating plastics pipe prior to bending. It is possible to devise several different types of heating apparatus to be used with heating appliances which give greater flexibility to their use (see Figure 3.16). With the use of the equipment illustrated and a little care it is possible to heat the plastics pipe fairly quickly, yet at the same time avoid the risk of overheating and charring it.

Radiated heat This form of heating is the safest method, but because of this it is very slow. It is also difficult to control the area being heated. It is therefore seldom used.

Direct heat This is perhaps the most commonly used method. It is fast and effective but it must be stressed that it is very easy to destroy the nature of the plastics by overheating or charring it. As already stated, plastics are poor conductors of heat and this means that if a direct flame is played on to one point, even for a very short time, the result will be charring and irretrievable damage to the pipe. It cannot be overemphasised how important it is to keep the flame continually on the move, at the same

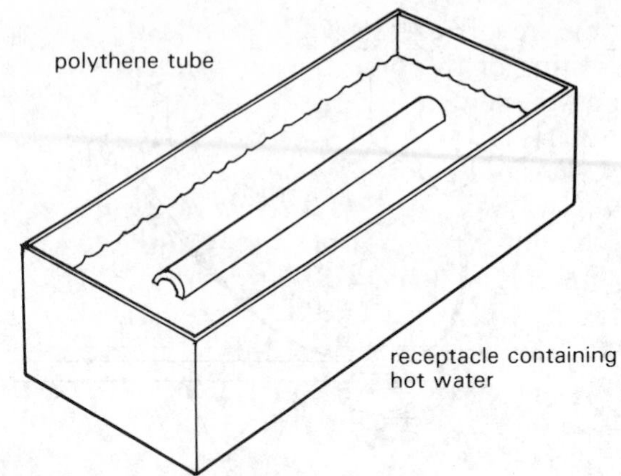

Figure 3.15 *Bending of pipes – methods of heating*

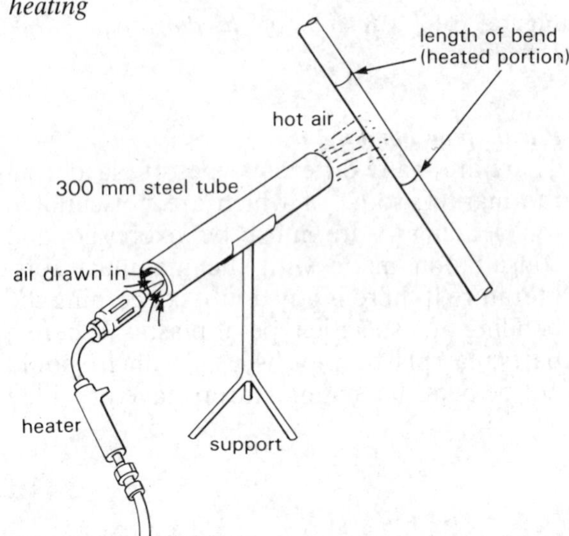

Figure 3.16 *Heating plastic pipe*

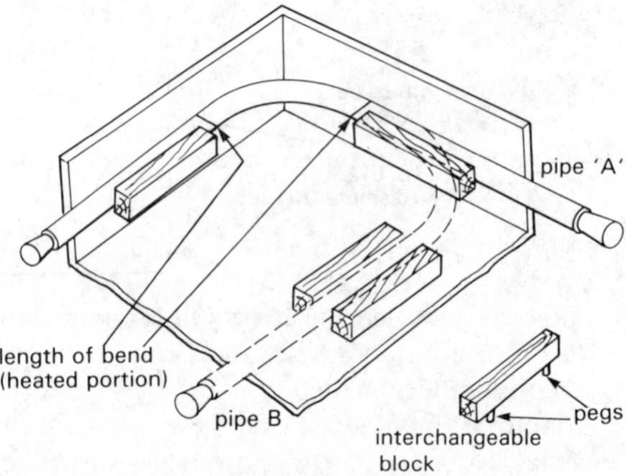

Figure 3.17 *Purpose-made jigs*

time revolving the pipe to ensure a slow heating of the pipe throughout its thickness and the length of bend.

Method of bending plastic pipes

The walls of the pipe must be fully supported both during the heating and bending process. This can be achieved by the use of a rubber insert which performs the function of a traditional steel-bending spring used for the bending of copper pipes. An alternative method is to support the walls of the pipe with a dry loose in-fill such as dry sharp sand. The application is fully described on p. 107 for the bending of copper tube.

Bending operation

1–4 As for copper tube (see pp. 107 and 108).

5 Heat the pipe very slowly, revolving it continuously, and constantly moving the torch flame backwards and forwards along the length of the bend.

Note This is a slow procedure and must not be rushed. Plastics have a very low conductivity and will very easily char. The pipe is ready for bending when it becomes pliable and floppy. A slight change in surface appearance may be detected: it takes on a more shiny look.

6 To form the bend place the heated pipe into a purpose-made jig (see Figure 3.17) or simply bend by hand. No force is required.

7 Cool the bend by applying cold water. This will set the bend in the required position.

8 Remove the bung and return the sand to its receptacle for reuse.

Polythene pipe This is a plastic material used for conveying cold or at least cool liquids. It is a clean, lightweight, smooth bore pipe, unaffected by corrosion. It has a low conductivity which means it is a poor conductor of heat. Plastics have a high rate of thermal expansion which must always be taken into account when jointing and fixing plastic pipes. They also have considerable elasticity, particularly when warmed above a temperature of 60 °C.

Bending polythene pipe Polythene can be bent cold as long as the bend has an easy radius of approximately 12 × dia. of pipe. A bend with a large radius is not usually acceptable. It is therefore necessary to heat the tube so as to produce neat bends with smaller radii of approximately 4 × dia. of pipe.

Polyvinyl chloride tube This type of tube is known simply as PVC. It is light in weight, smooth of bore, resistant to corrosion, all of which make it a very useful conduit for the conveying of cold or hot water.

Unplasticised polyvinyl chloride This is PVC without additives. In this form it is more rigid and can be fractured if subjected to a severe blow. It is more rigid than plasticised polyvinyl chloride but when fixed, supported and used in the correct manner, should give trouble-free service. It is used for above- and below-ground drainage discharge systems. This type of plastic should not be bent, any change of direction being accommodated by purpose-made bends.

Plasticised polyvinyl chloride This is the same plastic, manufactured with the addition of rubber and thus changing the rigid PVC

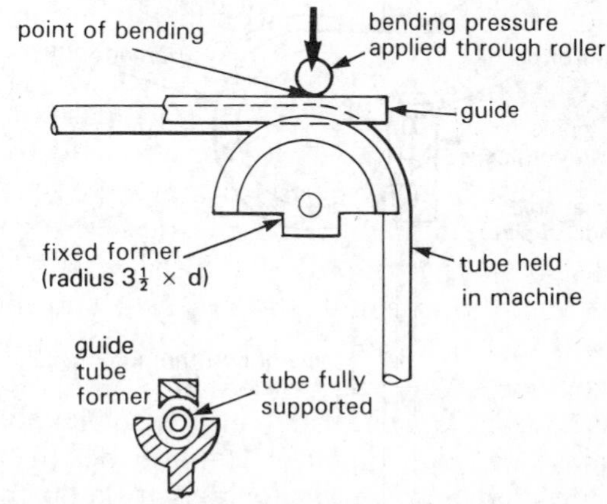

Figure 3.18 *Compression bending*

into a more flexible material which makes it suitable for bending and more resistant to impact.

Machine bending

Bending machines are supplied in various forms suitable for bending all types of metal pipes both ferrous and non-ferrous, thin- and thick-walled varieties. They come under one of the following headings:

1 Compression bending,
2 Draw bending,
3 Push bending.

Compression bending This method is used when bending thin-walled pipes as it gives the greatest support to the pipe at the point of bending. In this type of machine the centre former is fixed; the pipe is fitted into the groove of the former, and is surrounded and held in position by the guide. The former and guide together support the walls of the pipe and prevent it from collapsing during the bending operation (see Figure 3.18).

The pressure for bending is transmitted by a rotating lever arm and roller, positioned adjacent to the guide. The position of the roller is vital to the quality of the bend: too little pressure will result in a wrinkled bend; too much pressure will give a bend with excessive throating.

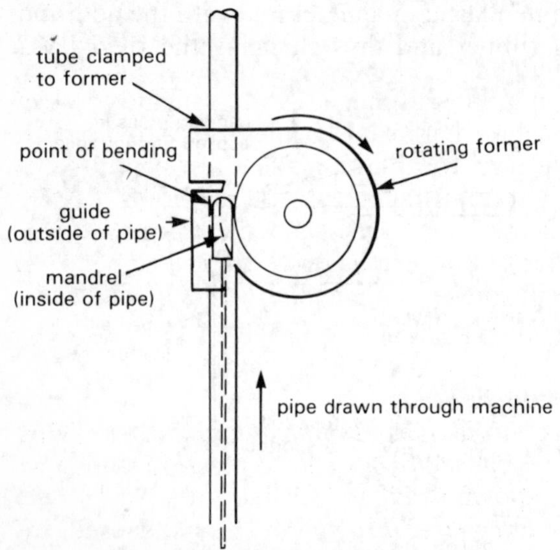

Figure 3.19 *Draw bending*

Draw bending Although very effective, the type of machine used for draw bending is of a more specialist nature (see Figure 3.19). It can produce bends to a much smaller radius than those required in normal domestic work. This type of machine is not commonly used on site work. Due to the accuracy and perfection of the bending and the necessary tooling provided, these bending machines are also expensive to purchase.

Operation In this type of machine the pipe is clamped to a centre-rotating former which when operated pulls the pipe forward, so forming the bend. The pipe is again fully supported by means of an internal mandrel and an external guide, the adjustment and position of the mandrel being very critical.

Push bending This type of bending is the simplest and requires the least skill or knowledge of pipe bending on the part of the operator. It is sometimes called 'centre point bending' because the bending pressure is applied at a single point in the centre of the bend.

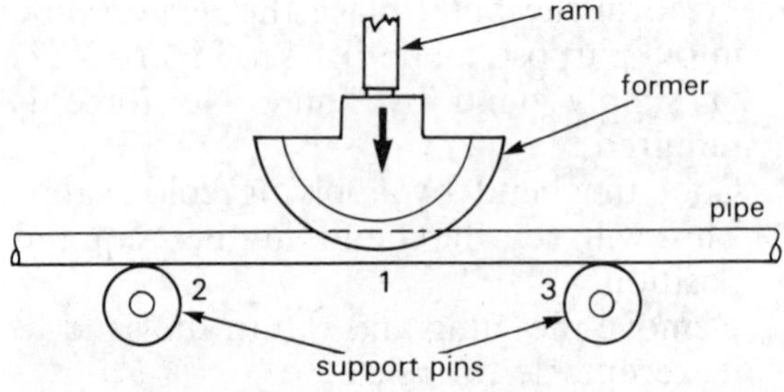

Figure 3.20 *Push bending*

This type of machine (see Figure 3.20) is also known as the three point bender because of its two support points in addition to the centre bending point. It is very satisfactory for bending heavy gauge (thick-walled) steel pipe, but unsuitable for light gauge (thin-walled) pipe, unless the walls of the pipe are

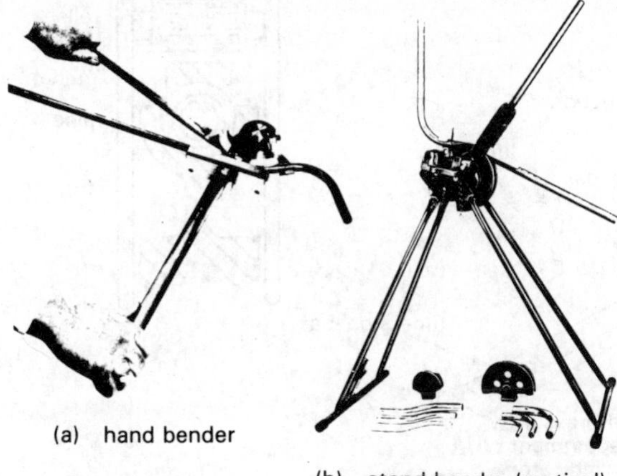

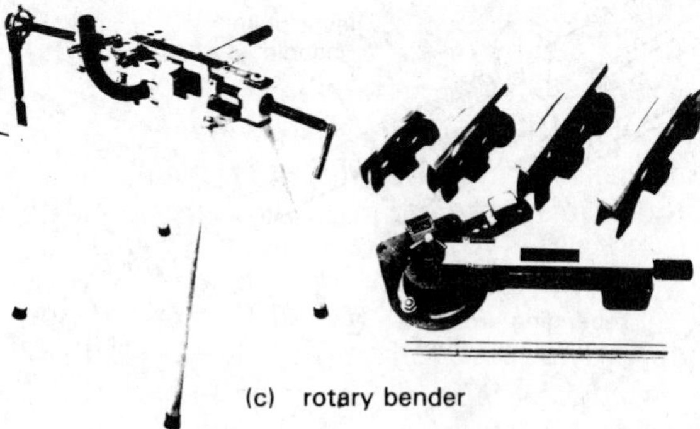

Figure 3.21 *A range of machines in common use*

Figure 3.22 *Rothenberger bender (copper and steel tube)*

again supported by the use of a suitable loose fill material.

Types of bender

Light gauge copper tube can be easily bent by machines of which there are several different types to choose from.

Bending machines Benders (a) and (b) in Figure 3.21 are by far the most commonly used, and are similar in their set-up and use. They can be used in either vertical or horizontal positions. Proficiency in their use is a must for all practising plumbers. Figure 3.23 shows an exploded view of a bending machine, detailing its component parts.

Note When bending, the pipe is completely encircled by former and guide, thereby supporting the pipe wall and preventing deformation.

Setting the pressure indicator Some of the smaller types of machines have fixed formers and guides with no pressure adjustment yet still give satisfactory performance, particularly when new. The better machines have incorporated in their design an adjustable pressure indicator and the correct positioning of this is very important if perfect bends are to be produced. Figure 3.24 indicates the correct position and also the reduced and increased pressure position.

The information given forms the basic guidelines about where to commence pipe bending, although with slightly worn machine parts and various grades of pipe it is always advisable to make a test bend first as slight adjustments may be required.

Figure 3.24(A) indicates the correct setting. This setting should give a perfect bend on light gauge copper pipe. Pressure indicator should be parallel with pipe.

Faults in bends

1 *Throat of bend rippled, heel flattens* This is caused by a reduced pressure setting as shown in Figure 3.24b.
2 *Excessive throating* This is caused by increased pressure as shown in Figure 3.24c.

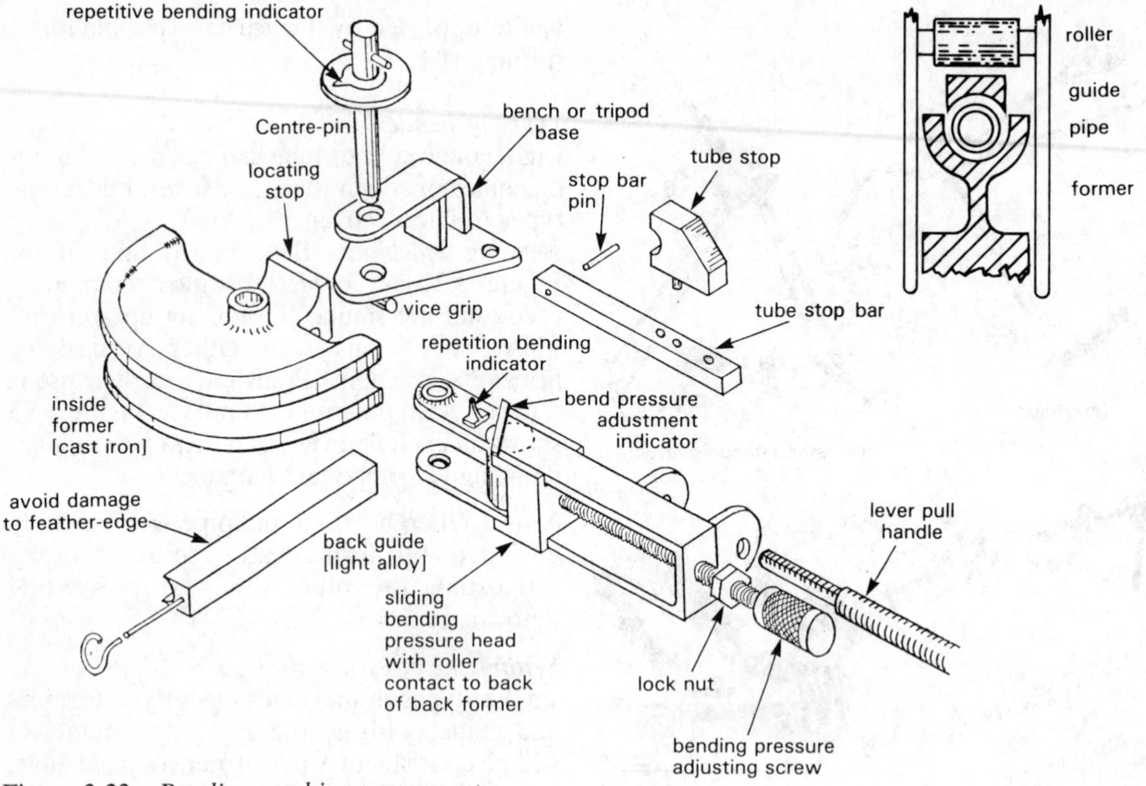

Figure 3.23 *Bending machine components*

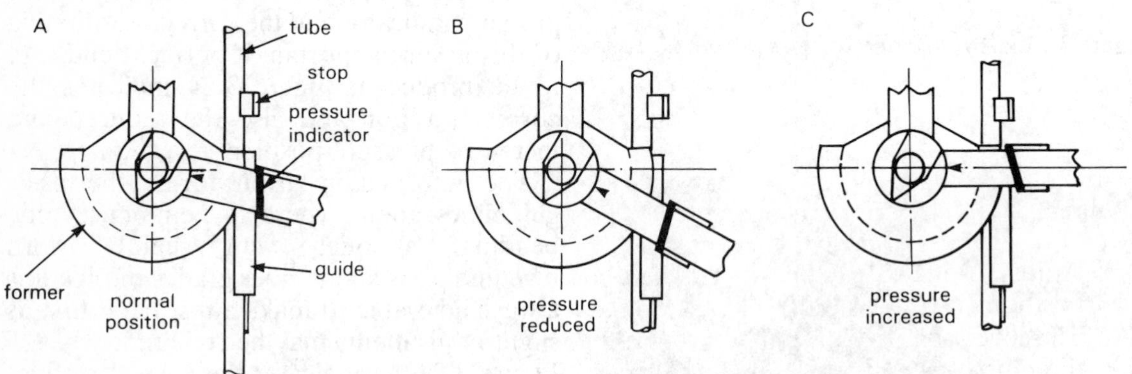

Figure 3.24 *Pressure adjustment*

Making a square bend on light gauge copper pipe

In this example we will be setting the bend to the outside of the former.

Method

1 Mark off pipe to the required distance;
2 Insert measured distance in the backside of the machine;
3 Ensure that the pipe fits right into the former and on to the stop;
4 Place alloy guide around the pipe, tighten pressure slightly to hold pipe in position;
5 Place square against mark on pipe, adjust pipe until the square touches the outside of the former;

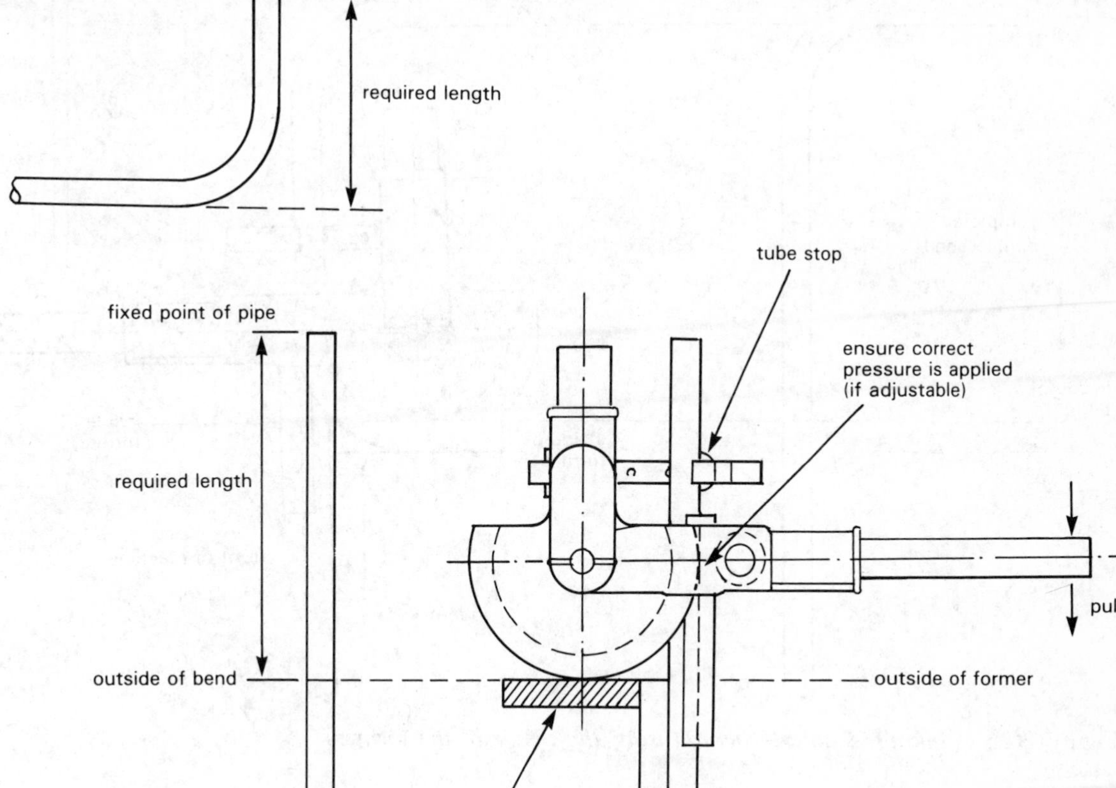

Figure 3.25 *Making a square bend*

6 Adjust pressure to the correct bending position (if of the adjustable pattern);
7 The lever arm is then pulled around and the pipe bent to the required 90° angle;
8 The bend will need to be very slightly over-pulled to counteract the spring back in the bend.

In the example shown in Figure 3.26 the procedure is almost identical to that explained previously and shown in Figure 3.25, except that in this case we work to the inside of the bend and the inside of the former. Both methods are equally correct and will produce identical bends.

Method

1–4 As for previous bend;
5 Adjust pipe until the square touches the inside of the former;
6–7 As for previous bend.

Note It must always be remembered that all bends *must* be placed at the *back* of the machine or errors in distances will occur.

Making an offset

An offset is also known as a double set, an ordinary single bend being known as a set. The method of setting up and operating the machine is as previously described.

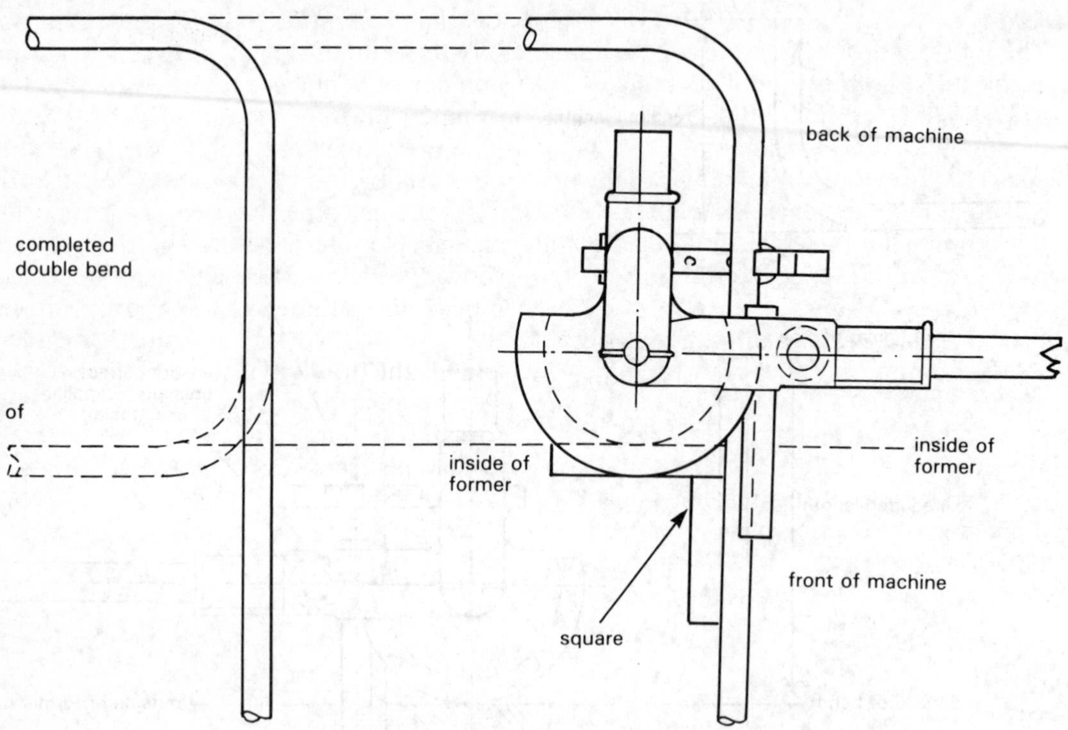

back of machine
completed double bend
inside of bend
inside of former
inside of former
front of machine
square

Figure 3.26 *Making a double bend (using the inside of the former)*

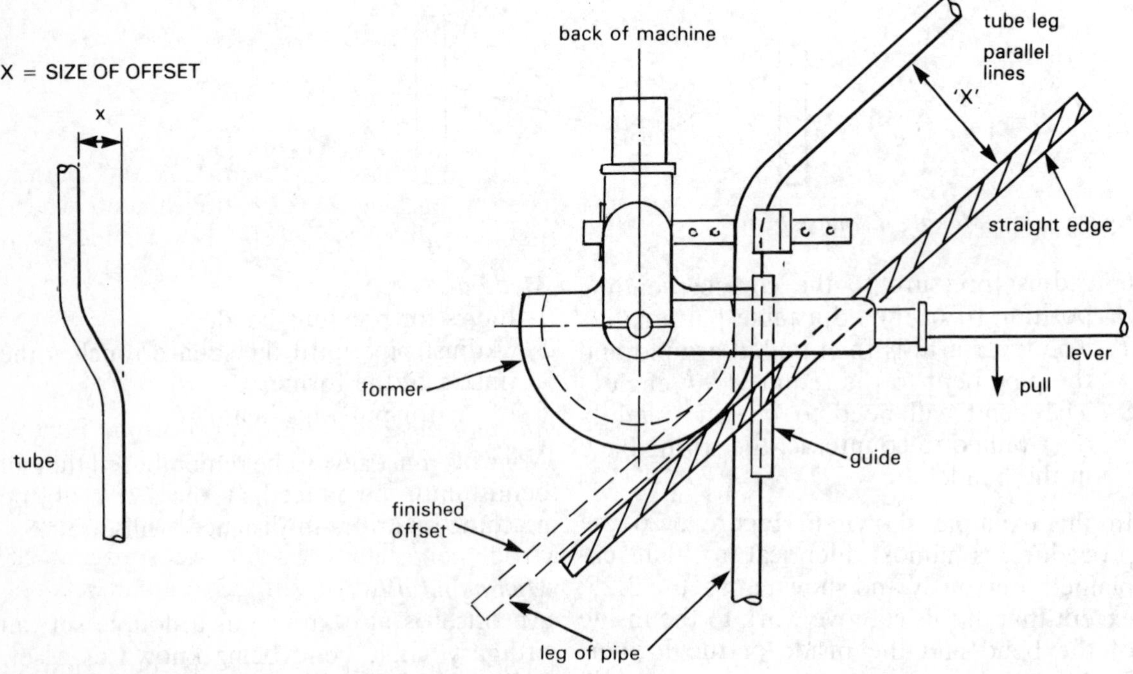

Figure 3.27 *Making an offset*

Method

1 The first bend or set on the pipe is made at the required position. The angle of the bend is not critical but 45° is usually recommended as satisfactory.
2 As stated previously all the bends must be placed at the back of the machine. Adjust the pipe in the machine, holding a slight pressure on the lever handle to hold the pipe in place.
3 Place straight edge against the outside of the former and parallel with the pipe.
4 Adjust the pipe in the machine until the required measurement is obtained.
Note To increase the size of the offset, push the pipe further through the machine. To decrease the size pull the pipe towards the front of the machine.
5 Apply pressure to lever arm and bend pipe until the legs are parallel.

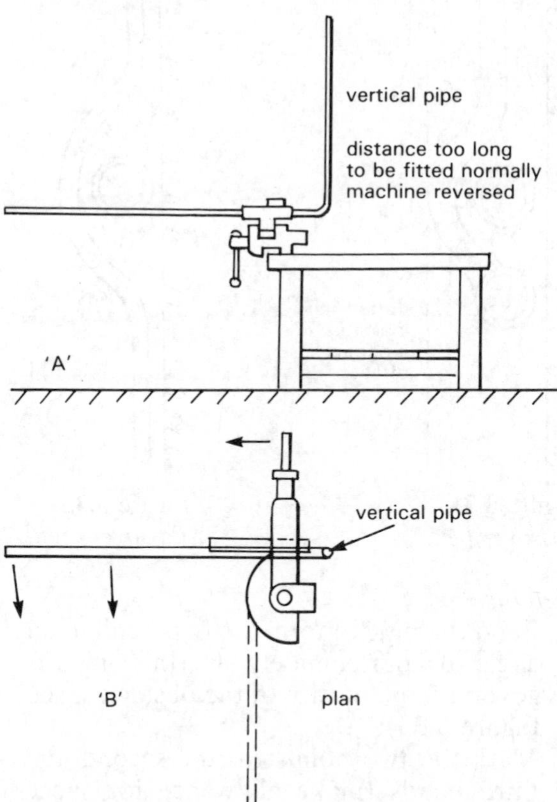

Figure 3.28 *Reverse bending*

Left-hand bending It is sometimes necessary to form left-hand bends, such as when a number of bends are required on one length of pipe or it may be impossible to locate the pipe in the machine due to the length of the pipe fouling the bench or the floor. In these cases the machine stop bar and lever arm are changed to operate from the reverse side as shown in Figure 3.28. The long length of the pipe is now situated in a vertical position. In the case of a vertical positioned bender on tripods the problem is solved by reversing the stop and placing the pipe in the bottom of the former and pulling upwards instead of downwards in the normal way.

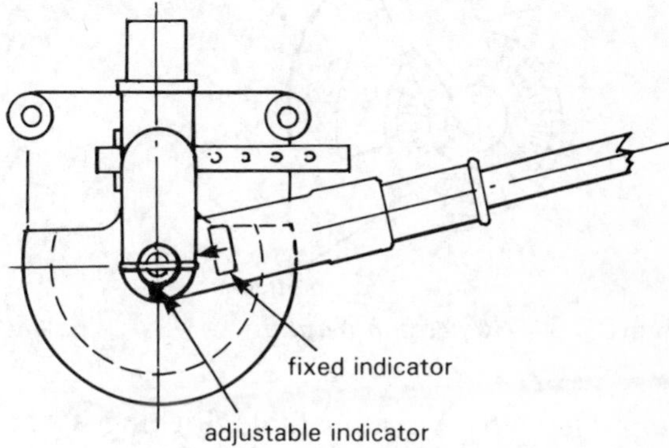

Figure 3.29 *Repetition bending*

Repetition bends On some of the better types of machine there are attachments which enable bends to be accurately repeated such as:

1 Indicators,
2 Stops,
3 Graduated protractors.

Indicator method This is as shown in Figure 3.29 where both fixed and adjustable indicators are an integral part of the bending machine. First, it is always necessary to make a trial bend, allowing for the spring back which will differ for different diameter pipes and machines.

1 Set and bend the pipe to the required angle;

2 Adjust the bend indicator to coincide with the fixed indicator on the handle mechanism;
3 Remove the pipe, ensuring the indicator is not altered. The machine is now set for repeat bends.

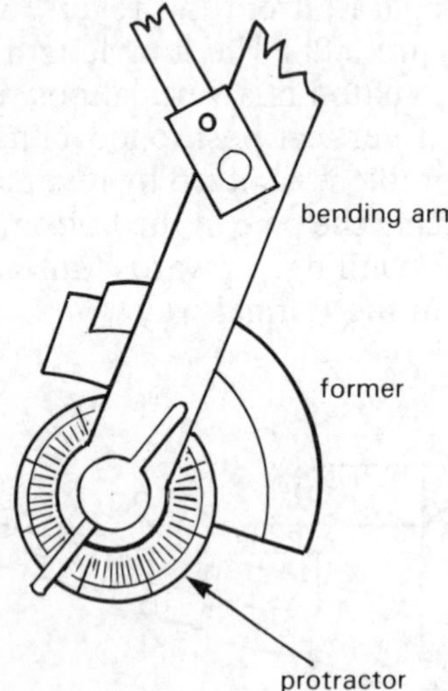

Figure 3.30 *Repetition bending*

Stop method An adjustable locking stop, which is situated in the bending quadrant of certain machines, is an alternative method of making repeat bends. The method of bending is as described in the indicator method. When the trial bend is correct the stop is then locked in that position; all repeat bends will now be bent to the same degree.

Protractor method On some machines of the larger type the angles are shown on a protractor. On the top of the former in this instance the pipe is bent until the bending arm indicates the required angle on the protractor, as in Figure 3.30.

Passover bend This is a double bend and another variation of the offset. This type of bend is used where a branch pipe is required to pass over another parallel pipe as shown in Figure 3.31. The method of making a '*passover*' is the same as for an ordinary offset as shown by the dotted lines in the sketch. The size of the offset (C–C) is indicated by distance 'X'.

1 Set out passover bend to required dimensions;
2 Bend first bend to approximately 45°;
3 Place pipe in machine as previously shown to obtain second bend;
4 This second bend is over-pulled to give the required angled bend.

Crank Another variation of the passover bend is shown in Figure 3.32 and is known as a 'crank'. This type of bend requires both skill and practice to ensure that it fits evenly to wall and passes uniformly over the obstruction (pipe). The size of the former of the bending machine will govern the length of the 'crank'.

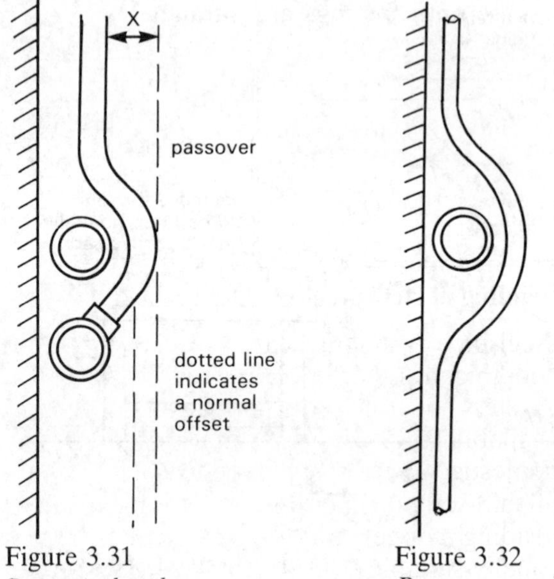

Figure 3.31
Passover bend

Figure 3.32
Passover crank

Method

1 Bend the pipe to form the first bend to an angle of approximately 90° (this angle is governed by the size of the obstacle) (see Figure 3.33);
2 Mark the two points for the second and third bends (make allowance for space between pipe and obstacle);
3 Line up mark with outside of former and

pull the bend to the required angle (use template);

4 Reverse the pipe and repeat operation for the third bend; check alignment of pipe and clearance of obstacle by the use of a straight edge (see Figure 3.34).

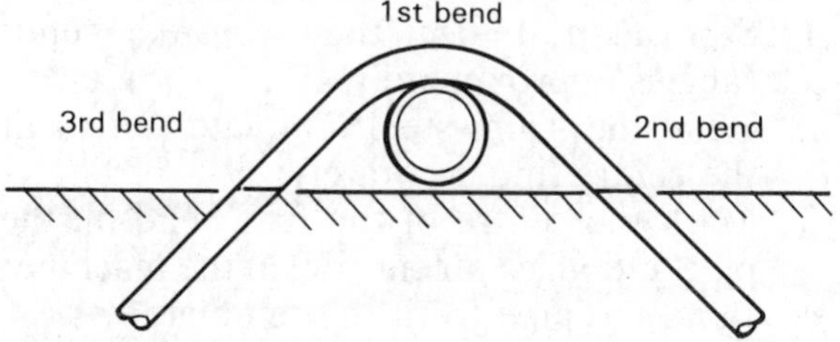

Figure 3.33 *Forming a crank*

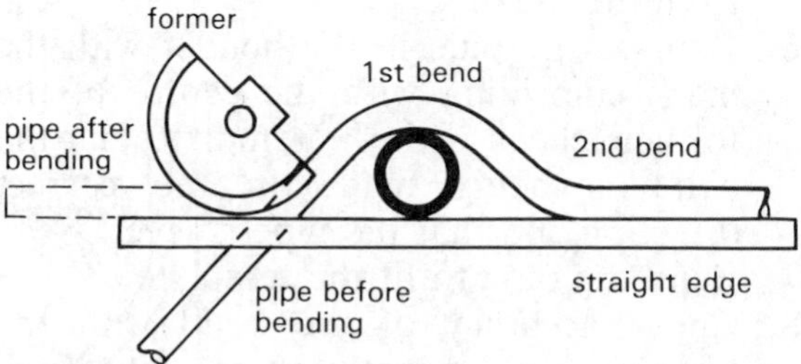

Figure 3.34 *Forming a crank*

Bending of steel pipe by machine

Steel pipe is manufactured in several grades and thicknesses depending on its use. Regardless of thickness there are machines available to bend the pipes used in any domestic situation satisfactorily. In the case of thin-walled pipes the machine shown for bending copper pipe (see Figure 3.23) is equally suitable; for the thicker-walled pipes the machine commonly used is the centre point push bender (see Figure 3.35). Although the actual working of the hydraulic bender is very simple there are one or two points which the operator must observe.

1 Select the correct former for the pipe to be bent. (Different formers are used when bending copper pipe with the same bender.)

2 Locate the pins in the correct holes. (This is most important or serious damage may result.)

3 The vent or breather valve is in the open position for most machines during the operation.

4 Should the ram fail to operate, check the oil level. If it is low, replenish, using only the correct type of oil. Do not overfill.

5 If the pipe is over-pulled replace the former with a flattener (supplied with most benders, see Figure 3.36); reverse the pipe, relocate pins in the flattening position, then apply pressure in the normal manner.

6 The use of a 4 mm steel template rod is strongly recommended.

7 During the bending operation the pipe often becomes wedged in the former; do *not* hammer the former to remove it. Completely remove the pipe and the former from the machine and strike the end of the pipe sharply on a wooden block. Hold the former if possible to prevent damage when it is released from the pipe (see Figure 3.37).

Making a square bend The method of making a square bend with a centre point bender will present no problem except when working from a fixed end of pipe. Even this is relatively simple once you appreciate that

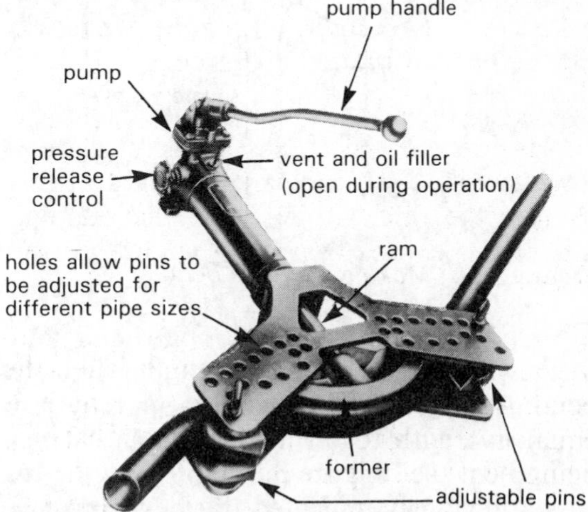

Figure 3.35 *Hydraulic bender*

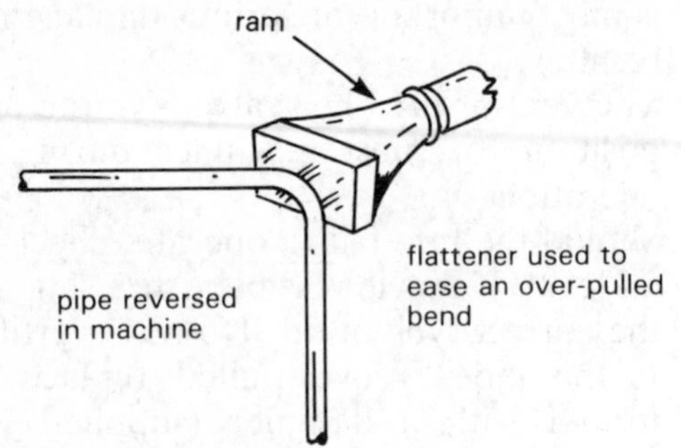

Figure 3.36 *Easing a bend*

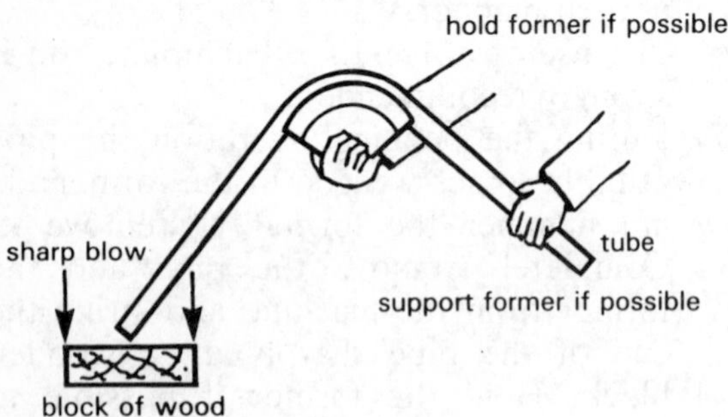

Figure 3.37 *Method of removing former*

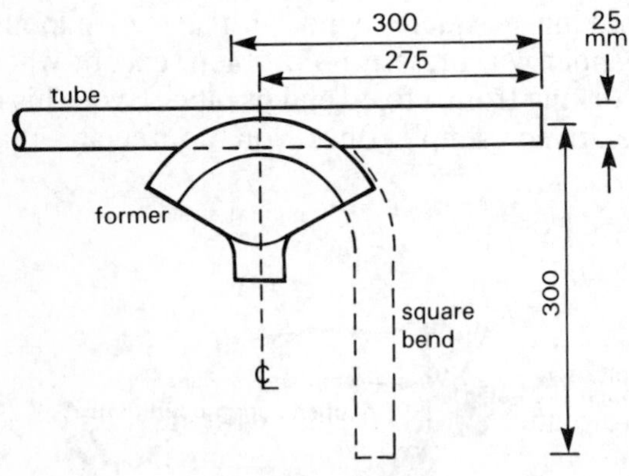

Figure 3.38 *Making a square bend*

there will be an increase in length when the bend is pulled. This increase in length is equal in length to the diameter of the pipe being bent (see Figure 3.38).

As previously indicated the bend must be slightly over-pulled to allow for spring back when the pressure is released on the ram. Check the bend with the aid of a square or a steel template before removing the pipe from the machine.

Offsets, passovers and cranks

Making an offset

1 Set out full size the required bends (angles approximately 45°);
2 Bend the 4 mm steel template rod to fit the centre line accurately;
3 Mark the centre of the first bend on the pipe and place it centrally in the machine;
4 Bend the pipe to fit the template;
5 Mark the centre of the second bend on the pipe (either from the template or from the drawing);
6 Replace the pipe in the bender with the mark coinciding with the centre of the former; check with the template that the bend is being made in the correct direction and that the pipe is level;
7 Bend the pipe to fit the template;
8 Check accuracy of the bend with the template before removing the pipe from the machine; if satisfactory remove and give final check with the drawing, straight edge and rule.

Passover It will easily be recognised that the passover bend (see Figure 3.40) is very similar to that of the offset shown in Figure 3.39, the difference being that the second bend is continued past the 45° angle until the required passover is obtained.

The method of setting out and bending is as already described for the offset.

Crank/passover Differences in terminology have already been pointed out and here again we find two different names given to an object.

Figure 3.41 is a further progression from the passover shown in Figure 3.40. Once again the actual bending procedure is as described for Figures 3.39 and 3.40, the centres of the bends being marked, then lined up with the centre of the former.

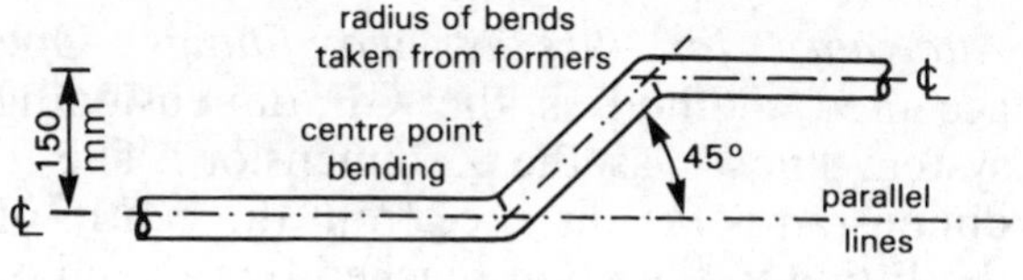

Figure 3.39 *Offset*

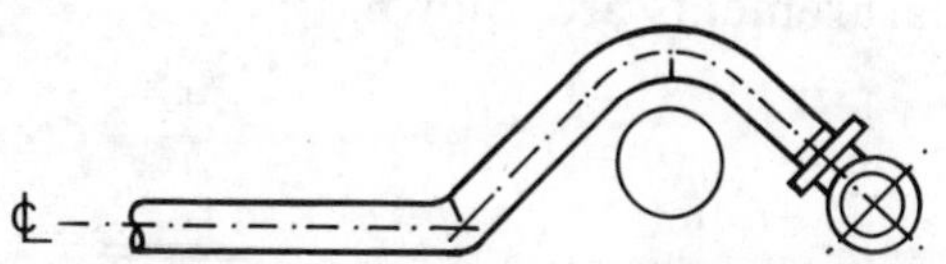

Figure 3.40 *Passover*

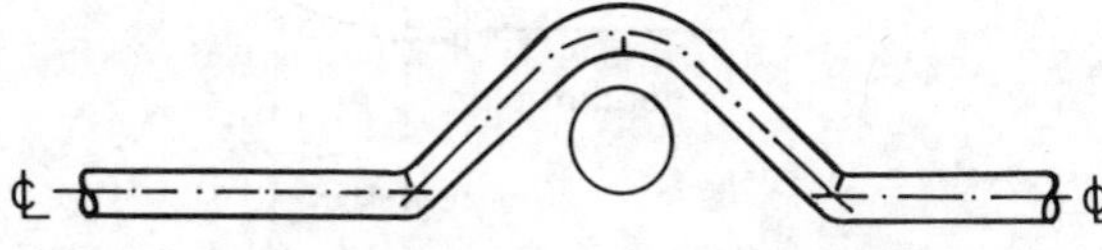

Figure 3.41 *Crank*

Emphasis on the use of templates cannot be over-stressed particularly when the bending becomes a little involved, i.e. in two planes or when the bending operation is to be performed some distance from the fixing location.

The GF dimensioning system

Malleable iron fittings Screwed malleable iron fittings are manufactured in accordance with BS 1256, with threads conforming to BS 21.

The specifying of the fittings is covered by BS 143 and 1256 which state:

1 *Equal sized fittings* Specify by reference to size, irrespective of the number of outlets, i.e. (½) 15 mm tee;
2 *Unequal fittings with two outlets* Specify the larger outlet size first, i.e. (¾ × ½) 20 × 15 mm elbow;
3 *Unequal fittings with more than two outlets* Specify outlet sizes in the following sequence (see Figure 3.42):

Note Where outlets 1 and 2 are equal, both are described by a single reference to size, i.e.

(A) 22 × 15 mm tee (¾ × ½)
(B) 22 × 15 mm cross (¾ × ½)
(C) 22 × 15 × 15 mm tee (¾ × ½ × ½).

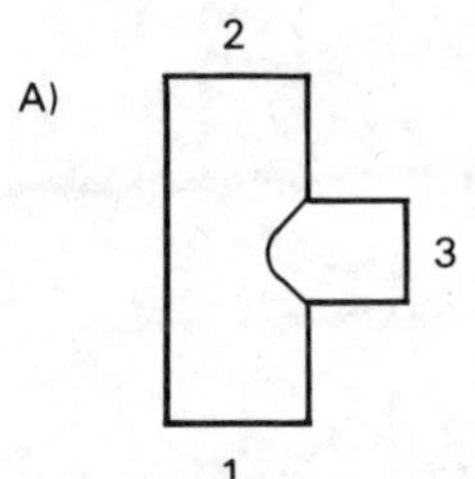

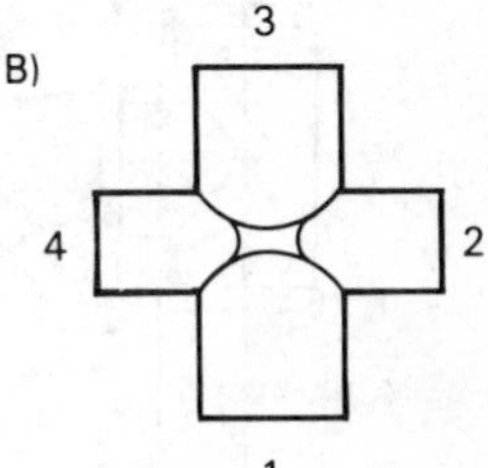

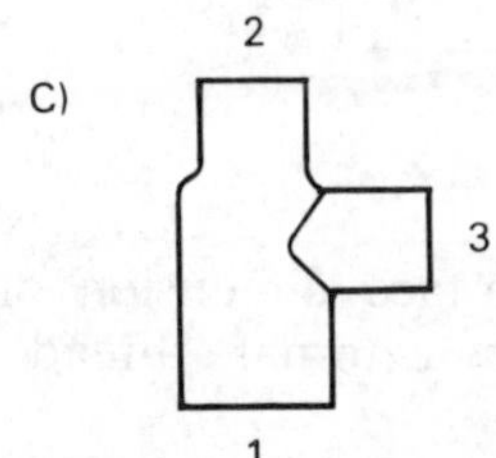

Figure 3.42

Sizes and dimensions Malleable iron fittings are manufactured for use with steel pipe and are identified by reference to the nominal size expressed in imperial terms. This is common in both the United Kingdom and in Europe, and the use of the term 'inches' will eventually disappear. The dimensions of fittings–as distinct from the nominal sizes–are expressed in millimetres in accordance with current metric practice. All the pipes and fittings are available in black or hot dipped galvanising.

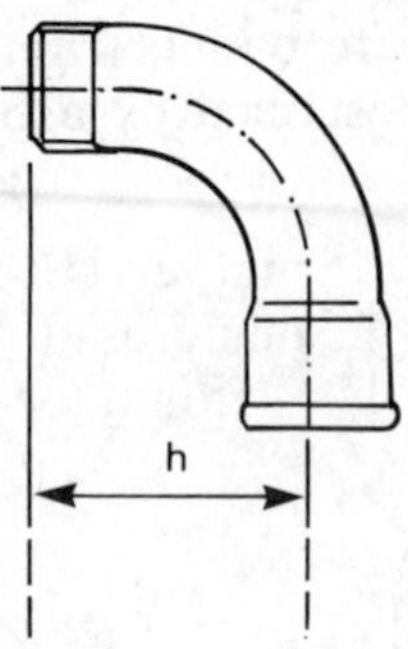

Figure 3.43

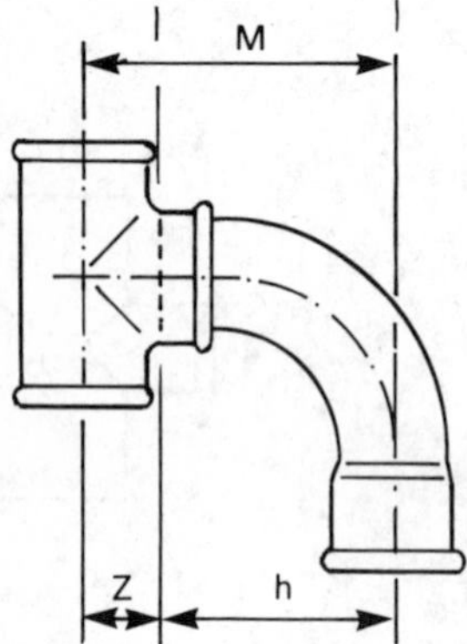

Figure 3.44

h is the centre-to-face dimension of the fitting with an external thread (see Figure 3.43);

Z is the dimension which remains after subtraction of the length of engagement of the thread from the centre-to-face or face-to-face dimension of a fitting (see Figure 3.43);

M is the centre-to-centre dimension of a pipe combination (see Figure 3.43). It is equal to the h and Z dimensions added together. M may also include the length of pipe between the two fittings (see Figure 3.46);

L is the cutting length of the pipe.

Note To determine the Z dimension of ancillary equipment, e.g. valves, take half the face-to-face dimension less one-third thread engagement length.

Allowance for threads and fittings One accepted method is the GF dimensioning system known as the Z dimension. The Z dimension is the distance from the centre of the fitting to the point reached by the end of the pipe when screwed the standard distance into the fitting (see Figure 3.45). This is a simple and reliable means of calculating the *actual length* of pipe when centre-to-centre measurements are known.

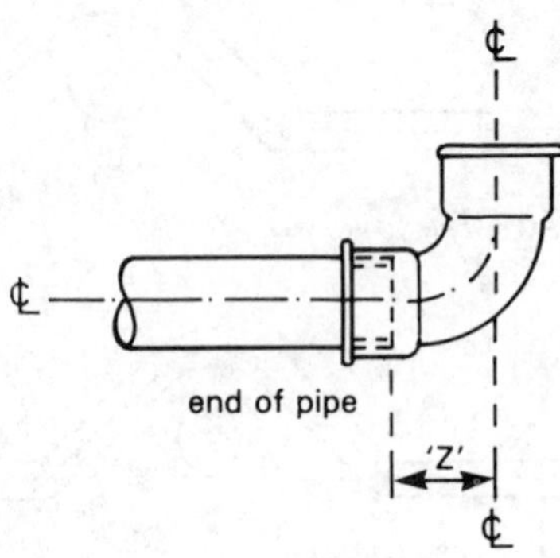

Figure 3.45

To obtain the length of pipe L subtract the two Z dimensions from the centre-to-centre dimension M. The length obtained will be the actual length of pipe required to be threaded to produce the assembly (see Figure 3.45).

Note Z measurements are read from the GF Tables (book of fittings).

Table 3.1 *Standard lengths of pipe threads*

Length of pipe thread			
Nominal pipe size	*Length of thread*	*Nominal pipe size*	*Length of thread*
8 mm	10 mm	32 mm	19 mm
10 mm	10 mm	40 mm	19 mm
15 mm	13 mm	50 mm	24 mm
20 mm	15 mm		
25 mm	17 mm		

The lengths of thread given are the standard engagement length of thread screwed into a fitting.

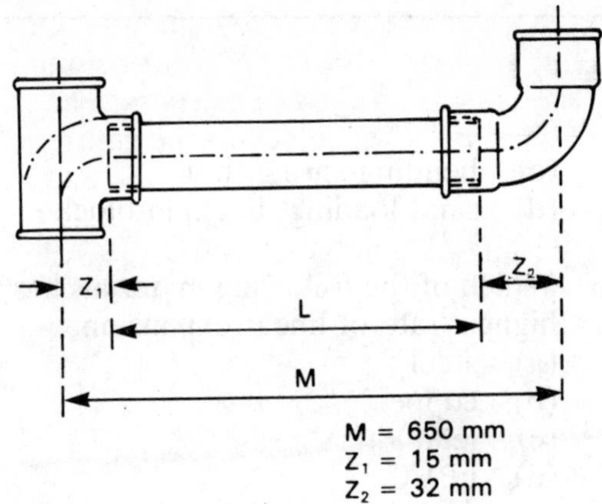

Figure 3.46

Example Calculate the length of 15 mm pipe required for Figure 3.46.

$L = M - (Z_1 + Z_2)$
$L = 650 - (15 + 32)$
$L = 650 - 47$
$L = 603$ mm (actual length of pipe)

Note For more detailed reading of this subject see GF literature.

Self-assessment questions

Working processes

1 Galvanising, the protective coating given to metals, is obtained by dipping the metal in:
(a) zinc
(b) tin
(c) solder
(d) aluminium

2 When a metal is annealed it is made:
(a) softer
(b) harder
(c) brittle
(d) less malleable

3 When removing a bending spring from a sharp set the correct procedure is to:
(a) twist to tighten the coil of the spring
(b) twist to loosen the coil of the spring
(c) straighten the pipe as far as possible
(d) pull it straight out

4 Throat rippling on machine-made bends on copper pipe is most probably caused by:
(a) an uneven pull on the machine handle
(b) the machine not being level
(c) pulling the bend too quickly
(d) an incorrectly adjusted or badly fitted guide

5 To obtain a 300 mm radius bend on 50 mm diameter low-carbon steel pipe, the most suitable method would be to use:
(a) sand loading, cold, bending ring
(b) rotary-type hydraulic bending machine
(c) bending spring, hot
(d) sand loading, hot, pin block

6 Which of the following materials has the highest rate of linear expansion:
(a) steel
(b) copper
(c) lead
(d) uPVC

7 An alloy is a:
(a) solution of metals
(b) mixture of metals
(c) compound
(d) an oxide of metals

8 To ensure that a hydraulic bending machine gives maximum thrust it is essential that the:
(a) correct former is used
(b) oil level is maintained
(c) machine is on a level base
(d) oil pressure is adjusted

9 Corrosion is the destruction of a metal by:
(a) physical change
(b) chemical change
(c) erosion
(d) incrustation

10 Pressure should never be exerted on the side of a grindstone wheel because:
(a) the stone may shatter
(b) the motor will slow down
(c) uneven wear will occur
(d) redressing will be required more frequently

4 Calculations

Introduction

After reading this chapter you should be able to:

1. Calculate areas and volumes.
2. Use simple formulae in calculations.
3. Complete calculations related to the cost of materials.
4. Complete calculations involving salaries.
5. Calculate perimeter lengths.
6. Use an electronic calculator to assist with basic mathematical functions.
7. Understand different methods of recording statistical information.
8. Apply square root calculations.
9. Calculate the capacity of pipes and vessels.
10. Understand and apply the theorem of Pythagoras to calculations related to building work.

Surface area and volume

The area of basic square and rectangular figures and the volume of cuboid figures have been dealt with in Chapter 9 of Book 1, where it was stated that area is a measure of the amount of surface and is measured in square units and that the volume of a body is a measure of the space it occupies and is measured in cubic units. It is necessary in this chapter to consider the area and volume of other shapes and figures, and also to calculate the perimeter length of these shapes.

Table 4.1 (on the next page) gives the area and perimeter formulae of some simple geometrical shapes.

Rectangles

A rectangle is a common shape in construction work and many calculation problems are based on plane figures. Often the shape in question is a combination of rectangular or square figures and there are several different methods of calculating their area as the following examples show.

Example 1 Calculate the total area of the shape in Figure 4.1 (on the next page). The shape is broken down into rectangles A, B and C. The area of each of these shapes is calculated and then added together to give the area of the whole shape.

$$\text{Area} = A + B + C$$

Table 1

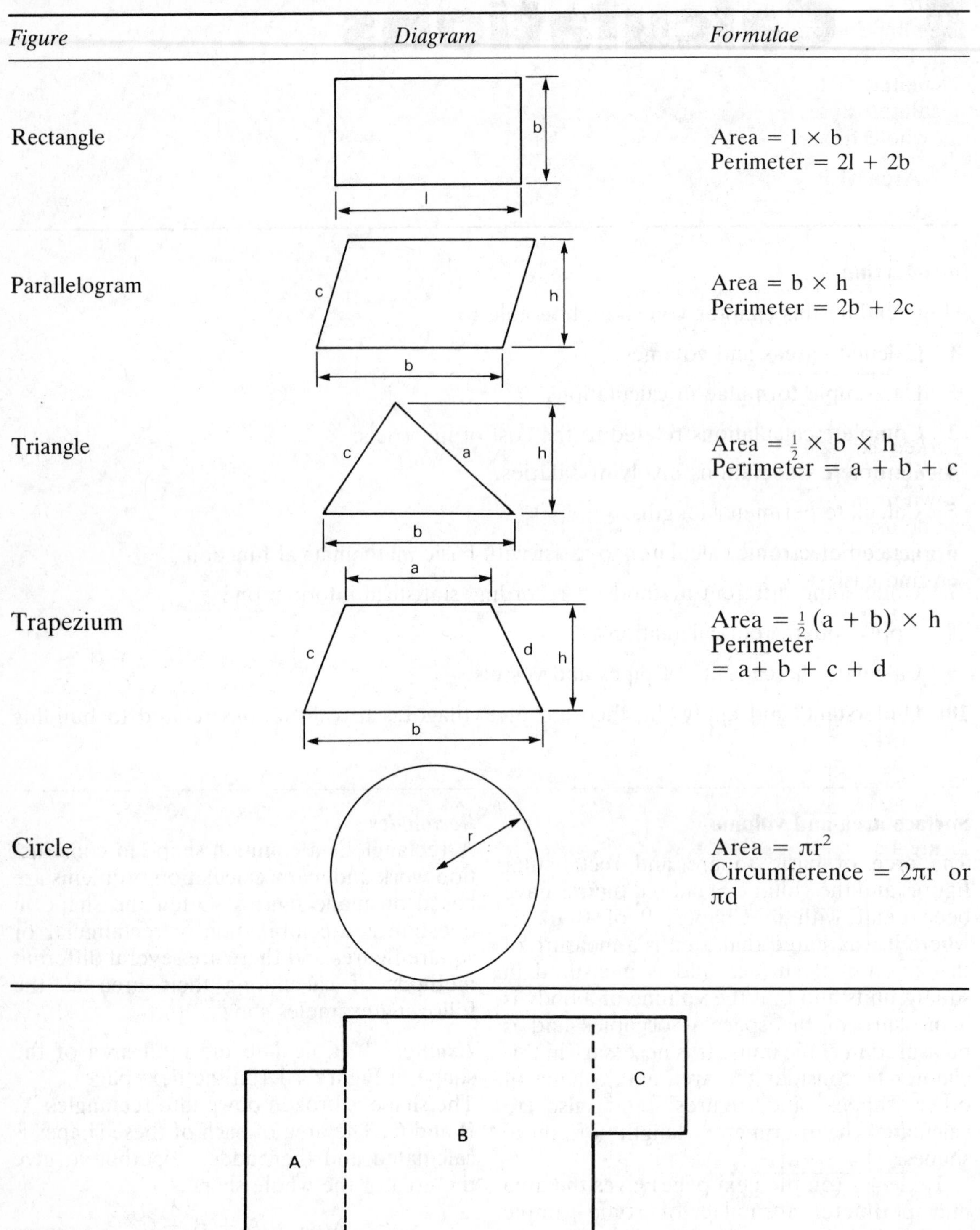

Figure	*Diagram*	*Formulae*
Rectangle		Area = $l \times b$ Perimeter = $2l + 2b$
Parallelogram		Area = $b \times h$ Perimeter = $2b + 2c$
Triangle		Area = $\frac{1}{2} \times b \times h$ Perimeter = $a + b + c$
Trapezium		Area = $\frac{1}{2}(a + b) \times h$ Perimeter $= a + b + c + d$
Circle		Area = πr^2 Circumference = $2\pi r$ or πd

Figure 4.1

Example 2 Calculate the area of shape B in Figure 4.2.
The shape is amended to include shapes D and E. The area of the whole figure is calculated. The areas of shapes D and E are calculated and subtracted from the area of the whole figure.

Area of B = total area − D and E

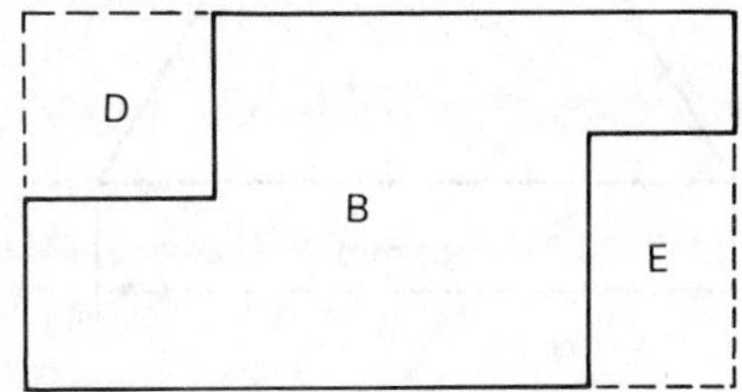

Figure 4.2

Worked examples
Calculate the surface area of the shapes shown in Figures 4.3 and 4.4. Either method can be used to calculate the areas, it is useful to practice both and then choose the method you find easiest.

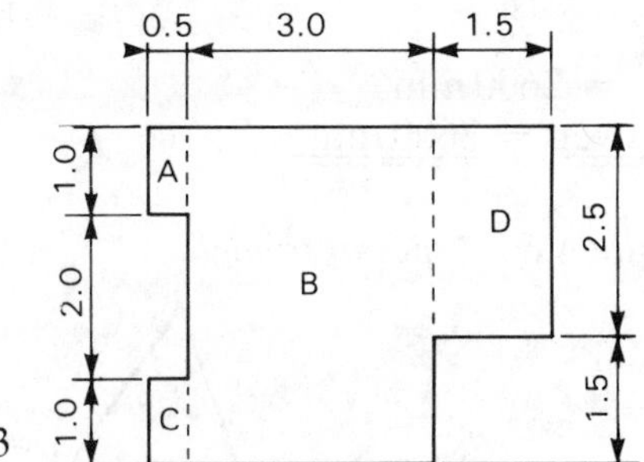

Figure 4.3

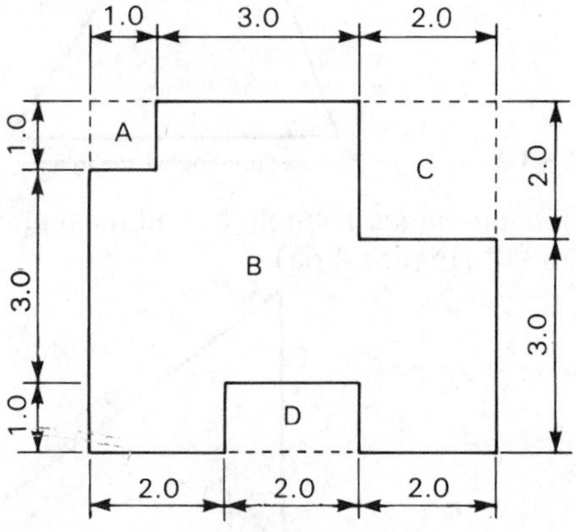

Figure 4.4

working margin

Example 3
Total surface area of Figure 4.3
= Sum of the areas of A + B + C + D

Area of A
= 1×0.5
= $\underline{0.5\,m^2}$

Area of B
= 3×4
= $\underline{12.0\,m^2}$

Area of C
= 1×0.5
= $\underline{0.5\,m^2}$

Area of D
= 2.5×1.5
= $\underline{3.75\,m^2}$

Total area
= $0.5 + 12.0 + 0.5 + 3.75\,m^2$
= $\underline{16.75\,m^2}$

Answer = $\underline{\underline{16.75\,m^2}}$

Working margin: $1.0 \times 0.5 = 0.5$; $3 \times 4 = 12$; $1.0 \times 0.5 = 0.5$; $2.5 \times 1.5 = 3.75$; $0.5 + 12.0 + 0.5 + 3.75 = 16.75$

Example 4
Surface area of part B of Figure 4.4
= Total area − areas of A, C, & D
= $6 \times 5\,m^2$
= $\underline{30\,m^2}$

Surface area of B
= 30 − areas of A, C, & D

Surface area of B
= $30 - (1 \times 1 + 2 \times 2 + 2 \times 1.0)$
= $30 - (1.0 + 4 + 2.0)$
= $30 - 7.0\,m^2$
= $23.0\,m^2$

Answer = $\underline{\underline{23.0\,m^2}}$

Working margin: $6.0 \times 5.0 = 30.0$; $2.0 \times 1.0 = 2.0$; $1.0 \times 1.0 = 1.0$; $1.0 + 4.0 + 2.0 = 7.0$; $2.0 \times 2.0 = 4.0$; $30.0 - 7.0 = 23.0$

Parallelograms

Parallelograms are four sided figures in which the sides opposite each other are equal in length and are also parallel to each other. The area of a parallelogram is calculated by multiplying the length of the base by the vertical or perpendicular height as shown in Example 5.

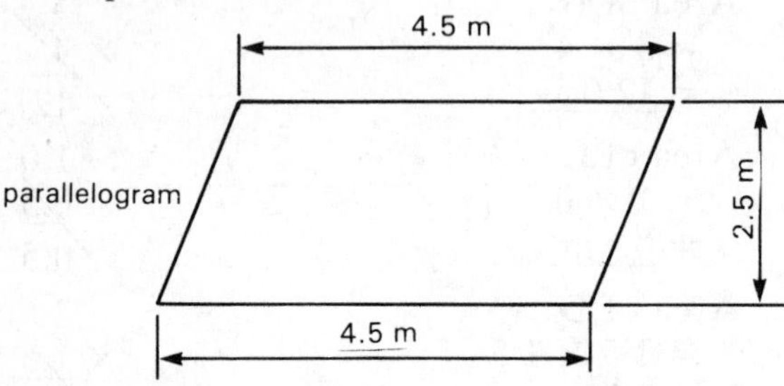

Figure 4.5a

Example 5 Calculate the surface area of the figure shown in Figure 4.5a.

Surface area of Figure 4.5a

= Base × vertical height

= $4.5 \times 2.5\,m^2$

= $11.25\,m^2$

Answer = $11.25\,m^2$

$$\begin{array}{r} 4.5 \\ \times\ 2.5 \\ \hline 225 \\ 90 \\ \hline 11.25 \end{array}$$

Triangles

A triangle is a three sided figure. There are several different types or forms of triangles as shown in Figure 4.6. There are also several different methods that may be used for calculating the area of a triangle, the most common method for most of the triangular shapes which will be encountered by the plumber is:

Area of triangle

$$= \frac{\text{Length of base} \times \text{perpendicular height}}{2}$$

$$= \frac{\text{Base} \times \text{height}}{2}$$

As with the area of a parallelogram it is the vertical height or perpendicular height which is used.

Example 6 Calculate the area of the triangle shown in Figure 4.5b.

Figure 4.5b

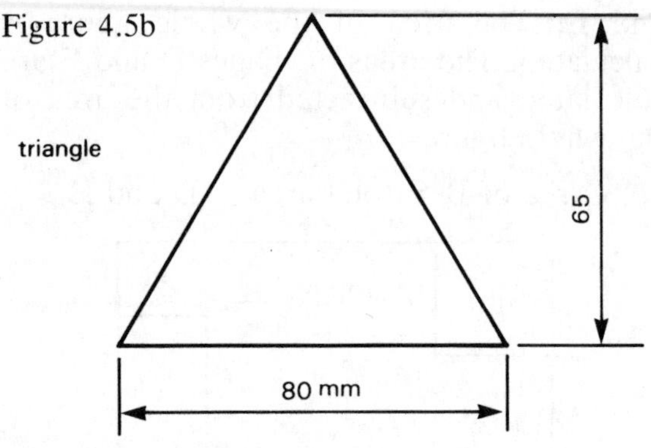

Formula

Area of the triangle

$$= \frac{\text{Base} \times \text{perpendicular height}}{2}$$

$$= \frac{80 \times 65}{2}\,mm^2$$

$$= \frac{5200}{2}\,mm^2$$

$$= 2600\,mm^2$$

Answer = $2600\,mm^2$

$$\begin{array}{r} 80 \\ \times\ 65 \\ \hline 400 \\ 480 \\ \hline 2)5200 \\ \hline 2600 \end{array}$$

Figure 4.6 *Types of triangle*

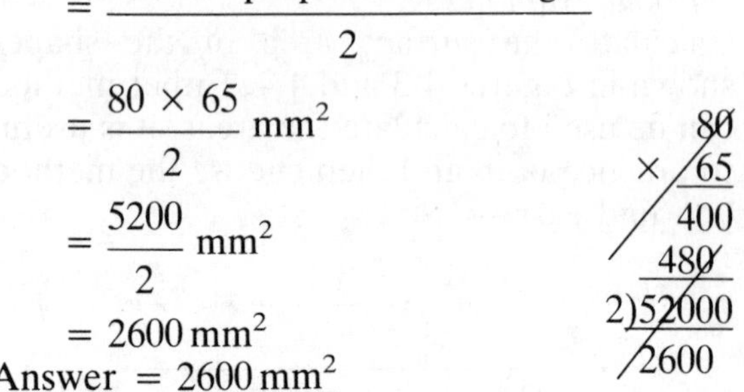

Figure 4.6a

(1) An *acute-angled* triangle has all its angles less than 90° (Figure 4.6a).

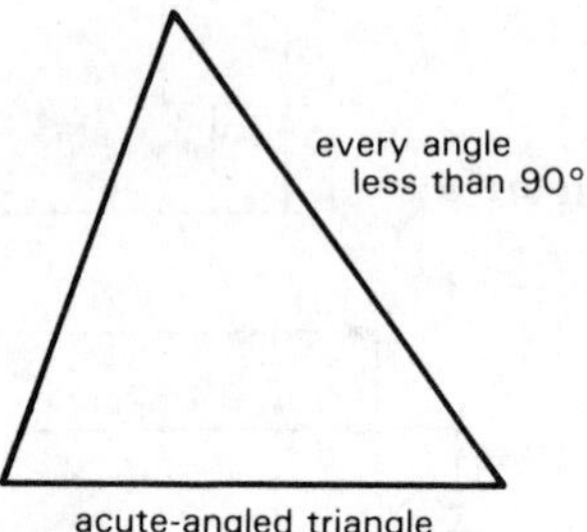

Figure 4.6b

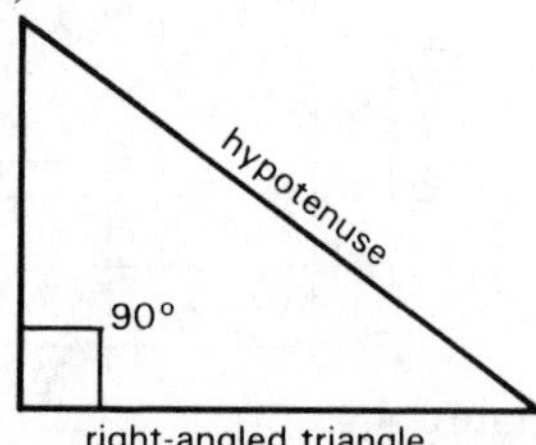

(2) A *right-angled* triangle has one of its angles equal to 90°. The side opposite to the right-angle is the longest side and it is called the hypotenuse (Figure 4.6b).

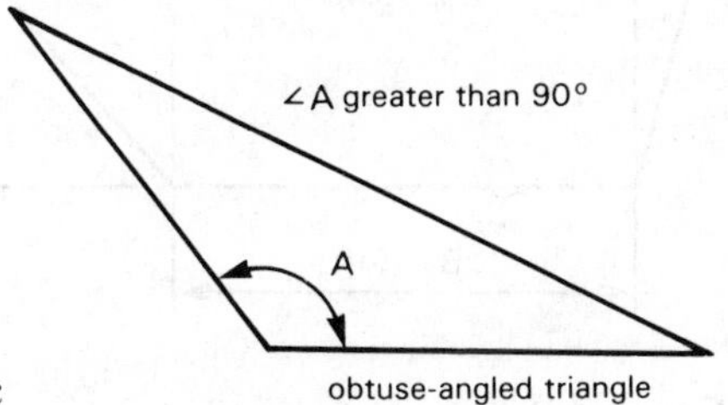

Figure 4.6c

(3) An *obtuse-angled* triangle has one angle greater than 90° (Figure 4.6c).

(4) A *scalene* triangle has all three sides of different length.

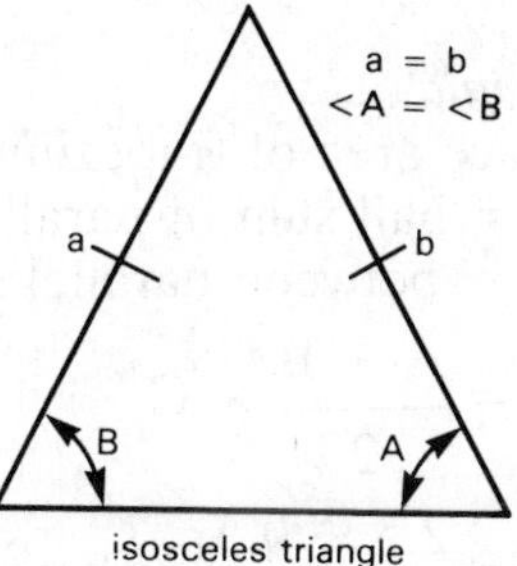

Figure 4.6d

(5) An *isosceles* triangle has two sides and two angles equal. The equal angles lie opposite to the equal sides (Figure 4.6d).

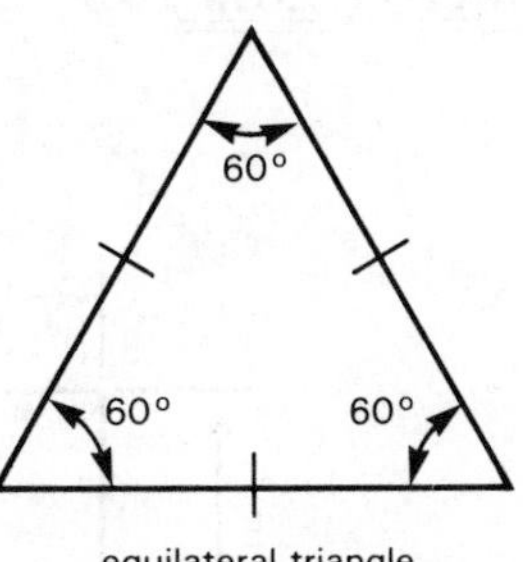

Figure 4.6e

(6) An *equilateral* triangle has all its sides and angles equal. Each angle of the triangle is 60° (Figure 4.6e).

When calculating the area of a right-angled triangle (a triangle in which one corner has an angle of 90°) it is often easier to extend the triangle into a square or rectangle as shown in Figure 4.7, then calculate the area of the figure and halve the answer for the triangle.

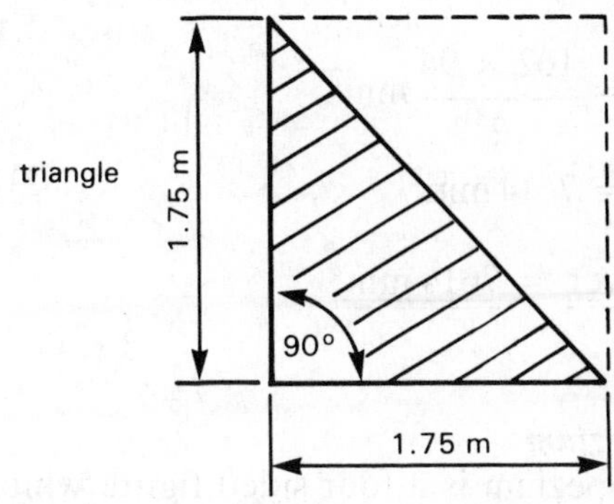

Figure 4.7

Example 7 Calculate the surface area of Figure 4.7.

Formula

Surface area of triangle

$$= \frac{\text{Base} \times \text{height}}{2}$$

$$= \frac{1.75 \times 1.75}{2}$$

$$= \frac{3.06}{2}\,\text{m}^2$$

$$= 1.53\,\text{m}^2$$

Answer = 1.53 m²

$$\begin{array}{r} \times\ 1.75 \\ \times\ 1.75 \\ \hline 875 \\ 1225 \\ 175 \\ \hline 2)3.0625 \\ 1.5312 \end{array}$$

Example 8 Calculate the surface area of Figure 4.8.

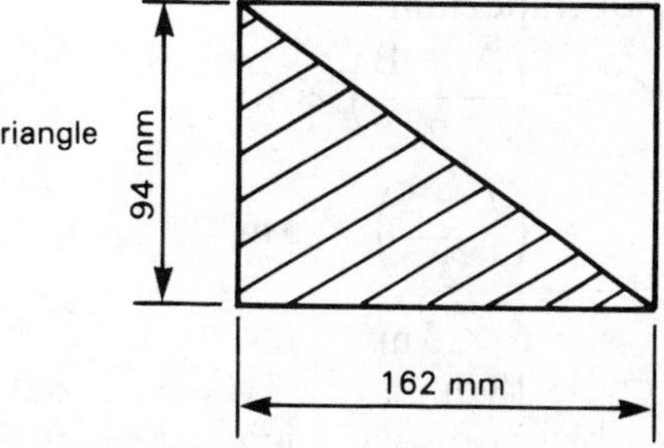

Figure 4.8

Formula

Surface area of triangle

$$= \frac{\text{Base} \times \text{height}}{2}$$

$$= \frac{162 \times 94}{2}\,\text{mm}^2$$

$$= 7614\,\text{mm}^2$$

$$\begin{array}{r} 162 \\ \times\ 94 \\ \hline 648 \\ 1458 \\ \hline 2)15228 \\ \hline 7614 \end{array}$$

Answer $= \underline{\underline{7614\,\text{mm}^2}}$

Trapezium

A trapezium is a four sided figure which has one or both pairs of sides unequal. The area of a trapezium can be viewed as the sum of the areas of two triangles and a rectangle as shown in Figure 4.9. However there is a more convenient method using the following formula.

Formula

Area of a trapezium = half the sum of the parallel sides × distance between them.

Usually written as

$$\text{Area} = \left(\frac{A + B}{2}\right) \times C$$

Where A & B = Length of parallel sides
C = Distance between sides

Note: See Figure 4.9

Example 9 Calculate the area of the trapezium shown in Figure 4.9.

Area of trapezium

$$= \left(\frac{A + B}{2}\right) \times C$$

$$= \left(\frac{7 + 5}{2}\right) \times 3\,\text{m}^2$$

$$= 6 \times 3\,\text{m}^2$$

$$= 18\,\text{m}^2$$

$$\begin{array}{r} 7 \\ +\ 5 \\ \hline 2)12 \\ \hline 6 \\ \times\ 3 \\ \hline 18 \end{array}$$

Answer $= \underline{\underline{18\,\text{m}^2}}$

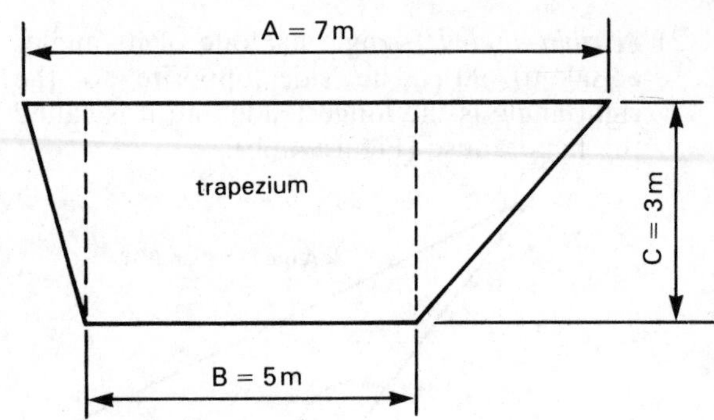

Figure 4.9

Example 10 Calculate the surface area of the side cheek of the dormer window shown in Figure 4.10.

Formula

Surface area of trapezium

= half sum of parallel sides × distance between parallel sides

$$= \frac{A + B}{2} \times C$$

$$= \frac{2.5 + 0.5}{2} \times 2.\,\text{m}^2$$

$$= 1.5 \times 2\,\text{m}^2$$

$$= 3.0\,\text{m}^2$$

$$\begin{array}{r} 2.5 \\ +\ 0.5 \\ \hline 2)3.0 \\ \hline 1.5 \\ \times\ 2 \\ \hline 3.0 \end{array}$$

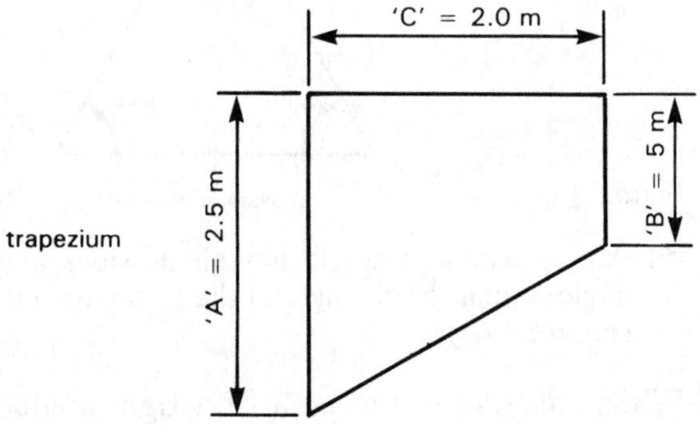

Answer $= \underline{\underline{3.0\,\text{m}^2}}$

Figure 4.10

The circle

Before attempting to calculate the area of a circle it is necessary to understand the names used to identify the different parts of a circle, as shown in Figure 4.11. The line which gives the circle its shape and marks the outer edge is called the *circumference*. The distance from the centre of a circle to any point on its circumference is called its *radius* and is represented by its symbol r. A straight line which extends from the circumference and passes through the centre of the circle to the circumference on the other side of the circle is twice the length of the radius and is called the *diameter*, represented by the symbol d.

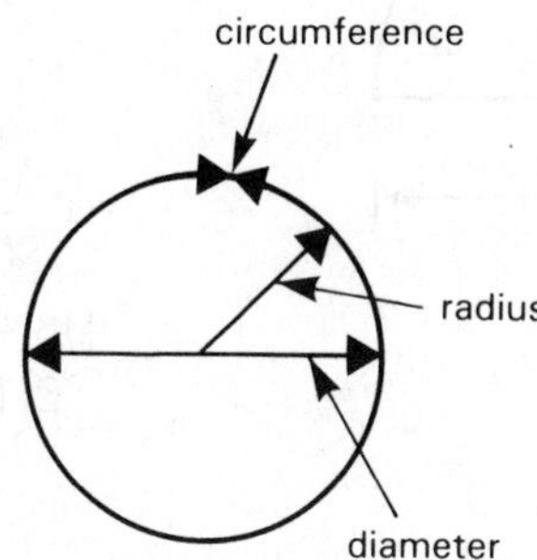

Figure 4.11 *Basic parts of a circle*

An important mathematical constant is used when solving area and perimeter calculations of a circle. This constant is called 'Pi' and is represented by the symbol π. The numerical value of the constant is 3.142 and is the number of times that the diameter of the circle will go into the circumference length of the same circle. This constant applies to all circles, whatever their size. See Book 1 for more detailed information.

The area of a circle is calculated using the formula πr^2:

Area of a circle $= \pi \times r^2$

where $\pi = 3.142$

$r =$ radius of circle

The index (small) 2 indicates that the radius is to be squared.

Note: To square a figure you have to multiply it by itself

i.e. $4^2 = 4 \times 4$

$4^2 = 16$

$\therefore$ The surface area of a circle

$= \pi \times r^2$

$= \pi \times$ radius $\times$ radius

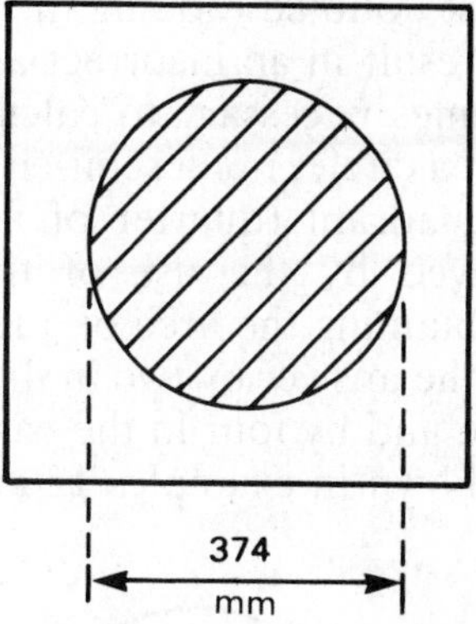

Figure 4.12

Example 11 Calculate the surface area of a circular sheet of metal as shown in Figure 4.12 with a diameter of 374 mm (note radius is half the diameter).

Formula

Surface area of circle

$= \pi \times r^2$

$= 3.142 \times 187 \times 187\,\text{mm}^2$

$= 109872\,\text{mm}^2$

Answer $= 109872\,\text{mm}^2$

2)374
187
× 187
1309
1496
187
34969
×3.142
69938
139876
34969
104907
109872.598

In some calculations the radius length may be given instead of the diameter (as in the previous example), it is always necessary to make sure that you use the radius length when solving circular areas. It must also be remembered to 'square' the radius (multiplying the number by itself) should the number be just doubled (adding the number to itself) will result in an incorrect answer.

It is sometimes necessary to calculate the area of part of a circle, i.e. a semi-circle (half a circle) or quadrant (quarter of a circle). This is achieved by the use of the same formula for obtaining the area of a full circle then dividing the answer by two in the case of the semi-circle and by four in the case of the quadrant as shown in examples 12 and 13.

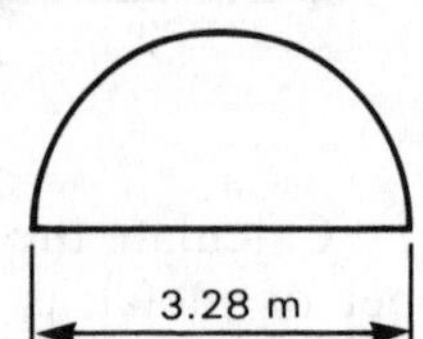

Figure 4.13

Example 12 Calculate the area of the semi-circular roof shown in Figure 4.13.

Formula

Area of a circle $= \pi r^2$

Area of semi-circle $= \dfrac{\pi \times r^2}{2}$

$= \dfrac{3.142 \times 1.64 \times 1.64}{2}\,\text{m}^2$

$= 4.225\,\text{m}^2$

Answer $= 4.225\,\text{m}^2$

2)3.28
1.64
1.64
656
984
164
2.6896
3.142
53792
107584
26896
80688
2)8.4507232
4.2253616

Example 13 Calculate the area of the quadrant shown in Figure 4.14.

Formula

Area of a circle $= \pi r^2$

Area of a quadrant $= \dfrac{\pi r^2}{4}$

$= \dfrac{3.142 \times 1.64 \times 1.64}{4}\,\text{m}^2$

$= 2.112\,\text{m}^2$

Answer $= 2.112\,\text{m}^2$

1.64
1.64
656
984
164
2.6896
3.142
53792
107584
26896
80688
4)8.4507232
2.112

1.64m

Figure 4.14

It is very easy to see and to appreciate that a semi-circle is half of a circle and that the area is obtained by simply dividing the area of the circle by two, also that a quadrant is a quarter of a circle and the area is obtained by dividing the area of the circle by four. It must be understood that to obtain these divisors all we are doing is to use the number of degrees in the circle as a divisor and the number of degrees in that part of the circle, i.e. semi-circle or whatever, as the numerator as the following examples will clearly show.

1 The number of degrees in a full circle is 360°.
2 The number of degrees in half a circle (semi-circle) is 180°.
3 The number of degrees in a quarter of a circle (quadrant) is 90°.

Example 14 The area of a semi-circle

$= \pi r^2 \times \dfrac{180}{360}$

$$= \pi r^2 \times \frac{1}{2}$$

Example 15 The area of a quadrant

$$= \pi r^2 \times \frac{90}{360}$$

$$= \pi r^2 \times \frac{1}{4}$$

It must now be easy to understand that to obtain the surface area of *any sector* of a circle it is a simple matter of dividing by 360 (the number of degrees in a circle) and multiplying by X (the number of degrees in the required sector).

The area of the sector of a circle

$$= \pi r^2 \times \frac{X}{360}$$

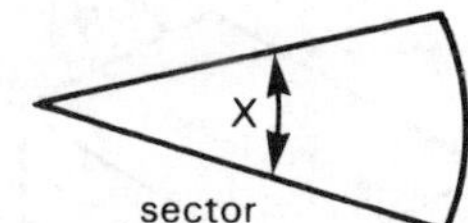

Figure 4.15

Alternative method of obtaining the area of a circle

The formula πr^2 is perhaps the one universally used for obtaining the area of a circle and has been fully explained in the previous examples. Plumbers always refer to the size of a pipe by its diameter so it therefore follows that a formula using the diameter dimension must be of considerable advantage. Therefore for calculations involving the size of pipes, i.e. their diameter, the formula for the *area of a circle* $\pi d^2/4$ is recommended and will be used later in the chapter when dealing with pipe sizing problems.

Perimeter length

The perimeter length of any flat shape or figure is the total length of its outside boundary line or lines. With straight sided figures such as squares, rectangles, parallelograms and triangles it simply means adding the lengths of all the sides together to obtain the perimeter length, as shown in the examples (see Table 4.1, p. 126).

Example 16 Calculate the perimeter of the trapezium shown in Figure 4.16.

Formula Perimeter of trapezium

= Sum of the length of all the sides.

= 1.423 + 1.206 + 0.980 + 0.940 m

= 4.549 m

Answer

= 4.549 m

1.423
1.206
0.980
0.940
4.549

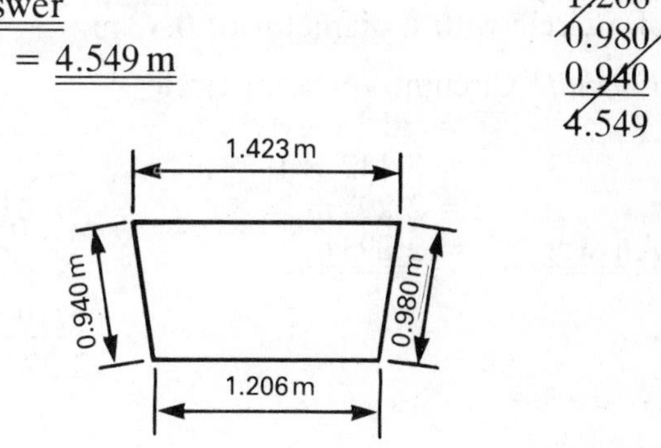

Figure 4.16

Example 17 Determine the perimeter of the triangle shown in Figure 4.17.

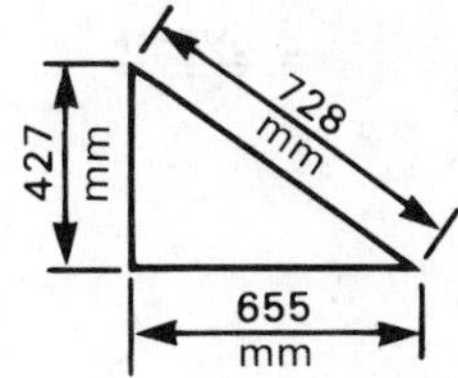

Figure 4.17

Formula Perimeter of triangle

= Sum of the lengths of all the sides.

= 728 + 655 + 427 mm

= 1810 mm

$$= \frac{1810}{1000}\,\text{m}$$

= 1.81 m

Answer

= 1.81 m

728
655
427
1810

Perimeter length of a circle
Calculations involving the perimeter length of a circle require a different method to that for straight-sided figures. The curved perimeter length of a circle is called its circumference.

When calculating the perimeter length of a circle (circumference) the distance is obtained by multiplying the diameter of the circle by π(3.142).

Formula Circumference of a circle = (a) πd
Alternative (b) 2πr

Example 18 Determine the circumference of a circle with a diameter of 0.73 m.

Formula Circumference of circle
= πd
= 3.142 × 0.73 m
= 2.293 m
Answer = 2.293 m

3.142
× 0.73
9426
21994
2.293 66

Alternative formula
Circumference of circle
= 2πr
= 2 × 3.142 × 0.365 m
= 2.293 m
Answer = 2.293 m

3.142
2
6284
0.365
31420
37704
18852
2.293 660

Surface area
The surface area is the amount of surface which covers a shape or figure. Plumbers have to work with cuboid shapes such as hot water tanks and/or cold water cisterns, or circular storage vessels such as hot storage cylinders.

With a closed cuboid shape as shown in Figure 4.18 the surface area may be calculated as in the following example.

Example 19 Calculate the total surface area of the closed cold water storage tank shown in Figure 4.18.

Formula Total surface area
= The sum of the area of all six sides
$= (1.25 \times 0.85 \times 2) + (1.25 \times 0.9 \times 2) + (0.9 \times 0.85 \times 2)\ m^2$
$= (2.125 + 2.25 + 1.53)\ m^2$
$= 5.905\ m^2$
Answer = $5.905\ m^2$

1.25
× 0.85
625
1000
×2
2.1250

1.25
×0.9
1.125
×2
2.250

0.9
× 0.85
45
72
0.765
×2
11.530

2.125
2.250
1.53
5.905

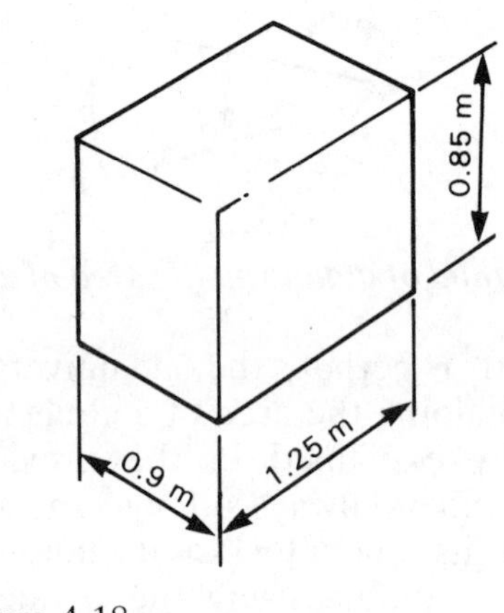

Figure 4.18

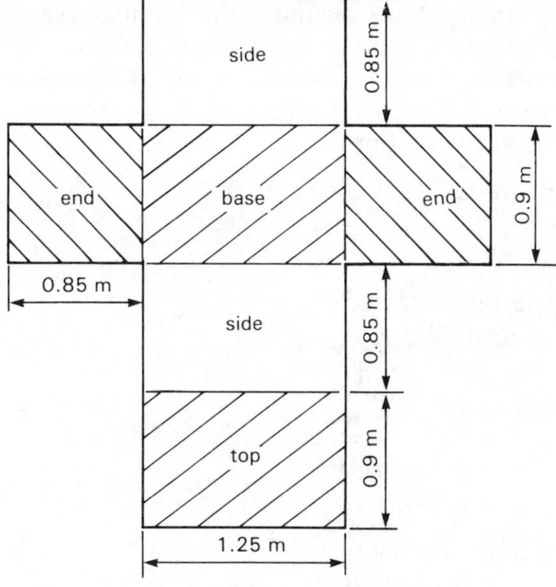

Figure 4.19

There are alternative methods of calculating the surface area. One such method is shown in Figure 4.19. Open out the tank and then treat each of the six parts as separate sections, then add all the sections together to obtain the total. Another method is to add together certain sections – shown hatched to form various rectangles – and to obtain the area of the enlarged sections, then adding each area together to obtain the total area.

Surface area of flat ended cylinders (cylindrical vessels)

The surface area of a cylinder as shown in Figure 4.20 really represents two circles for the ends and a rectangular shape for the side (when cut and opened). See development of cylinders in Book 1.

Example 20 Calculate the surface area of the flat ended cylinder as shown in Figure 4.20. The height is 930 mm and the diameter 350 mm.

Formula

Total surface area of cylinder

= Area of both ends + area of curved side

$= (\pi r^2 \times 2) + (\pi dh)$

$= (3.142 \times 175 \times 175 \times 2) + (3.142 \times 350 \times 930)\ \text{mm}^2$

$= 192447.5 + 1022721\ \text{mm}^2$

$= 1215168.5\ \text{mm}^2$

<u>Answer</u> = <u>$1215168.5\ \text{mm}^2$</u>

3.142
×175
549.850
175
96223.75
×2
192447.50

3.142
350
1099.70
×930
1022721.0
192447.5
1215168.5

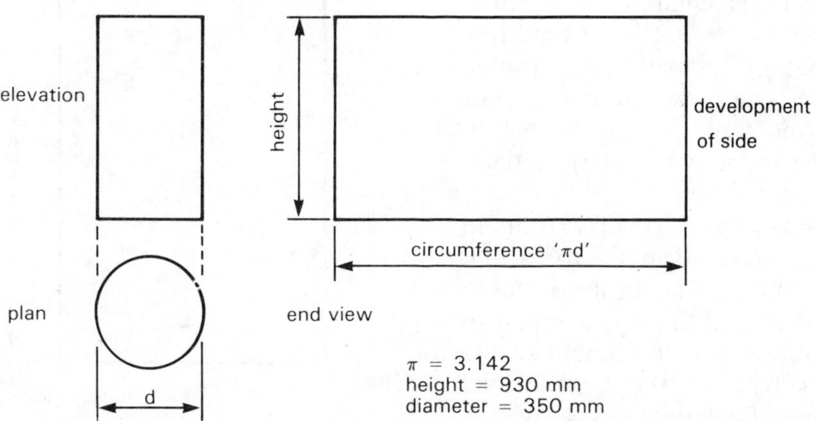

Figure 4.20 *Development of side*

To convert the answer to metres^2 divide by 1 000 000
(1 000 000 square millimetres in 1 square metre.)

$$\text{Total surface area} = \frac{1\,215\,168.5}{1\,000\,000}$$
$$= 1.215\,\text{m}^2 \quad \text{move the decimal point 6 places}$$

Answer $= 1.215\ \text{m}^2$

An alternative method is to convert mm to m first by dividing by 1000.
Total surface area of cylinder

$$= \left(3.142 \times \frac{175}{1000} \times \frac{175}{1000} \times 2\right) + \left(3.142 \times \frac{350}{1000} \times \frac{930}{1000}\right)\text{m}^2$$
$$= (3.142 \times 0.175 \times 0.175 \times 2) + (3.142 \times 0.350 \times 0.93)\,\text{m}^2$$
$$= 0.192 + 1.022\,\text{m}^2$$
$$= 1.214\,\text{m}^2$$

Answer $= 1.214\,\text{m}^2$

Volume
The volume of a body is the measure of the space it occupies. Basic volume calculations related to cubical figures have been dealt with in Book 1. Readers will remember that volume is measured in cubic units such as mm^3, m^3 etc. and calculated in a three dimensional manner, i.e. length × breadth × depth = units^3. When calculating the volume of curved shapes such as circular water storage cisterns and tanks it is necessary to make use of π (Pi) and area calculations dealt with earlier in this chapter.

Hot water storage cylinders have curved tops (domed) and bases (see Figure 4.21). For volume calculation purposes it is usual to assume that the curve in the top is equal to the curve in the base, and the height of the cylinder from its base to the start of the domed top is the dimension used for the height of the cylinder.

Example 21 Calculate the volume of a cylinder 1.25 m high with a diameter of 700 mm.
Note All dimensions must be in the same unit (1000 mm = 1 m).

Formula Volume of cylinder
$$= \text{Area of base} \times \text{height}$$
$$= \pi r^2 \times h$$

The radius in metres
$$= \frac{700}{2} \times \frac{1}{1000}$$
$$= 350 \times \frac{1}{1000}\,\text{m}$$
$$= 0.35\,\text{m}$$

Volume of cylinder
$$= 3.142 \times 0.35 \times 0.35 \times 1.25\,\text{m}^2$$
$$= 0.49\,\text{m}^3$$

Answer $= 0.49\,\text{m}^3$

A length of pipe, duct or any circular vessel can be dealt with in the same manner as shown in Example 21.

Example 22 Calculate the volume of air contained in any empty drainage pipeline

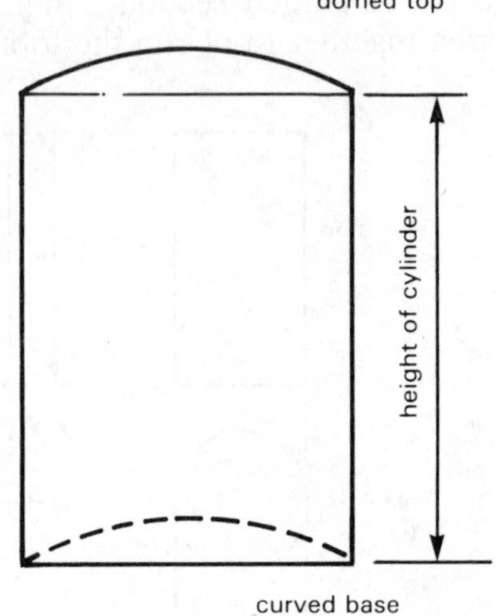

Figure 4.21

which is 16.5 m long and has a diameter of 100 m.

Note All dimensions must be in the same unit.

Formula Volume of pipeline
= Area of pipe × length of pipe
= $\pi r^2 \times l$ (length)

Radius of pipe in m
$= \frac{100}{2} \times \frac{1}{1000}$ m
= 0.05 m

Volume of pipeline
= $3.142 \times 0.05 \times 0.05 \times 16.5\,m^3$
= $0.1296\,m^3$
= $0.13\,m^3$ (nearest two places of decimal)

Answer = 0.13 cubic metres of air.

Note The rule used to obtain the answer to the nearest two places of decimal is to take the last number and if that number is five or greater you add one to the preceding figure.

Example (a) 0.1346
= 0.135 to the nearest 3 places
= 0.14 to the nearest 2 places

(b) 2.3061
= 2.306 to the nearest 3 places
= 2.31 to the nearest 2 places

Capacity

The measurement of liquids held in a vessel or pipeline is referred to as its capacity. The unit used for recording capacity is the litre. When calculating the capacity of pipes or storage vessels it is usual to calculate the volume of the pipe or vessel as shown in previous examples and then to multiply the volume answer by 1000 which will convert the volume figure to litres. This rule will apply if the volume is in cubic metres (m^3).
Note 1 m^3 of volume is equal to 1000 litres.

Example 23 Calculate the capacity in litres of the hot water storage vessel shown in Figure 4.22.

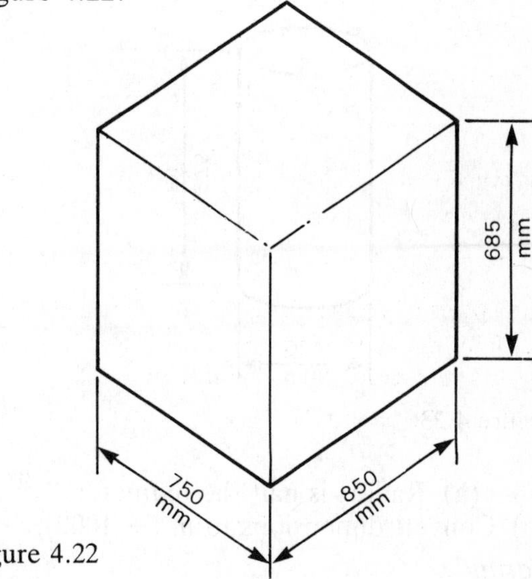

Figure 4.22

Note 1 m^3 = 1000 litres.

Formula
Volume of cubical tank
= length × breadth × depth
= L × B × D × 1000 litres

Convert dimensions to metres by dividing by 1000

Volume of cubical tank
$= \frac{850}{1000} \times \frac{750}{1000} \times \frac{685}{1000} \times 1000$ litres
= 0.85 × 0.75 × 0.685 × 1000 litres
= 436.68 litres

Answer = 436.68 litres

```
     0.85
     0.75
      425
     595
   0.6375
   ×0.685
    31875
   51000
  38250
0.4366875 ×1000
   436.68
```

Example 24 Calculate the capacity in litres of the cylinder shown in Figure 4.23.

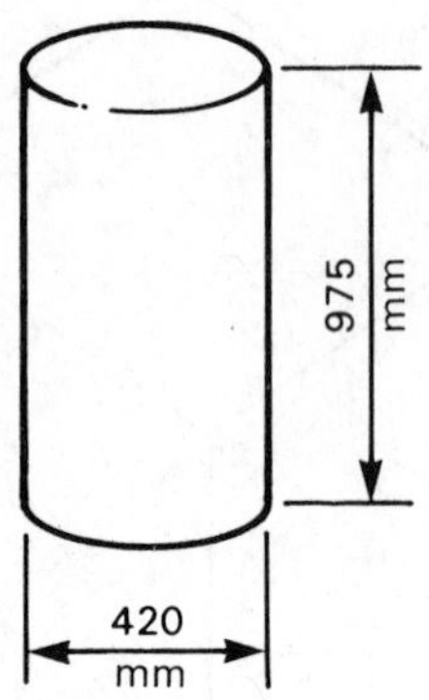

Figure 4.23

Note (a) Radius is half the diameter.
(b) Convert dimensions to m (÷ 1000).

Formula

Volume of cylinder

= Area of base × height m^3

= $\pi r^2 \times h \times 1000$ litres

$= 3.142 \times \frac{210}{1000} \times \frac{210}{1000} \times \frac{975}{1000} \times 1000$

= 3.142 × 0.21 × 0.21 × 0.975 × 1000 litres

= 135.09 litres

Answer = 135.09 litres

3.142
×.21
0.65982
×.21
0.1385622
×.975
0.1350981
×1000
135.0981

Example 25 Calculate the quantity of water in litres contained in a pipe which has a diameter of 38 mm and a length of 7.5 m.

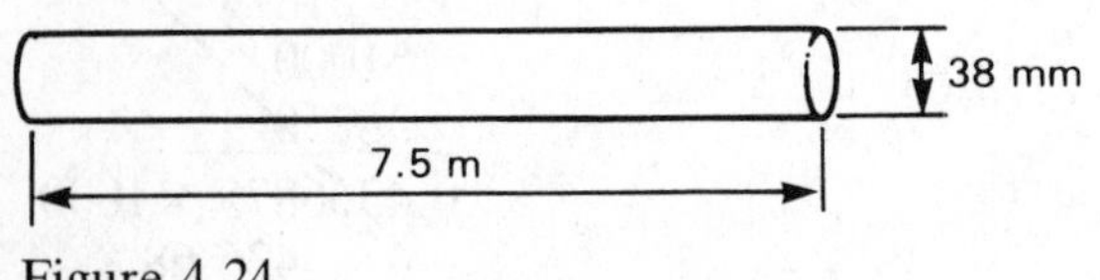

Figure 4.24

Formula

Volume of pipe

= Cross-sectional area of pipe × length m^3

= $\pi r^2 \times l \times 1000$ litres

$= 3.142 \times \frac{19}{1000} \times \frac{19}{1000} \times 7.5 \times 1000$ litres

= 3.142 × 0.019 × 0.019 × 7.5 × 1000 litres

= 8.506 litres

Answer = 8.506 litres

3.142
0.019
0.059698
×.019
0.0011342
×7.5
0.0085065
×1000
8.5065

Weight of water

For normal circumstances we assume that one litre of water weighs 1 kilogram (kg). It is useful for plumbers to know the weight (mass) of vessels or pipelines they are fixing so that they can ensure that these heavy components are adequately supported. From a calculation point of view it is quite easy to establish the weight of a volume of water. The volume should be calculated in cubic metres (m^3), if the figure is then multiplied by 1000 the volume will be converted into kilograms; this is the same process that is used for changing volume into litres (capacity) and reinforces the fact that one litre of water weighs 1 kg.

Example 26 Calculate the quantity and weight of water contained in a rectangular vessel with the following dimensions: 800 mm long, 750 mm wide, and 640 mm deep.

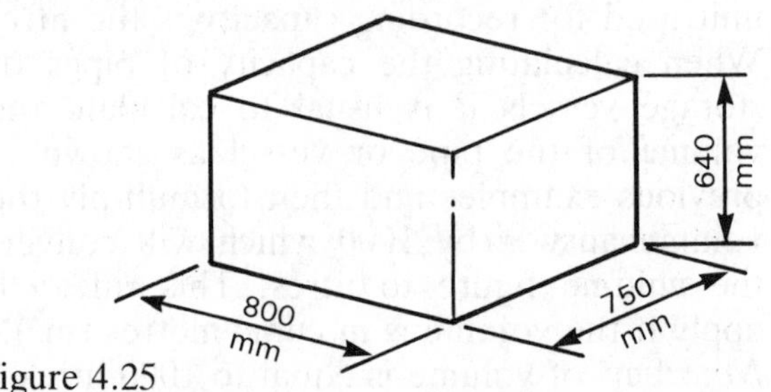

Figure 4.25

Note (a) 1 litre of water weighs 1 kg.
(b) 1 m^3 of water weighs 1000 kg.

Formula

Volume of vessel
= length × breadth × depth

Volume in litres
= L × B × D × 1000 litres.

Volume in weight

$= \frac{800}{1000} \times \frac{750}{1000} \times \frac{640}{1000} \times 1000\,\text{kg}$

= 0.8 × 0.75 × 0.64 × 1000 kg

= 384 kg

Answer = 384 kg

```
  0.8
  0.75
  0.6
  0.64
  0.0384
 ×1000
  384
```

Example 27 Calculate the quantity and the weight of water contained in a circular vessel 1.12 m in diameter and 950 mm high. The water level is 100 mm below the top edge of the cistern.

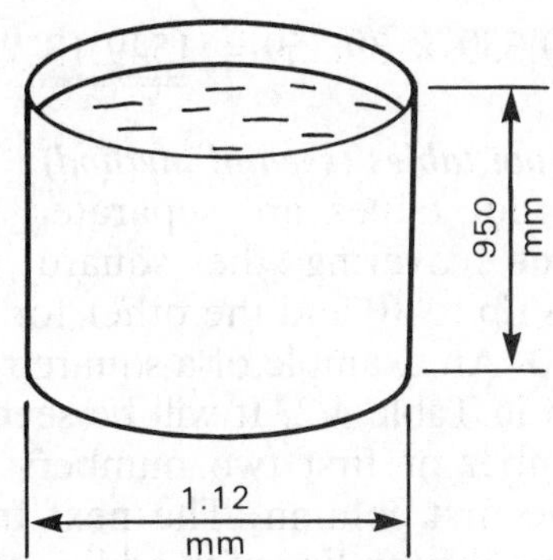

Figure 4.26

Procedure

(a) Pictorial presentation
(b) Convert dimensions to the same unit, i.e. metres
(c) State formula
(d) Each operation on a separate line

Volume of vessel
= Area of base × height

Volume of water
= Area of base × (height − 100 mm)

$= \pi r^2 \times \frac{(950 - 100)}{1000} \text{m}^3$

= 3.142 × 0.56 × 0.56 × 0.85 m

Note 1 m^3 water weighs 1000 kg

Weight of water
= 3.142 × 0.56 × 0.56 × 0.85 × 1000 kg
= 837.53 kg

Answer = 837.53 kg

```
   3.142
  ×0.56
   1.75952
  ×0.56
   0.9853312
  ×0.85
   0.8375315
  ×1000
   837.5315
```

Note In the practical situation the actual weight of the vessel must be either known or calculated and then added to the weight of water to obtain the actual weight that will need to be supported by the bearers or fixings.

Square roots

In a great many of our problems it is required to obtain the *square root* of a given number, i.e. the number which when multiplied by itself equals that given number.

(a) The square root of 9 which is written as $\sqrt{9} = 3$
Proof 3 × 3 = 9

(b) The square root of 16 $\sqrt{16} = 4$
Proof 4 × 4 = 16

(c) The square root of 144 $\sqrt{144} = 12$
Proof 12 × 12 = 144

Using these simple numbers the square root can easily be obtained. For larger numbers and numbers including decimal points (fractions) the answer is not so easily obtained and will require a special method or the use of a *square root table* or calculator. All three methods will be explained in this chapter.

The long hand method has to a large extent been superseded by the use of tables and/or calculators, but it is still useful, if not imperative, to be able to perform this task by this method. Let us therefore take a very simple example.

Example 28 What is the square root of 144?

```
         1 2.        Answer on this line
   1  | 1,44.00
      | 1
  22  | 044
  ×2  |  44
      |  00
```

Answer = 12

Note Follow carefully the following procedure.

Procedure:

(a) Set down the number as shown in example.
(b) Place the decimal point.
(c) Mark off the numbers in pairs from the *decimal point* both ways.
(d) Find a number which when multiplied by itself will give you the first number or near to it (it must not be greater). In this case the number is 1.
(e) Place a 1 in the answer also under the 1 in the original number and subtract.
(f) Bring down the next pair of numbers, i.e. 44 and place with the zero.
(g) Double the answer in this case 1 and place the 2 on the same line as the 44.
(h) You now need to find a number that when placed with the 2 and multiplied by the same number will give you the number 44. In this case the number required is 2.

Note Continue the process until you have the correct answer or until you have two or three decimal places, depending on the accuracy of the work.

Example 29 Find the square root of 1549.2.

Note (a) A zero must be added to make the 2 into 20 to make a pair of numbers.
(b) Add additional pairs of zeros to give the required number of decimal places.

```
                           3 9. 3 5 9
                     3  | 15,49.20,00,00
                        |  9
double answer       69  |  649
                    ×9  |  621
double answer      783  |   2820
                    ×3  |   2349
double answer     7865  |    47100
                    ×5  |    39325
                 78709  |     777500
                    ×9  |     708381
                        |      69119
                        |        remainder
```

Answer = 39.359 correct to three decimal places

Proof 39.359 × 39.359 = 1549.13

Square root tables (second method)

Square root tables are separated into two parts one covering the square roots of numbers up to 10 and the other for numbers up to 100. An example of a square root table is shown in Table 4.2. It will be seen that the first number or first two numbers are read down the first column. The next fraction is read along the top line with additional figures being read from the mean difference columns as shown in the following examples.

Example 30 Find the square root of 18.5

Procedure

(a) Locate 18 in the extreme left hand column.
(b) Place a straight edge (rule) along this horizontal line of figures.

Table 4.2 *Square roots*

	0	1	2	3	4	5	6	7	8	9	1	2	3	4	5	6	7	8	9
10	1000	1005	1010	1015	1020	1025	1030	1034	1039	1044	0	1	1	2	2	3	3	4	4
	3612	3178	3194	3209	3225	3240	3256	3271	3286	3302	2	3	5	6	8	9	11	12	14
11	1049	1054	1058	1063	1068	1072	1077	1082	1086	1091	0	1	1	2	2	3	3	4	4
	3317	3332	3347	3362	3376	3391	3406	3421	3435	3450	1	3	4	6	7	9	10	12	13
12	1095	1100	1105	1109	1114	1118	1122	1127	1131	1136	0	1	1	2	2	3	3	4	4
	3464	3479	3493	3507	3521	3536	3550	3564	3578	3592	1	3	4	6	7	8	10	11	13
13	1140	1145	1149	1153	1158	1162	1166	1170	1175	1179	0	1	1	2	2	3	3	3	4
	3606	3619	3633	3647	3661	3674	3688	3701	3715	3728	1	3	4	5	7	8	10	11	12
14	1183	1187	1192	1196	1200	1204	1208	1212	1217	1221	0	1	1	2	2	3	3	3	4
	3742	3755	3768	3782	3795	3808	3821	3834	3847	3860	1	3	4	5	7	8	9	11	12
15	1225	1229	1233	1237	1241	1245	1249	1253	1257	1261	0	1	1	2	2	3	3	3	4
	3873	3886	3899	3912	3924	3937	3950	3962	3975	3987	1	3	4	5	6	8	9	10	11
16	1265	1269	1273	1277	1281	1285	1288	1292	1296	1300	0	1	1	2	2	3	3	3	4
	4000	4012	4025	4037	4050	4062	4074	4087	4099	4111	1	2	4	5	6	7	9	10	11
17	1304	1308	1311	1315	1319	1323	1327	1330	1334	1338	0	1	1	2	2	3	3	3	3
	4123	4135	4147	4159	4171	4183	4195	4207	4219	4231	1	2	4	5	6	7	8	10	11
18	1342	1345	1349	1353	1356	1360	1364	1367	1371	1375	0	1	1	1	2	2	3	3	3
	4243	4254	4266	4278	4290	**4301**	4313	4324	4336	4347	1	2	3	5	6	7	8	9	10
19	1378	1382	1386	1389	1393	1396	1400	1404	1407	1411	0	1	1	1	2	2	3	3	3
	4359	4370	4382	4393	4405	4416	4427	4438	4450	4461	1	2	3	5	6	7	8	9	10
20	1414	1418	1421	1425	1428	1432	1435	1439	1442	1446	0	1	1	1	2	2	2	3	3
	4472	4483	4494	4506	4517	4528	4539	4550	4561	4572	1	2	3	4	5	7	8	9	10
21	1449	1453	1456	1459	1463	1466	1470	1473	1476	1480	0	1	1	1	2	2	2	3	3
	4583	4593	4604	4615	4626	4637	4648	4658	4669	4680	1	2	3	4	5	6	8	9	10
22	1483	1487	1490	1493	1497	1500	1503	1507	1510	1513	0	1	1	1	2	2	2	3	3
	4690	4701	4712	4722	4733	4743	4754	4764	4775	4785	1	2	3	4	5	6	7	8	9
23	1517	1520	1523	1526	1530	1533	1536	1539	1543	1546	0	1	1	1	2	2	2	3	3
	4796	4806	4817	4827	4837	4848	4858	4868	4879	4889	1	2	3	4	5	6	7	8	9
24	1549	1552	1556	1559	1562	1565	1568	1572	1575	1578	0	1	1	1	2	2	2	3	3
	4899	4909	4919	4930	4940	4950	4956	4970	4980	4990	1	2	3	4	5	6	7	8	9
25	1581	1584	1587	1591	1594	1597	1600	1603	1606	1609	0	1	1	1	2	2	2	3	3
	5000	5010	5020	5030	5040	5050	5060	5070	5079	5089	1	2	3	4	5	6	7	8	9
26	1612	1616	1619	1622	1625	1628	1631	1634	1637	1640	0	1	1	1	2	2	2	2	3
	5099	5109	5119	5128	5138	5148	5158	5167	5177	5187	1	2	3	4	5	6	7	8	9
27	1643	1646	1649	1652	1655	1658	1661	1664	1667	1670	0	1	1	1	2	2	2	2	3
	5196	5206	5215	5225	5235	5244	5254	5263	5273	5282	1	2	3	4	5	6	7	8	9
28	1673	1676	1679	1682	1685	1688	1691	1694	1697	1700	0	1	1	1	1	2	2	2	3
	5292	5301	5310	5320	5329	5339	5348	**5357**	5367	5376	1	2	3	**4**	5	6	7	7	8
29	1703	1706	1709	1712	1715	1718	1720	1723	1726	1729	0	1	1	1	1	2	2	2	3
	5385	5394	5404	5413	5422	5431	5441	5450	5459	5468	1	2	3	4	5	5	6	7	8
30	1732	1735	1738	1741	1744	1746	1749	1752	1755	1758	0	1	1	1	1	2	2	2	3
	5477	5486	5495	5505	5514	5523	5532	5541	5550	5559	1	2	3	4	4	5	6	7	8
31	1761	1764	1766	1769	1772	1775	1778	1780	1783	1786	0	1	1	1	1	2	2	2	3
	5568	5577	5586	5595	5604	5612	5621	5630	5639	5648	1	2	3	3	4	5	6	7	8
32	1789	1792	1794	1797	1800	1803	1806	1808	1811	1814	0	1	1	1	1	2	2	2	2
	5657	5666	5675	5683	5692	5701	5710	5718	5727	5736	1	2	3	3	4	5	6	7	8

The first significant figure and the position of the decimal point must be determined by inspection.

(c) Locate 5 on the top line and drop a vertical line down until it corresponds with the horizontal line obtained previously.
(d) Read off this number of 4.301, this being the square root required.

Therefore the square root of 18.5 = 4.301
Answer = 4.301

Proof 4.301 × 4.301 = 18.5

Example 31 Find the square root of 28.74.
Procedure
(a) Locate 28 in the extreme left hand column.
(b) Locate .7 in the top line to obtain 5.357.
(c) Locate .4 in the difference column to obtain .4.
(d) Add 5.357 and .4 to give 5.361.

Therefore the square root of 28.74 = 5.361
Answer = 5.361
Proof 5.361 × 5.361 = 28.74

Square root by calculator (third method)
This is by far the easiest, quickest and of course the most reliable method, provided the operator of the calculator takes care to depress the correct keys. All that is required is to insert the required number which will be indicated on the display panel (screen) then to depress the appropriate square root key, usually indicated by the square root sign $\sqrt{\ }$. The required square root will now appear on the screen.

Example 32 What is the square root of 339.3?
Procedure
(a) Record the number 339.3 on the display screen.
(b) Actuate the square root key.
(c) Answer will now appear on the display panel 18.42.

Therefore the square root of 339.3 = 18.42
Answer = 18.42
Proof 18.42 × 18.42 = 339.3

The Pythagoras Theorem
This theorem is very useful in certain calculations where it is necessary to find the lengths of the sides of right-angled triangles. Pythagoras' Theorem states that in any right-angled triangle the *square* of the *hypotenuse* is equal to the sum of the squares of the other two sides.

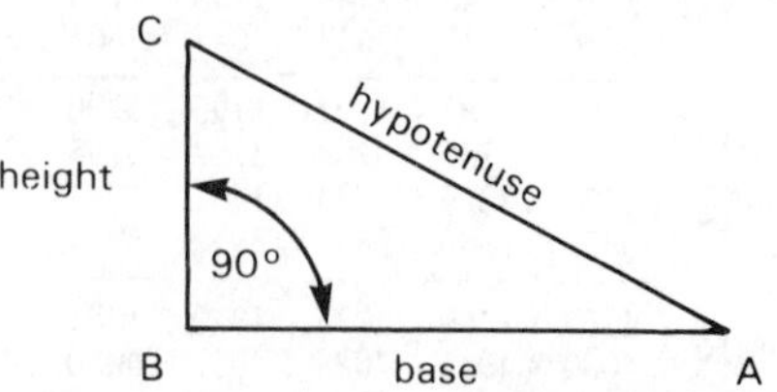

Figure 4.27a *Definition of a right-angled triangle*

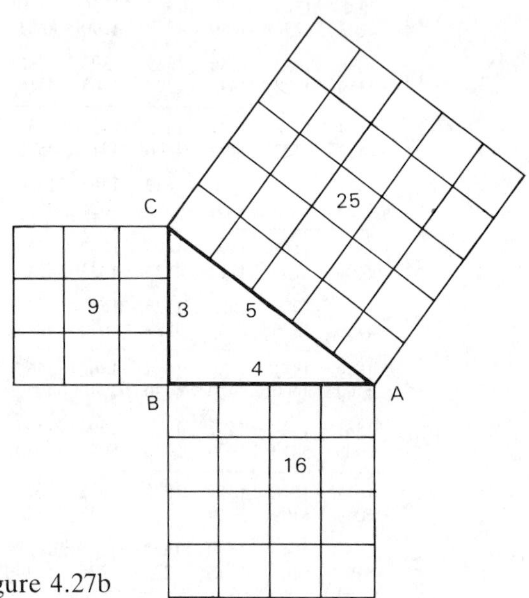

Figure 4.27b

Figure 4.27b illustrates a right-angled triangle drawn in the ratio of 3, 4, 5. It can be seen that the square of the sides A–B and B–C when added together equal the square of the side A–C. Right-angled triangles often occur in construction and plumbing work; in practice this means that when any two sides of a right-angled triangle are known the third side can be easily calculated using the Theorem of Pythagoras.

Example 33 Calculate the length of the third side of a right-angled triangle where the base is 3.5 and the height is 2.75 m.

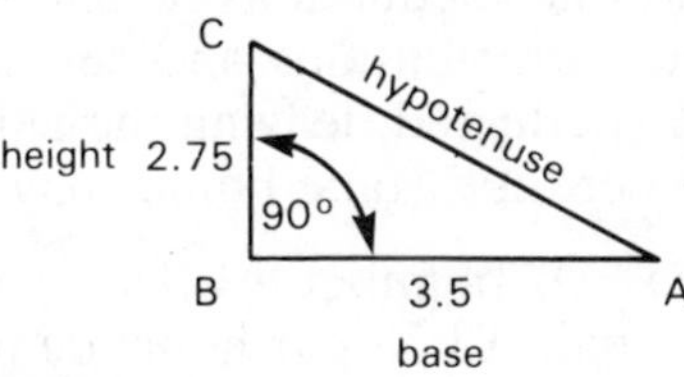

Figure 4.28a

Formula The Theorem of Pythagoras

The square of the hypotenuse =
sum of the squares of the other two sides.

$$\text{(hypotenuse) } AC^2 = AB^2 + BC^2$$

Taking the square root of both sides of the equation gives:

$$AC = \sqrt{AB^2 + BC^2}$$
$$AC = \sqrt{3.5^2 + 2.75^2}$$
$$AC = \sqrt{12.25 + 7.56}$$
$$AC = \sqrt{19.81}$$
$$AC = 4.45\,\text{m}$$
$$\underline{\underline{\text{Answer}}} = \underline{\underline{4.45\,\text{m}}}$$

Example 34 A 6.25 m ladder is placed against the eaves of a building. If the foot of the ladder is 3.5 m from the building calculate the height of the eaves.
Note It is always advisable to sketch the problem and to identify the known dimensions.

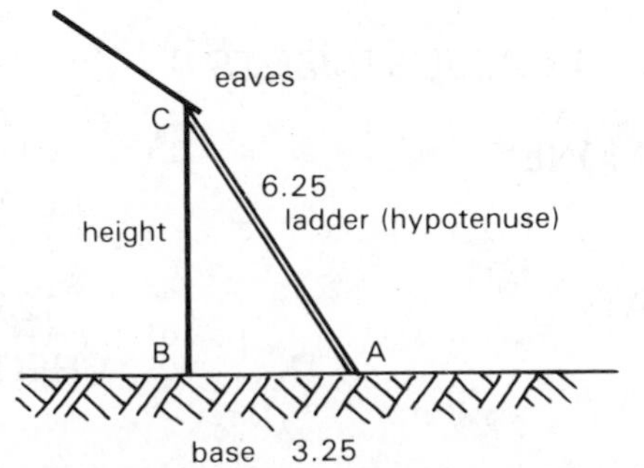

Figure 4.28b

Formula The Theorem of Pythagoras.

The square of the hypotenuse =
sum of the squares of the other two sides.

$$\text{(hypotenuse) } AC^2 = AB^2 + BC^2$$
$$6.25^2 = 3.5^2 + BC^2$$
$$(6.25^2)-(3.25)^2 = BC^2$$
$$39.06-12.25 = BC^2$$
$$26.81 = BC^2$$

Taking the square square root of both sides of the equation gives:

$$\sqrt{26.81} = BC$$
$$5.18 = BC$$

Answer Height of eaves = 5.18 m

Salaries

Everyone who works for an employer receives a wage or salary in return for their labours. Wages may be paid at so much per hour, so much per week or so much per month. Most employees work a basic week of so many hours and it is this basic week which determines the wages of the employee. Overtime is additional time worked over and above the basic week and it is often paid for at plus or enhanced rates of pay, which is extra to the basic wage.

Manual workers such as plumbers are usually paid an hourly rate. The Joint Industry Training Board for Plumbing Mechanical Engineering Services in England and Wales prescribes a wages structure based on three grades of craftsperson:

1 Trained plumber,
2 Advanced plumber,
3 Technician plumber.

Basic wages per week (gross) are calculated by multiplying the agreed hourly rate by the total number of hours worked during the basic week. For example, a plumber being paid £3.58 an hour and working a 40 hour week can expect to earn £143.20 per week before deductions (tax, pensions, etc.)

$$£3.58 \times 40 = £143.20$$

Apprentices and trainees are usually paid a percentage of the craftspersons' rate depending on their age or year of traineeship.

Salary deductions

Income tax Taxes are levied by the Chancellor of the Exchequer in order to produce income for the government. Every person who has an income above a certain minimum amount has to pay part of that income to the government in income tax. Tax is not paid on the whole income – certain allowances are made as follows:

1 An allowance, the amount of which varies according to whether the taxpayer is a single person or married;
2 Allowances for dependent relatives, etc.;
3 Allowances for superannuation contributions;
4 Allowances for payment of subscription fees to trade unions and professional bodies.

The residue left after the allowances have been deducted from the gross income or wage is called the taxable income; tax is then deducted at a standard percentage rate from this figure. Other deductions such as National Insurance contributions and superannuation are also deducted, leaving the employee with his or her 'net' (take home) pay.

Example 35 A plumber works a 46 hour week and is paid £3.58 per hour; deductions such as income tax, National Insurance, etc. amount to 32% of the gross wage. Calculate the plumber's net (take home) pay.

$$\begin{aligned} \text{Gross wage} &= £3.58 \times 46 \\ &= £164.68 \\ \text{Deductions} &= 32\% \text{ of } £164.68 \\ &= £52.69 \\ &= £164.68 - £52.69 \\ \text{Net wage} &= \underline{\underline{£111.99}} \end{aligned}$$

In addition to the four allowances listed previously tax should not be paid on items such as travelling expenses and tool money. Figure 4.29 illustrates a typical employee's payslip and shows how net wage or pay is arrived at.

COMPANY L B CONSTRUCTION SERVICES			WEEK
NAME M. R. BOYCE		DATE	NUMBER
GROSS PAY		*DEDUCTIONS*	
HOURS 40 AT 3.50	140.00	NATIONAL INSURANCE	8.50
HOURS 5 AT 4.25	21.25	INCOME TAX	37.20
TOTAL	161.25	TOTAL	45.70
TAXABLE GROSS		*AFTER TAX ADJUSTMENTS*	
TAX TO DATE		EXPENSES	7.65
		TOOL MONEY	1.20
TAX CODE		TOTAL	8.85
TOTAL N.I.		NET PAY	124.40
N.I. NUMBER			CHEQUE

Figure 4.29 *Employee's payslip*

Example 30 Calculate the net pay of an employee who is paid £4.18 an hour for a 40 hour week. Deductions from salary total £42.63 and after tax additions are £14.37.

Gross wage 40 × £4.18 =	167.20
Less £42.63 deductions =	124.57
Plus £14.37 additions =	138.94
∴ Net pay of employee =	£138.94

Cost calculations

Cost calculations are a very important part of the builder's work and the success of a business will depend greatly on accurate estimating and costing of work. People who try to price a job by guesswork will usually find that their estimate is either too high for the client to accept or so low that a reasonable profit cannot be made. The examples in this section cover basic calculations relating to time and materials.

Example 36 A trench excavated for a new drainage pipeline is 23 m long, 450 mm wide and averages 1.5 m in depth. How many cubic metres of earth are removed and what would be the excavation cost at £17.56 per cubic metre?

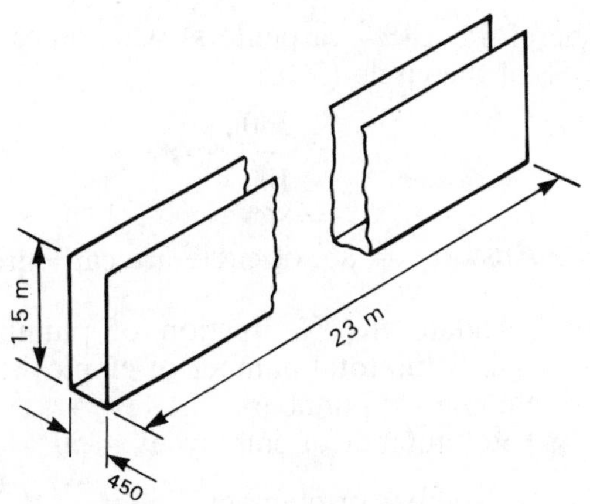

Figure 4.30a

Formula Volume of excavation = length × breadth × depth

Note Convert all dimensions to the same unit.

(a) 450 mm to metres.
(b) 1000 mm in 1 metre.

Volume of excavation $= L \times B \times D$

$$= 23 \times \frac{450}{1000} \times 1.5\,\text{m}^3$$

$$= 23 \times .45 \times 1.5\,\text{m}^3$$

$$= 15.525\,\text{m}^3$$

The cost of the excavation is £17.56 per cubic metre.

Cost of excavation = 15.525 × £17.56
= £272.61

Answer = £272.61

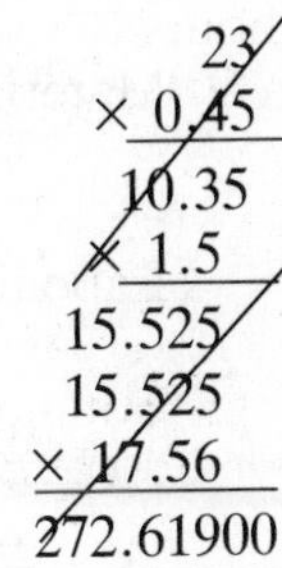

Example 37 Calculate the cost of covering the flat roof shown in Figure 4.30b. The material cost is £9.16 per square metre and the plumber's labour charge is £7.80 per hour; the work takes 32 hours to complete.

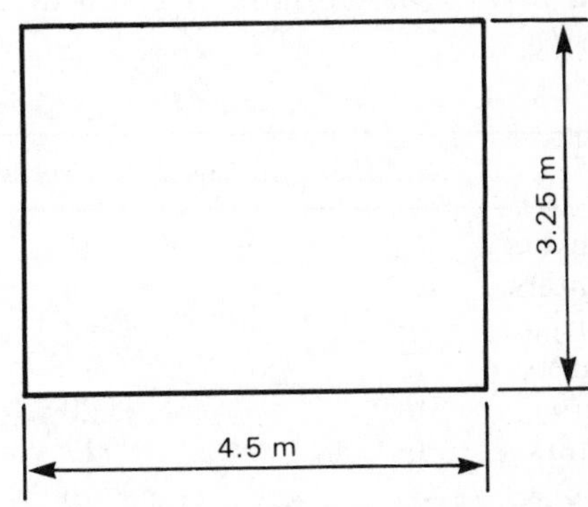

Figure 4.30b

Formula Area of roof = length × breadth
= $4.5 \times 3.25\,m^2$
= $14.625\,m^2$

Cost of material = 14.625 × £9.16
= £133.96

Cost of labour = 32 × £7.80
= £249.6

Total cost of job = £249.6 + £133.96
= £385.56

Answer = £383.56

Example 38 Calculate the cost of replacing a cold water cistern. The new material cost is

Cistern = £31.70
Cistern lid = £8.65
Fittings = £9.30
Insulating cover = £7.86

The labour charge is £8.30 an hour for 6 hours.

Material cost = £31.70 + £8.65 + £9.30 + £7.86
= £57.51

Labour cost = £8.30 × 6
= £49.80

Total cost = £57.51 + £49.80
= £107.31

Answer = £107.31

Statistics and recording information

A useful method of presenting information is by the use of a chart and these can be represented pictorially in several ways.

For example, suppose that on a certain building site the number of persons employed in various crafts is as given in Table 4.3.

Table 4.3

Craft	*Number employed*	*Percentage*
Bricklayers	8	16
Carpenters	12	24
Electricians	4	8
Labourers	6	12
Painters	10	20
Plasterers	6	12
Plumbers	4	8
Total	50	100

The information in the table can be represented pictorially in several ways:

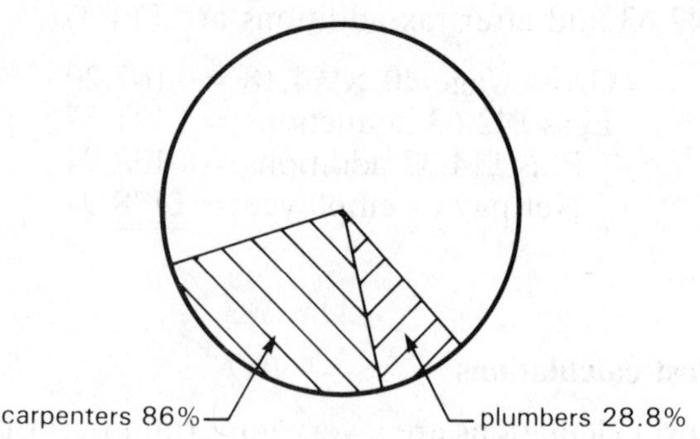

Figure 4.31

Method 1 'The pie chart'. Figure 4.31 displays the proportions as angles (or sector areas), the complete circle representing the total number of persons employed and containing 360°. Thus for carpenters the angle is obtained as follows:

(a) Calculate the % fraction of carpenters against the total number of employees

Number of carpenters = 12
Total number of employees = 50

% Number of carpenters $= \frac{12}{50} \times 100$
= 24

Answer = 24% carpenters.

Therefore 24% carpenters will represent 24% of the circle (360°)

$= \frac{360}{100} \times 24$
= 86°

Answer = 86° of circle are carpenters.

(b) Calculate the % fraction of plumbers against the total number of employees

Number of plumbers = 4
Total number of employees = 50

% Number of plumbers $= \frac{4}{50} \times 100$

Answer = 8% plumbers.

Therefore 8% plumbers will represent 8% of the circle (360°)

$$= 360 \times \frac{8}{100}$$

$$= 28.8\% \text{ of circle are plumbers.}$$

The pie chart
Note: The full circle contains 360 degrees.

Method 2 'The bar chart' Figure 4.32 relies on heights (or areas) to convey the proportions, the total height of the diagram representing 100%. Figure 4.32 uses the same work-force as in the previous method.

Individual percentage of craftsmen

(a) Bricklayers $\frac{8}{50} \times 100 = 16\%$

(b) Carpenters $\frac{12}{50} \times 100 = 24\%$

(c) Electricians $4 \times 2 = 8\%$
(d) Labourers $6 \times 2 = 12\%$
(e) Painters $10 \times 2 = 20\%$
(f) Plasterers $6 \times 2 = 12\%$
(g) Plumbers $4 \times 2 = 8\%$

Method 3 'The horizontal bar chart' Figure 4.33 gives a better comparison of the various types of personnel employed on the site. It also displays the total number.

Frequency distributions

Suppose we measure and record the lengths of 40 radiators fitted in a building and these lengths are as in Table 4.4.
These figures do not mean very much as they stand and so we rearranged them into what is called a frequency distribution. To do this we collect all the 1.1 m readings and so on. A tally chart (see Table 4.5) is the best way of

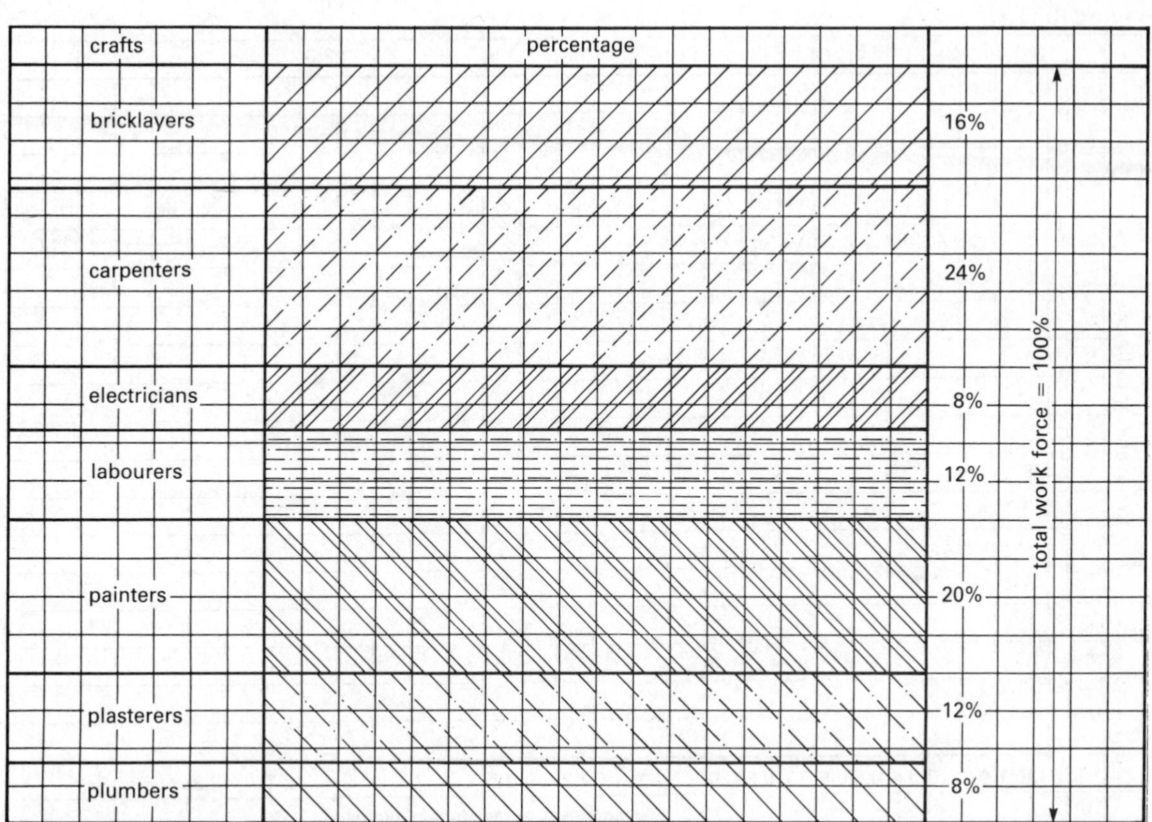

Figure 4.32

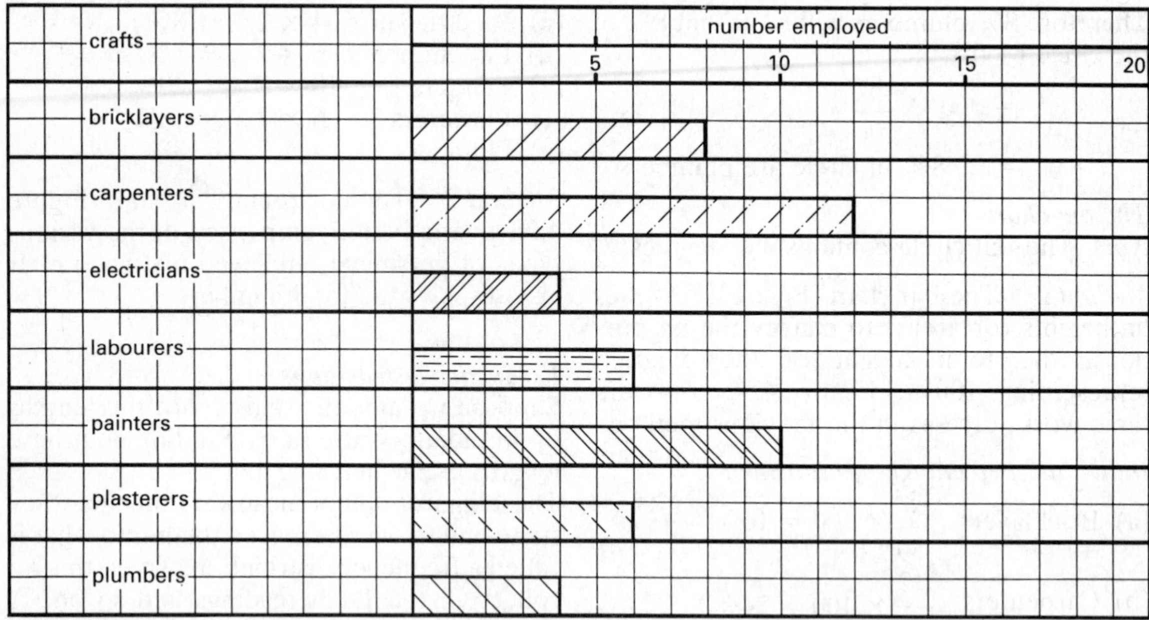

Figure 4.33

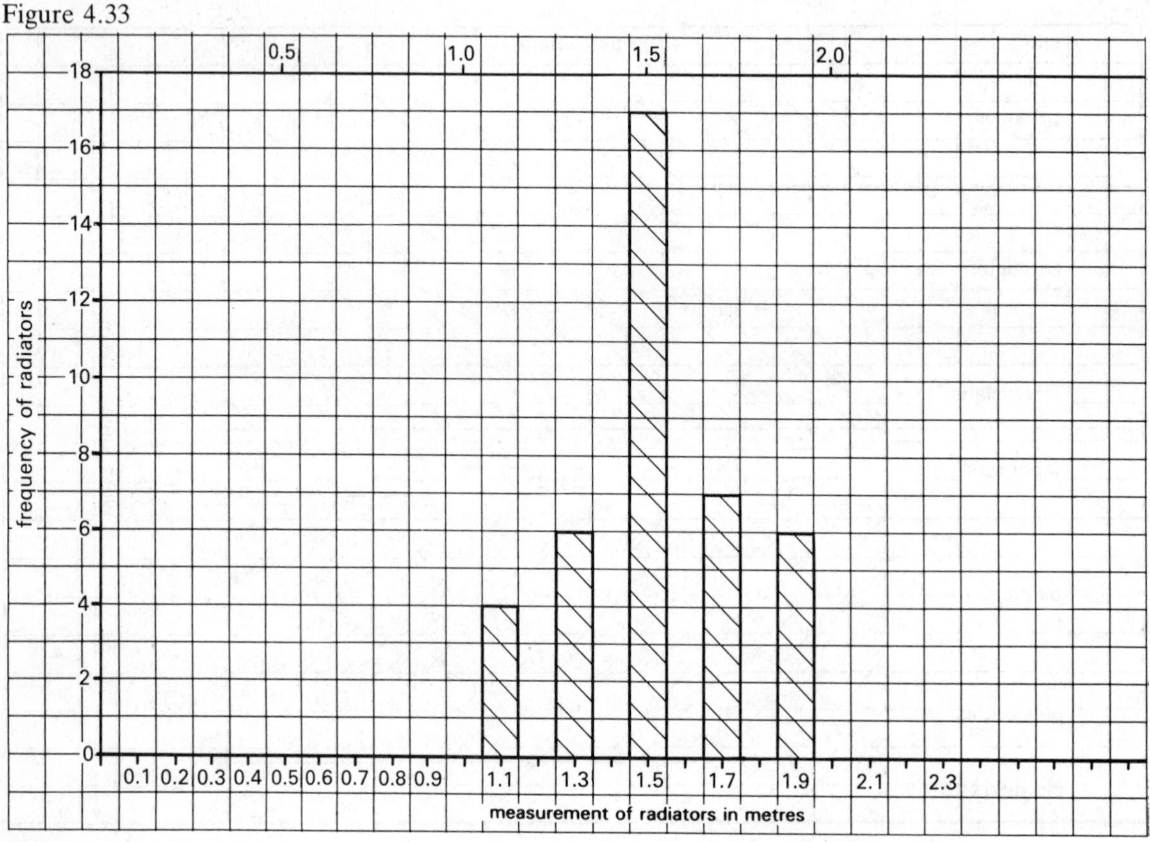

Figure 4.34 *Histogram*

Table 4.4 *Radiator lengths (m)*

1.5	1.3	1.5	1.1	1.5
1.7	1.1	1.9	1.7	1.3
1.5	1.3	1.1	1.5	1.7
1.9	1.5	1.1	1.5	1.5
1.7	1.5	1.3	1.3	1.5
1.5	1.7	1.5	1.9	1.3
1.5	1.7	1.5	1.9	1.7
1.5	1.9	1.5	1.5	1.9

doing this. Every time a measurement arises a tally mark is placed opposite the appropriate measurement. The fifth tally mark is usually made in an oblique direction, thus binding the tally marks into bundles of five to make counting easier. When the tally marks are complete the marks are counted and the numerical value recorded in the column headed 'frequency'. The frequency is the number of times each measurement occurs. From Table 4.5 it will be seen that the measurement 1.1 m occurs four times (that is it has a frequency of 4), the measurement 1.3 m occurs 6 times (a frequency of 6), and so on.

Table 4.5 *Tally chart*

Measurement in metres	*Number of radiators with this measurement*	*Frequency*
1.1	IIII	4
1.3	~~IIII~~ I	6
1.5	~~IIII~~ ~~IIII~~ ~~IIII~~ II	17
1.7	~~IIII~~ II	7
1.9	~~IIII~~ I	6

The histogram

The frequency distribution becomes even more understandable if we draw a diagram to represent it. The best form of diagram is the histogram (see Figure 4.34) which consists of a set of rectangles, in this case each of the same width, whose heights represent the frequencies. On studying the histogram the pattern of the variation is easily understood, most of the frequencies or values being grouped near the centre of the diagram with a few frequencies being more widely dispersed.

Electronic calculators

A great deal of time and effort can often be used in solving arithmetical problems, and much of this time and effort may be saved by the use of calculators. There are many types of calculator on the market, but for craft students a calculator with the functions of adding, subtracting, multiplying and division is adequate, although one with percentages, roots and powers will be even more useful.

Figure 4.35

The keyboard of a calculator has 10 number keys marked 0, 1, 2, 3, 4, 5, 6, 7, 8, 9, as shown in Figure 4.35, and a decimal point. On basic calculators there are four function keys such as +, −, × and ÷; in addition there is also an = key. There is always a cancel or clear key, usually marked C, which is used to clear the display. Before each calculation is worked it is advisable to depress the C key to ensure that the calculator is clear of previous exercises.

Note As calculators vary from one manufacturer to another you should read carefully the users' instructions before attempting to work any of these machines. Figure 4.35 illustrates a typical electronic calculator.

Example 39 A calculator showed that

$$\frac{0.563 \times 0.621}{0.0362} = 0.967$$

Is this a feasible answer?

Rough check: $\dfrac{0.5 \times 0.6}{0.03} = 10$

The answer in this case is *not feasible* and must be discounted. On performing the calculation again, the answer is found to be 9.67 which agrees with the approximation of the rough check.

5 Science

Introduction

After reading this chapter you should be able to:

1 State the common causes of corrosion.

2 Demonstrate knowledge of how to prevent corrosion occurring.

3 Name and recognise fluxes used by the plumber.

4 Describe the causes of condensation and state measures to prevent condensation forming.

5 State the properties of common insulating materials.

6 Describe the uses and application of insulating materials.

7 Define the terms 'anode' and 'cathode'.

8 Demonstrate knowledge of Archimedes' Principle.

9 Describe common examples of capillarity and surface tension related to building work.

10 Demonstrate knowledge of pressure in liquids and gases and relate this knowledge to plumbing situations.

Fluxes

A soldering flux is a liquid or solid material which, when heated, is capable of promoting or accelerating the wetting of metals with solder. Table 5.1 lists some of the fluxes commonly used by plumbers. The purpose of a soldering flux is to remove and exclude oxides and other impurities from the joint being soldered. Anything interfering with the attainment of uniform contact between the surface of the base metal and the molten solder will prevent the formation of a sound joint. An efficient flux removes tarnish films and oxides from the metal and solder, and prevents re-oxidation of the surfaces when they are heated. It is designed to lower the surface tension of the molten solder so that the solder will flow readily and adhere to the metal. The flux should be readily displaced from the metal by the molten solder.

Surfaces to be soldered are often covered with films of oil, grease, paint, heavy oxides or atmospheric grime which must be removed. When clean metal surfaces are exposed to the air, chemical reactions occur, depositing fresh surface films. These reactions are generally accelerated as the temperature is raised, and although nitrides, sulphides, and carbides are formed in some instances, the prevalent reaction is oxidation. The rate of oxide formation, its structure and

Table 5.1 *Fluxes used by the plumber*

Name	*Uses*	*State*	*Notes*
Zinc chloride (killed spirits)	All forms of copper bit soldering	Liquid	Produced by dissolving zinc in hydrochloric acid. The flux is actively corrosive and must be removed by washing the soldered joint following the jointing process.
Zinc-ammonium chloride	All forms of copper bit soldering	Liquid	This flux is active at lower temperatures than zinc chloride. This is helpful when soldering metals with a relatively low melting point. The flux is corrosive and must be removed by washing the soldered joint to remove all traces of the flux.
Tallow	Lead soldering	Solid	Tallow is obtained from the fat of animals. These organic fats contain glycerin, which makes them mildly acidic.
Resin	Tinning (brass and copper)	Powder or paste	Resin is obtained from pine tree bark. It is a gum-like substance. This flux is usually applied in powder form, and is sprinkled on the surfaces to be tinned. It is mildly corrosive at soldering temperature and non-corrosive when cold.

resistance to removal with a flux, varies with each base metal. Aluminium, stainless and high alloy steels, aluminium and silicon bronzes, when exposed to air, form hard adherent oxide films – highly active and corrosive fluxes are used to remove and prevent the reforming of these films during soldering. Lead, copper and silver on the other hand form less tenacious films and at a slower rate, so that mild fluxes remove them easily and prevent them from reforming.

Flux action

In most soldering operations the flux removes the oxide film from the base metal and solder by dissolving or loosening the film and floating it off into the main body of the flux. Because of the refractory nature of many oxide films it has been suggested that the flux wets, coagulates and suspends the oxide which has been loosened by a penetrating and reducing action. The molten flux then forms a protective blanket over the bare metal which prevents the film from reforming. Liquid solder displaces the flux and reacts with the base metal to form an intermolecular bond. The solder layer builds up in thickness and when the heat is removed it solidifies.

Types of flux

A functional method for classifying fluxes is based upon the nature of their residues. They are classified into three main groups:

1 *high corrosive*,
2 *intermediate*,
3 *non-corrosive* fluxes.

Good soldering practice requires the selection, from the three main groups, of the mildest flux that will perform satisfactorily in a specific application.

The *highly corrosive* fluxes consist of inorganic acids and salts. These fluxes are used to best advantage where conditions require a rapid and highly activated fluxing. They can be applied as solutions, pastes, or as powders. Corrosive fluxes are almost always required when using the higher melting temperature solders.

The corrosive fluxes have one distinct disadvantage in that the residue remains chemically active after soldering. This residue, if not removed, may cause severe corrosion at the joint.

Zinc chloride is the main ingredient in the majority of corrosive fluxes. It can be prepared by adding an excess of zinc to concentrated hydrochloric acid, or it can be purchased as fused zinc chloride which is readily available and more convenient to use.

A water solution of ammonium chloride may be used as a flux. It is less effective than zinc chloride.

A combination of one part of ammonium chloride to three parts of zinc chloride forms a eutectic flux mixture which is considerably more effective than either constituent when used alone.

The *intermediate fluxes*, as a class, are weaker than the inorganic powder types. They consist mainly of mild organic bases and certain of their derivatives. These fluxes are active at soldering temperatures but the period of activity is short because of their susceptibility to thermal decomposition. They are useful, however, in quick spot soldering operations and, when properly used, their residue is relatively inert and easily removed with water. Intermediate fluxes are useful where sufficient heat can be applied to decompose or volatilise fully the corrosive constituents contained within the flux.

Non-corrosive fluxes White resin dissolved in a suitable organic solvent is the closest approach to a non-corrosive flux.

The active constituent, abietic acid, becomes mildly active at soldering temperatures.

This form of flux is widely used in plastic form in cored solder wire.

Because of the slow fluxing action of resin on anything but clean or precoated metal surfaces, a group of stabilised and activated resin fluxes have been developed, aimed at increasing the fluxing action without altering the non-corrosive nature of the residue.

Paste fluxes In the plumbing trade it is sometimes convenient to have the flux in the form of a paste. Paste fluxes can be more easily localised at the joint and have the advantage of not draining off the surface or spreading to other parts of the work where flux may be harmful. The paste-forming ingredients may be water, petroleum jelly, tallow or lanolin, with glycerine or other moisture-retaining substances. If the pastes contain inorganic salts, such as zinc or ammonium chloride, they are classified as corrosive fluxes. Paste fluxes have been developed for universal application containing resins dissolved in butyl cellosolve and with plasticisers added to provide the flux activity. Resin pastes have also been developed which are non-corrosive and meet the requirements of the electrical industry.

Lead-free solders

To comply with the new water by-laws, lead-free solder must be used when soldering joints on copper pipes, where these pipes are to be used for conveying water for drinking and the preparation of food. A range of British Standard lead-free solders is available from stockists as sticks, bars and solid wire. The standard paste fluxes, such as Baker's Blue and Powerflow can be used with these lead-free solders.

Condensation

Principles involved in condensation

The amount of water vapour that air can contain is limited and when this limit is reached the air is said to be saturated. The saturation point varies with temperature – the higher the temperature of the air, the greater the weight of water vapour it can contain. Water vapour is a gas, and in a mixture of gases, such as when present in the air, it contributes to the total vapour pressure exerted by the mixture. The ratio of the vapour pressure of any mixture of water vapour and air to the vapour pressure of a saturated mixture at the same temperature is the relative humidity (RH), which is expressed as a percentage. Alternatively, relative humidity can be regarded as the amount of water vapour in the air expressed as a percentage of the amount that would saturate it at the same temperature.

In conditions of, for example, 20 °C and 80% RH all the moisture can be held in the air. If more water vapour is introduced into the air and the temperature remains constant, the relative humidity will increase; saturation point (100% RH) may be reached and thereafter any further vapour will be deposited as condensation. If, on the other hand the amount of water vapour remains constant but the temperature falls, because the colder air can support less moisture, the RH will rise until at about 15 °C it is 100% and any further cooling will cause water to condense. This is the dew-point of that air which at 20 °C had an RH of 80%.

Conditions producing condensation

It has been shown that changes in temperature or in moisture content can cause condensation to occur. These changes can occur naturally – by changes in atmospheric conditions, or artificially – by living habits or industrial processes.

Atmospheric conditions

When warm damp weather follows a period of cold, the fabric of a heavy structure which has not been fully heated will not warm up immediately but may remain comparatively cold for several hours. When the warm, moist, incoming air comes into contact with cold wall or pipe surfaces which are below its dew-point, water will condense upon them (see Figure 5.1) but as the walls warm up and eventually exceed the dew-point, condensation ceases and the condensed moisture evaporates. A building of light construction will warm more rapidly and is less likely to suffer condensation from this cause.

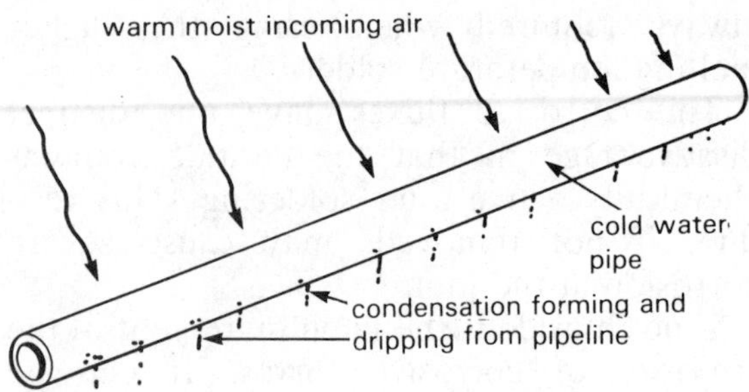

Figure 5.1 *Action of condensation*

A solid floor, with a non-insulating finish, has a surface that is slow to warm, and if there is a rise in temperature and humidity of the air above, it may suffer condensation for several hours. In general, the bigger the heat capacity of the structure, the longer will condensation persist on its surface in adverse conditions.

Artificial influences

The humidity inside an occupied building is usually higher than outside. People themselves and many of their activities increase the amount of moisture in the air. A person sitting down will breathe out more than a litre of water (see Figure 5.2) as vapour in

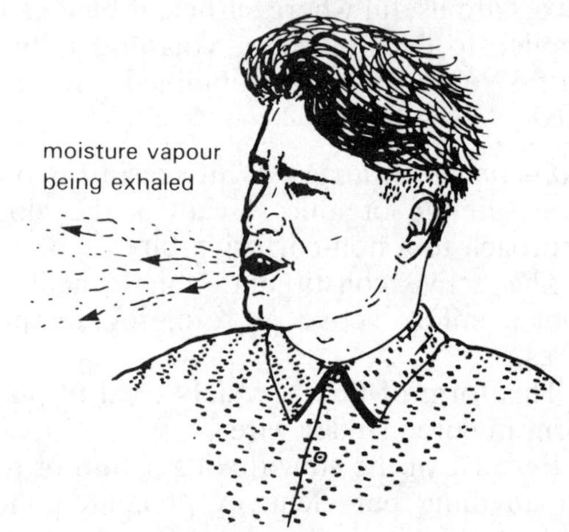

Figure 5.2 *Increasing moisture in the air*

twenty-four hours; physical exertion may raise this to four times the rate. Moisture vapour is released by cooking, by clothes-washing and drying and by the combustion of oil, gas or solid fuel (see Figure 5.3). A litre

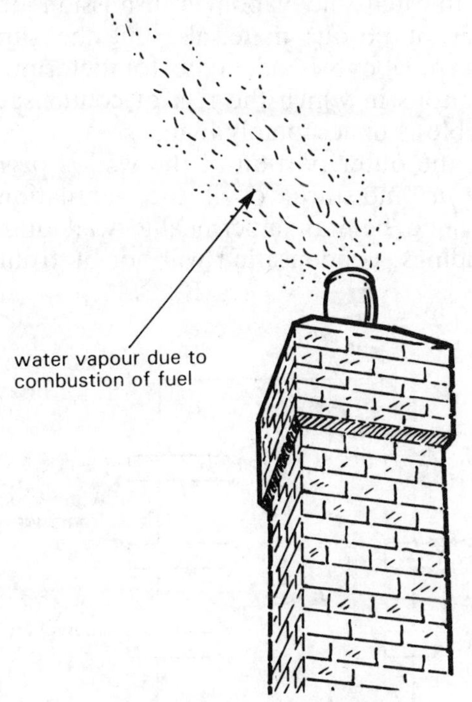

Figure 5.3 *Water-vapour plume from a chimney*

of oil burnt produces in vapour form the equivalent of about a litre of water and if burnt in a flueless appliance this vapour is emitted into the air within the building.

Many industrial processes require high humidities and temperatures and some release large quantities of steam. The risk of condensation is great when the RH and temperature of the air remain above 60% and 20 °C for long periods.

Condensation, particularly in dwellings, does not necessarily occur in the room where the water vapour is produced. A kitchen or bathroom in which vapour is produced may be warm enough to remain free from condensation except perhaps on cold single-glazed windows, cold-water pipes and other cold surfaces. But if this water vapour is allowed to disperse through the dwelling into colder spaces such as the stairwell and unheated bedrooms, condensation will occur on the cold surfaces of those rooms, which may be remote from the source of the moisture. Soft furnishings, including bedding and clothing, may become damp because of this.

Removal of the moisture-laden air from the building from a point near to the source of the moisture will greatly reduce the likelihood of condensation, for example, air movement caused by air bricks, open windows, ventilating fans and chimneys (see Figure 5.4).

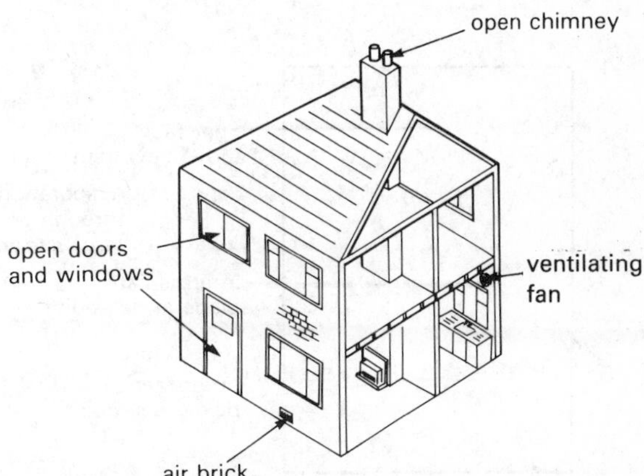

Figure 5.4 *Air passages in a building*

Absorbent surfaces and materials

Temporary or intermittent condensation which is clearly visible on a non-absorbent surface may pass unnoticed on an absorbent surface or material. Condensed water can be absorbed and held until conditions change and allow it to dry out, but condensation can only be accommodated in this way if the periods of condensation are short enough and drying periods long enough to avoid complete saturation of the absorbent material. This is the principle on which anti-condensation paint works.

When the material of a wall, roof or similar building element is permeable to water vapour – and this applies to nearly all building materials – a dew-point temperature is associated with each point within the material. In the same way that a temperature gradient exists through a structure, depending on the thermal properties of the component materials, so a dew-point gradient depending on their water vapour diffusion properties also exists. If at any point the actual temperature is below the dew-point, then condensation will occur at that point within the material or structure. For example, the temperature through the thickness of a wall may vary from the inner face being above dew-point to the outer face below dew-point; at some intermediate position the temperature will then be equal to the dew-point and condensation will begin at this plane (see Figure 5.5).

The exact processes taking place are complex and are further complicated by the fact that the condensed water changes both the thermal and vapour-transmission properties of porous materials, but the simple concept above is adequate for determining situations in which the risk of condensation trouble is unacceptably high.

If the outer portion of the wall is permeable to moisture, or if the ventilation is provided behind impermeable wall or roof claddings, condensation will not be trouble-

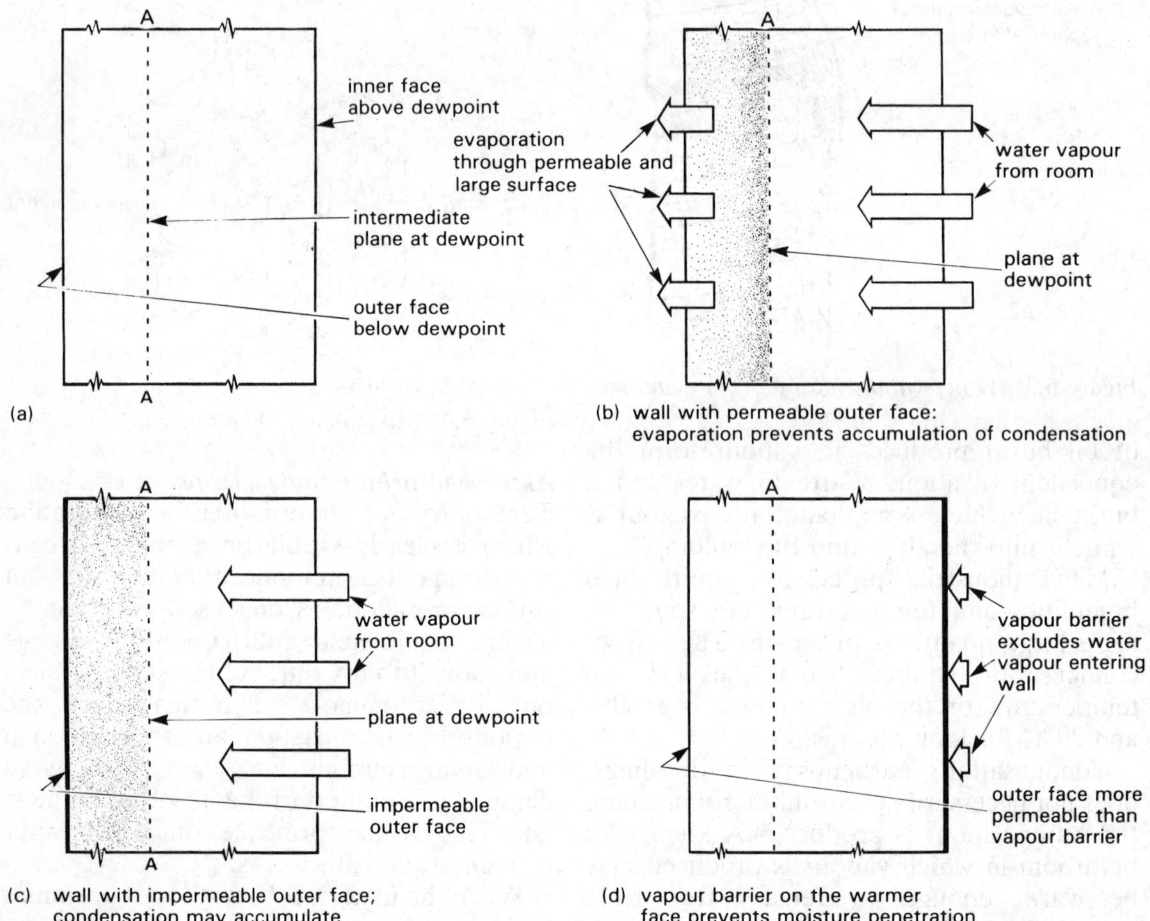

Figure 5.5 *Temperature conditions in wall which may lead to interstitial condensation*

some because the moisture can evaporate gradually to the outside air (see Figure 5.5b).

If the outer surface is impermeable, the condensed moisture tends to accumulate in the wall and may ultimately saturate the material (see Figure 5.5c). The situation will be most severe when the humidity of the indoor air is high.

A vapour barrier on the inner face of the wall (on the potentially warm side of any layer of insulating material) will prevent the passage of water vapour into the wall but only if it is undamaged and continuous. If the outer face of the wall is more permeable than the inner vapour barrier, any moisture contained in the wall can escape to the outside air (see Figure 5.5d). If, however, the outer face of the wall has an impermeable cladding, or if the cladding is of organic material that would suffer in prolonged damp conditions, a ventilated cavity should be formed between the cladding and the wall so that any moisture evaporating from the wall surface is removed.

Corrosion

In plumbing, examples of corrosion of metals are not difficult to find. When we consider that pipes carrying water (which may itself be corrosive) are laid in soil with strongly corrosive properties, i.e. clays, clinker or ash, it is not surprising that corrosion takes place. What is sometimes difficult to determine is the cause of corrosion. The chief causes of corrosion are:

1 The effects of air and water,
2 The direct effect of acids,
3 Electrolytic action.

When metals are exposed to the atmosphere they form a layer of oxide on their surface. The speed at which the oxide forms varies with each metal. For atmospheric corrosion to take place it is not necessary for the metal to be constantly wet or even exposed to rain for long periods.

The gases present in the atmosphere that have the greatest effect on metals are oxygen, carbon dioxide, sulphur dioxide and sulphur trioxide, together with the water vapour in the atmosphere.

Oxygen produces a film of oxide on metals. Carbon dioxide may mix with rainwater to form a weak solution of carbonic acid and in contact with metals it tends to promote the formation of carbonate films, as with sheet copper when it produces a basic carbonate film (patina). Sulphur dioxide is probably the greatest accelerator of atmospheric corrosion. It is a gas, ejected with flue gases, which when mixed with rainwater forms a weak sulphurous acid.

Sulphur trioxide also combines with water to form sulphuric acid. These gases are found in considerable concentrations in industrial areas and iron rusts 3–4 times as fast and zinc 6 times as fast in industrial areas as in rural areas. Seaside towns also suffer from atmospheric corrosion because of the sodium chloride (salt) particles present in the air and derived from the sea.

Copper

The bright lustre of copper sheet changes after a time to a greenish shade (patina). This discoloration is due to the combined effect of carbon dioxide, and sulphur dioxide which form carbonates and sulphates. Copper tubes that have been exposed to ashes containing sulphur exhibit the same hard green coating. The green covering on sheet copper is a hard and tough protective layer which prevents any further attack by the atmosphere.

Lead

Lead also forms a tough non-scaling film (lead oxide) on its surface when exposed to the atmosphere. If the atmosphere is an industrial one, the surface of the lead appears to be ingrained with sticky tar-like deposits.

Lead is also affected by the acids present in some timbers. Oak boards in particular contain acids that attack lead. The action of alkali in cement upon lead flashings or weatherings can also be detrimental in moist or permanently damp situations.

Zinc

This is a very stable metal in dry air, and in moist air forms a film of oxide and carbonate. The carbonate is formed from the carbon dioxide present in the air. It is liable to attack from sulphurous gases in the atmosphere – this results in the zinc being converted to zinc sulphate, a whitish compound. In addition, any solutions containing ammonia tend to attack the metal quite rapidly.

Aluminium

This metal also forms a hard and durable oxide film that protects the underlying metal from further attack. If exposed to alkali attack the material should be protected with bituminous paint.

Any weaknesses in the oxide film on aluminium are exploited by corrosive elements in the atmosphere. This results in 'pitting', which is a localised form of attack that is not serious: it does not remain as a steady rate of corrosion but falls off as the oxide film builds up. Alloys of aluminium are liable to attack by salt solution; seaside atmosphere is inclined to be aggressive because of its high salt content.

Electrolytic action

Almost any hot water system offers many examples of electrolytic action. The cold water cistern may be made of steel covered with zinc (galvanised), the hot water cylinder of copper, the pipework of lead, copper or steel, often with brass or gunmetal connections or wiped soldered joints. It has been found that certain types of water are capable of dissolving small quantities of copper in hot water systems. This copper-bearing water comes in contact with zinc coatings and some of the zinc is dissolved by electrolytic action between the two metals. Once the zinc coating has been perforated then the attack of the steel underneath proceeds at a rapid pace. The ability of the water to dissolve copper is important and it is known that any increase in the carbon dioxide content is liable to increase the copper-solvency of the water.

In hard water districts the formation of scale on pipes, boiler and cylinder, provided it is an unbroken film, often prevents attacks due to electrolytic action.

In general it is best therefore to construct the hot-water system of one metal only – unless previous experience in the district shows that corrosion problems do not arise from mixing metals in an installation. Before the introduction of the plastic flushing cistern, with nylon siphon and ball valve float, it was common to find a flushing cistern made of iron, with a brass ball valve; a copper ball float with a soldered seam and perhaps a lead-alloy siphon. It is not surprising that in this confined space the life of the soldered seam on the copper ball float was short, especially when the local water tended to be slightly acid.

The use of other metals in addition to lead as a roof covering have made it necessary for the plumber to be most careful when fixing roof coverings. An aluminium-covered roof provides a weathertight finish, but its efficiency would soon be affected if a copper or iron rainwater pipe discharged water on to it from an old lead- or zinc-covered roof. The aluminium roof would soon be pitted and perforated as a result of the electrolytic action between the dissimilar metals.

Corrosion in central heating systems

Steel is a man-made alloy and it is skill and knowledge that make it possible to convert the oxidic ore into iron and steel.

Common red rust is probably the best known of all the corrosion products of iron. Others are white, green and black. The black oxide, also known as ferrous oxide or magnetite, is most commonly found within central heating systems in the form of a black sludge (see Figure 5.6). Red rust requires moisture and generous supplies of free oxygen for its formation, while the black

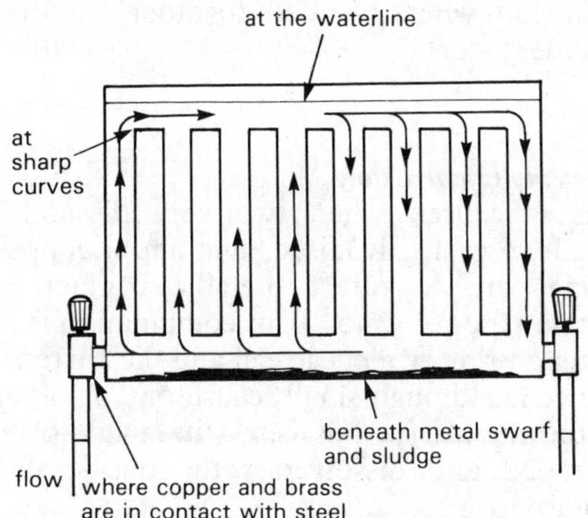

Figure 5.6 *Corrosion attack in a radiator*

oxide of iron has a lower oxygen content in its molecule and it will form when there is very little free and dissolved oxygen available in the water.

Hydrogen gas is a by-product of this corrosion process, and the frequent necessity to bleed the gas from radiators (see Figure 5.7) clearly indicates that corrosion is taking place.

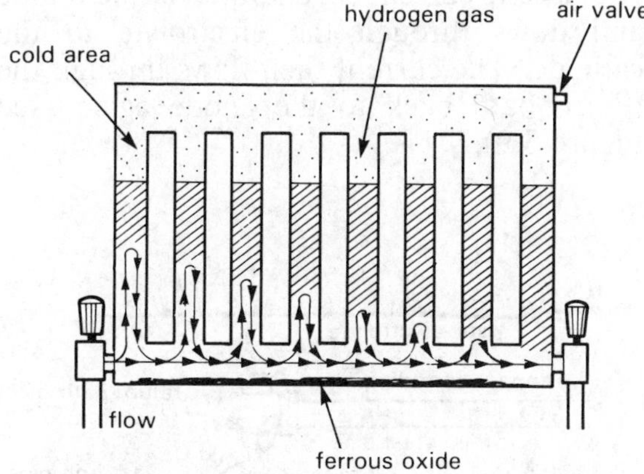

Figure 5.7 *Build-up of hydrogen gas and reduced hot water circulation due to corrosion*

A simple way to determine if the water in a central heating system is corrosive is to carry out the corrosion test as shown in Figure 5.8.

(1) fill a small jar with water drawn from a radiator vent.

(2) add a few clean steel nails (not galvanised nails) to simulate the steel of the radiators. Close the jar and leave for three days.

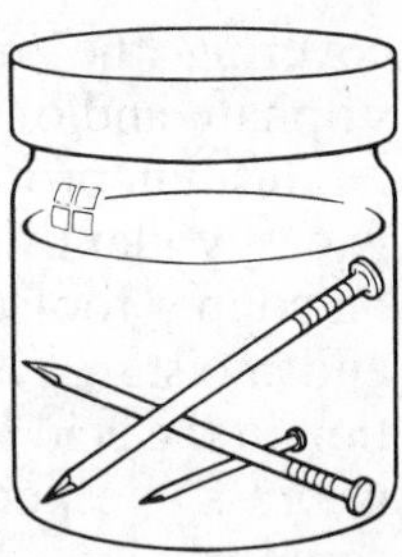

if the nails rust, you can be sure that all steel within the system is corroding. In addition, when the system is heated, corrosion processes become more aggressive.

Figure 5.8 *A simple test for corrosion*

Corrosion is the 'eating away' of a substance by an attacking influence, which is usually external. The term corrosion is often misapplied to cases of encrustation or deposition where no actual corrosion has occurred.

The more common corrosive agents with which we in the plumbing trade are concerned are:

1 Air containing moisture, carbon dioxide, sulphur dioxide, sulphuric acid, or combinations of these,
2 Water containing dissolved air, mineral or vegetable acids, alkalis and certain salts.

All acids and the strong alkalis are corrosive, but as plumbers we are mainly concerned with those agents likely to attack plumbing pipework, roof work, sanitary fittings, etc.

Many soils are slightly acid or alkaline, and with the inevitable moisture have detrimental effects on many metals.

Atmospheric corrosion

Pure air or pure water acting independently have practically no corrosive action. Moist air and water with dissolved air attack iron and steel very quickly, producing the familiar oxide known as 'rust'. If this corrosive action is unchecked the metal will be completely destroyed.

If sulphur dioxide or carbon dioxide are present in the air, copper is attacked, covering the metal with a film of basic sulphate and/or carbonate.

This film protects the underlying copper; it is easily identified as the green coating seen on copper roofs. Zinc, while withstanding air and moisture, is subject to quick deterioration in the acid air of industrial towns, so also is brass – especially brass with a high zinc content.

Lead and tin withstand atmospheric corrosion well. Aluminium is corroded by the atmosphere to the extent of surface dullness, but is seriously damaged by alkaline solutions. This also applies to tin and lead solders.

Corrosion by water

The corrosive effects of impure water are very important. The case of iron and steel have already been referred to. So called 'soft' waters have a pronounced action on lead. The strength of a lead pipe is not greatly affected by this minute corrosion, but since a very small quantity of lead in domestic water supplies (plumbo-solvency) is highly dangerous to health, the matter assumes great importance.

Very few waters attack copper, but with highly acidic waters, green staining of sanitary fittings may occur (cupro-solvency). Generally, in the case of neutral or hard waters, the tubes become coated internally with a thin protective film.

Where the attack, however, is persistent, no harmful effects will occur to people consuming the water. Tin is sometimes used as a coating for lead and copper pipes, but commercial pure tin must be used, otherwise the coating is useless.

Corrosion resistance

Gold and the rare metals of the platinum group can be regarded as incorrosible for all practical purposes. To a lesser degree, nickel and chromium, much used for ornamental finishes, are resistant to corrosion.

The alloys of the 'stainless' steel group (steel and nickel) are reasonably immune to corrosion when blended together to form 'stainless steel'.

Electrolytic corrosion

This is caused when two very dissimilar metals, e.g. a galvanised tube and a copper fitting, are in direct metallic contact in certain types of water. This combination is in effect a primary electric cell and the currents induced, although small, cause one or other of the metals (in this case, the zinc) to be corroded and dissolved with considerable rapidity.

Specks of iron rust resting in a brass tube may cause perforation of the tube in certain types of water. This form of corrosion can take place in water systems or in damp soil. Electrolysis or 'galvanic corrosion' requires four things in order for it to take place. These are:

1 an anode – the corroding area,
2 an electrolyte – the means of carrying the electric current (water or soil),
3 a cathode – the protected area,
4 a return path – for the corrosion currents.

An electric current is generated at the anode and flows through the electrolyte to the cathode. The current then flows through the return path back to the anode again (see Figure 5.9).

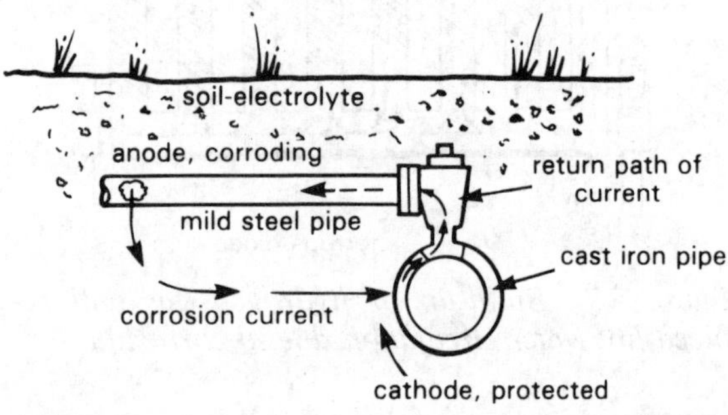

Figure 5.9 *Electrolytic corrosion, dissimilar metals*

The principal causes of electrolytic corrosion are:

1 Different metals joined together and both in contact with the electrolyte, for example, mild steel radiators and copper pipe in a wet central heating system. Or a steel service pipe connected to a cast iron main. In both examples the steel is the anode or corroding area. The metals do not have to be completely different to set up electrolysis. It is sufficient to have clean, pure metal at one point and scale, impurities or scarring at another. Corrosion can take place between iron and particles of graphite or carbon in the same metal.
2 Differences in the chemical environment of the metal. For example, in buried pipes, a lack of oxygen or a concentration of soil chemicals or bacteria at the anode point (see Figure 5.10).
3 Stray electric currents. This may occur where bonding is ineffective and a gas service pipe acts as an electrical earth return. It happens on gas mains when in contact with other authorities' plant or electrified railway systems (see Figure 5.11).

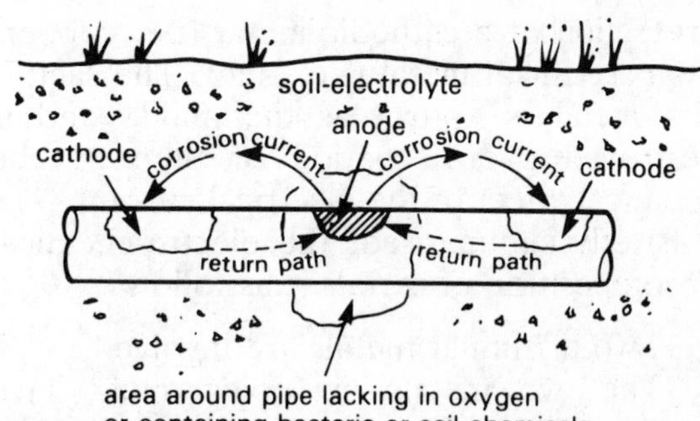

Figure 5.10 *Electrolytic corrosion, difference in environment*

The electro-chemical series

The farther apart two metals appear in the table below the more active the corrosion will be when they are placed in contact in a slightly aqueous solution.

Cathodic protection

Cathodic protection is a form of corrosion control designed and arranged to combat the chemical effect of electric current flows induced by electrolytic action. It is the

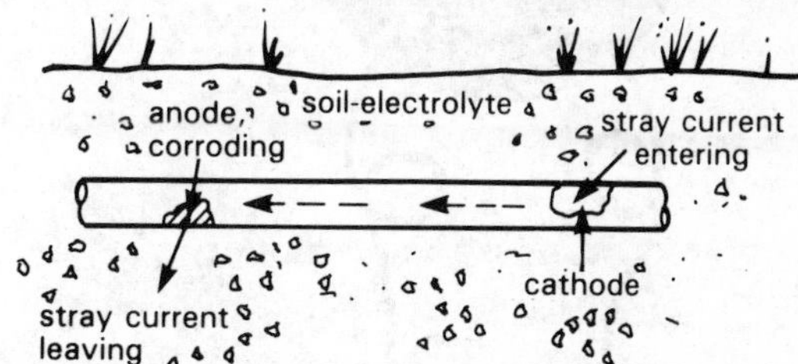

Figure 5.11 *Electrolytic corrosion, stray currents*

	Metal	*Chemical symbol*
Cathodic	1 Gold	Au
	2 Platinum	Pt
	3 Silver	Ag
	4 Mercury	Hg
	5 Copper	Cu
	6 Lead	Pb
	7 Tin	Sn
	8 Nickel	Ni
	9 Cadmium	Cd
	10 Iron	Fe
	11 Chromium	Cr
	12 Zinc	Zn
	13 Aluminium	Al
Anodic	14 Magnesium	Mg

protection of a cathodic metal (i.e. copper) by a sacrificial metal (i.e. zinc). The sacrificial metal is known as the anode and is destroyed over a period of time by the chemical effect of the electrical current.

Briefly summarised, the electro-chemical decomposition of metals is as follows:

1 Two dissimilar metals are involved,
2 The two metals or 'poles' must have contact so that current flow can take place,
3 The 'poles' must be immersed in an electrolyte, that is, a liquid or moist substance capable of conducting electricity.

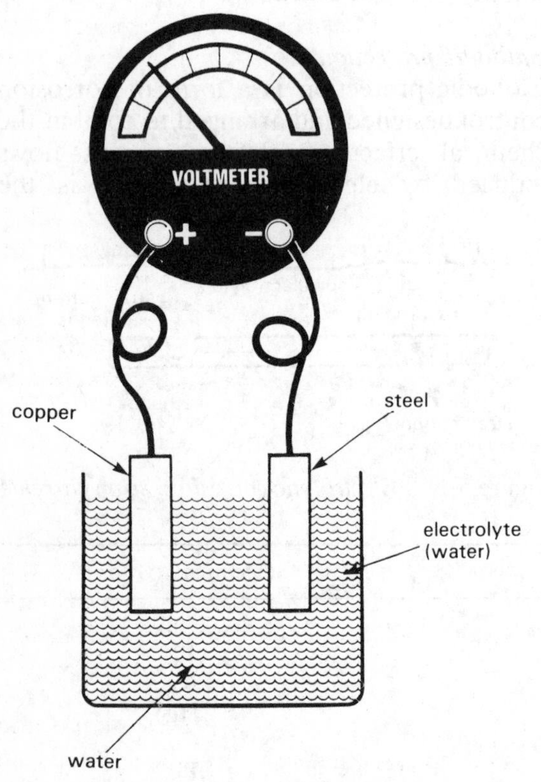

Figure 5.12 *Example of electrolysis. Copper and steel, immersed in water, and connected to a voltmeter confirm that electricity is being generated. Steel is the 'sacrificial' element in this instance. Hot water, or the addition of impurities to the water, will increase electrolytic corrosion and thus the voltage.*

The simplest example of electrolysis, or corrosion due to the chemical effect of electric current flow, is the voltaic cell. This comprises a jar to contain the electrolyte and the two 'poles' or dissimilar metals, say steel and copper. If the 'poles' are connected, by a wire or similar connector, outside the electrolyte then an electric current will flow around the circuit.

This simple arrangement (see Figure 5.12) produces an electric cell or 'battery' capable of producing electrical energy and this can be measured on suitable instruments. The current generated is very small, as also is the voltage, but if the current flow is allowed to continue, there will be evidence of the steel 'pole' seemingly being dissolved away. This electro-chemical decomposition is the form of corrosion which cathodic protection aims to inhibit or stop.

Zinc is said to be anodic to copper, or copper is cathodic to zinc, and a feature of electrolytic decomposition, or corrosion, is that the anodic metal is the one which is corroded by the chemical effect of the electric current passage as just outlined.

Different metals have varying capacities of current flow. These are referred to as their potential differences and enable a table to be drawn up to indicate the likely electro-chemical reaction one might expect when any two dissimilar metals are being considered. For example:

	Metal	*Potential difference*	*Volts*
Cathodic	Copper	0.35	positive
	Lead	0.13	negative
	Nickel	0.25	negative
	Chromium	0.71	negative
	Zinc	0.75	negative
	Aluminium	1.70	negative
Anodic	Magnesium	2.38	negative

from which it will be seen that each metal is cathodic to all those metals listed beneath it.

Corrosion resistance

Water pipes

1	Suitable for hard water	Copper and galvanised low carbon steel.
2	Suitable for soft water	Copper, stainless steel, galvanised low carbon steel.

Lead
Suitable for most waters except soft acid waters, where plumbo-solvency may occur. (WRC water supply by-laws prohibit the use of lead as a water supply pipe material.)

Galvanised steel
Galvanised low carbon steel pipes are given an additional protection against corrosion by the formation of a scale, 'calcium carbonate', from hard waters. However, some hard water with a high free carbon dioxide content will form a loose deposit (see Figure 5.13) which gives no protection.

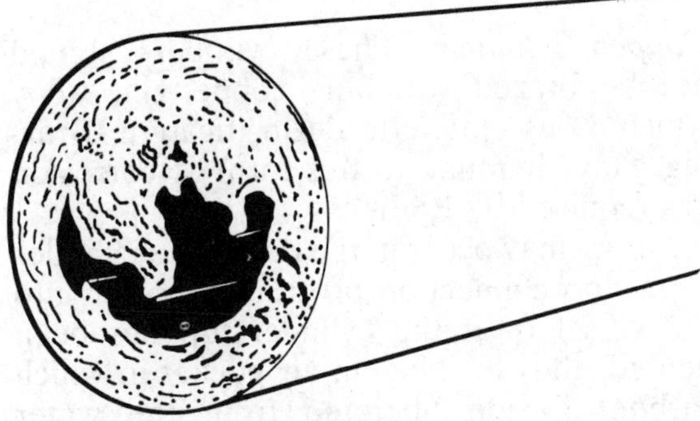

Figure 5.13 *Galvanised steel water pipe blocked by corrosion deposits*

Copper
Most types of water will dissolve minute particles of copper from new pipes. This may be sufficient to deposit green copper salts in fitments. Where this occurs (neutral and hard waters) it will usually cure itself as it forms a protective film over the internal surface. These small copper particles may also cause corrosion to galvanised steel cylinders, etc. and pitting in aluminium kettles.

Corrosive soldering flux residues can often cause pitting. No more flux than is necessary should be used in making capillary soldered joints, and all surplus flux should be removed on completion of jointing process.

Corrosion by building materials
Some types of wood have a corrosive action on lead, and latex cements and foamed concrete will effect copper. Some wood preservatives contain copper compounds and can cause corrosion of aluminium. Copper is not affected by cement or lime mortar, but should be protected from contact with magnesium oxychloride flooring or quick setting materials such as Prompt cement.

Lead is not affected by lime mortar, but must be protected from fresh cement mortar.

Galvanised coatings are not usually attacked by lime or cement mortars once they have set. Aluminium is usually resistant to dry concrete and plaster after setting, but it is liable to attack when they are damp.

Corrosion of water cisterns and tanks and the exterior of underground pipes may often be prevented by cathodic protection. This involves a natural small electric current passing through the water or soil between the metal to be protected and a suitable anode. If the anode is of magnesium or zinc alloy it is connected to the metal to be protected. The two metals, the pipe and the anode, act as an electric cell and a current passes between them, all the corrosion taking place at the anode (sacrificial metal), which has therefore to be replaced eventually. Permanent non-corrodible anodes are sometimes used, but the electric current has then to be provided from an external source, through a transformer and rectifier.

Fittings
Galvanised fittings should always be used with galvanised pipe. Brass fittings may be

cast or hot pressed. Cast fittings are usually of alpha brass. Hot pressed fittings are of 'duplex brass' which some waters will affect with dezincification.

Dezincification may cause:

1 Blockage by build up of corrosion products,
2 Mechanical failure due to conversion of brass to porous copper,
3 Slow seepage of water through the material.

Cisterns

Galvanised steel cisterns will be affected by soft water with a high carbon dioxide content or, where copper is present in the water, some waters are cupro-solvent. Where this is likely to occur the painting of the inside of the cistern with a suitable bituminous paint is recommended. They may also be protected by means of magnesium anodes (Figure 5.14) or other forms of cathodic protection.

Copper vessels should not be soft soldered as this may give rise to corrosion of the tin in the solder.

Figure 5.14 *Sacrificial anode fitted to a hot water tank*

Ball valves

In certain areas with hard water, especially those with a high free carbon dioxide content, ball valve seatings may become eroded as well as corroded by the action of the fast-flowing water. Where this occurs, phosphor-bronze or non-metallic seatings should be used.

Hot water tanks

Galvanised steel is a suitable material in hard water districts but may be affected by soft water which has a high carbon dioxide content. Failure of this material is usually due to:

1 High copper content in the water,
2 Excessively high water temperature,
3 Debris and metal filings, etc. on the bottom of the tank,
4 Damage to the protective galvanised coating.

Copper circulating pipes or cold feed services should not be used with galvanised steel tanks, although magnesium anodes may be used to provide protection in areas where failure might occur.

Copper cylinders These cylinders should not be brazed with any copper-zinc alloy which is susceptible to dezincification. Brazing alloys immune to this form of corrosion are required by BS 699.

Pitting may occur in the dome of a cylinder if the top connection protrudes too far into the vessel or if the cylinder top has been dented, thus forming an air pocket in which carbon dioxide liberated from the water during the heating process can accumulate. Fittings for cylinders should be manufactured from brass, or where dezincification may occur, of copper or gun metal.

Boilers

Domestic boilers are usually of cast iron, which resists attack from the water inside and combustion products outside. In soft-water areas, cast iron may be susceptible to attack.

In this case Bower Barffed boilers should be used.

Copper boilers will resist corrosion in waterways, but are more readily affected by combustion products. In soft-water areas, copper–steel boilers are sometimes used. These have an internal surface of copper and an external one of steel.

Aluminium bronze boilers are claimed to have a high resistance to both corrosive waters and combustion products.

Radiators

Corrosion to light-gauge steel radiators depends upon the presence of dissolved oxygen in the water. In closed systems this is soon reduced to a very low quantity and therefore these radiators may be safely used. They should not, however, be used in open circuits where fresh water containing oxygen is replenished.

Towel rails are usually of chromium-plated brass or copper. Copper or arsenical brass to BS 885 is satisfactory but non-arsenical brass rails should not be used as they may suffer dezincification.

Impingement attack

The rapid flow of turbulent water (see Figure 5.15) has a characteristic form of damage, small deep pitting, often of horseshoe shape. It occurs more often in heating systems than cold-water systems due to rapid pumping or local turbulence set up by partially opened valves or abrupt changes of pipe size, sharp elbows etc.

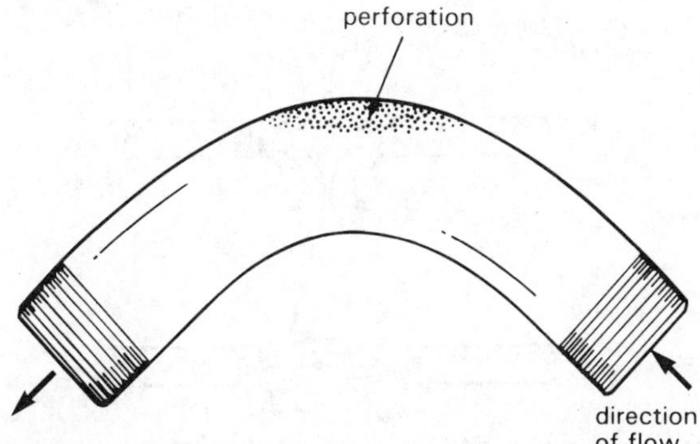

Figure 5.15 *Impingement damage to a low carbon steel bend*

Copper can suffer attack at speeds of above 2 m/sec. For marine work other materials have been developed to overcome this attack. Aluminium and brass will stand speeds of up to 3 m/s and cupro-nickels will stand still higher speeds.

Underground corrosion

Certain types of soil will affect pipes laid underground. Heavy clay may cause trouble, especially if waterlogged, as it may then contain sulphate-reducing bacteria which will corrode steel, lead and copper. Made-up ground containing cinders is very corrosive, and pipes laid in this or any other corrosive soil should be wrapped in one of the proprietary tapes sold for this purpose. Pipes should not be wrapped in hessian or similar material as this may lead to microbiological attack producing corrosive acids. Copper may be protected by the use of plastic covered tube.

Fittings

Dezincification of duplex brass fittings may be caused by some soils. In this case gun metal or copper fittings should be used. Cast iron may corrode in some soils by graphitisation. Connected to non-ferrous metals it will generally suffer initial corrosion, but after its surface has suffered graphitisation the corrosion potential between the two is reversed and the non-ferrous material becomes attacked.

Electric leakage

An electrical leakage from faulty apparatus earthed to the main water supply can cause severe corrosion to pipes.

Roofing

Lead will corrode if laid direct on oak or unseasoned timber. The corrosion will be on the underside in the form of a white powder and is due to the liberation of acetic acid from the wood. Lead should always be laid on bitumenised felt.

Copper

Once green patina – a form of corrosion on copper roofs – has formed, it affords protection against further attack.

Another form of corrosion may occur from water dripping from rusting steel, slates that contain pyrites, or lichen-covered slates or tiles as this water is highly corrosive.

Zinc

Zinc may suffer 'white rusting' on the underside if condensation can occur there. It is also readily attacked by a polluted industrial atmosphere.

Aluminium

Aluminium is corroded by the soluble corrosion products of other metals. It is important that no drippings be received from any copper or copper alloy structure, e.g. copper gutters, lightning conductors, etc.

Soot

Soot deposits from chimneys will cause corrosion on all types of metal roofing (a) because a corrosion cell is set up between soot particles and the metal; and (b) because the soot will contain sulphur acids formed by the combustion of the fuel.

Rainwater goods

Cast iron is generally satisfactory. Gutters should be cleaned and painted regularly both inside and outside.

Galvanised gutters and pipes have good resistance to corrosion, but should not be used in conjunction with copper roofing materials.

Aluminium gutters and pipes have good resistance except where allowed to receive drippings from copper.

Magnetism

A magnet is a piece of metal which can attract to itself, and hold, pieces of iron. The invisible force that enables magnets to attract other objects is called magnetism.

Magnetic materials

A material is said to be magnetic if it is attracted to a magnet. The main magnetic materials are the metals, iron, nickel and cobalt.

Alnico – an alloy containing aluminium, nickel, iron, cobalt and copper – is used to make permanent magnets.

Plumbers use magnets to clean swarf and steel filings out of steel cisterns and tanks, following the cutting of holes for pipe connections.

Magnets are also used in some types of automatic control and are a useful aid for holding objects in place or position during working processes.

Forces between magnetic poles are shown in Figure 5.16 which illustrates repulsion at (a) and attraction at (b). The lines of magnetic force create a 'magnetic flux' or total force between the poles. These lines always form complete loops and never cross each other. Figure 5.17 shows the lines of magnetic flux which form the magnetic 'field' around a bar magnet.

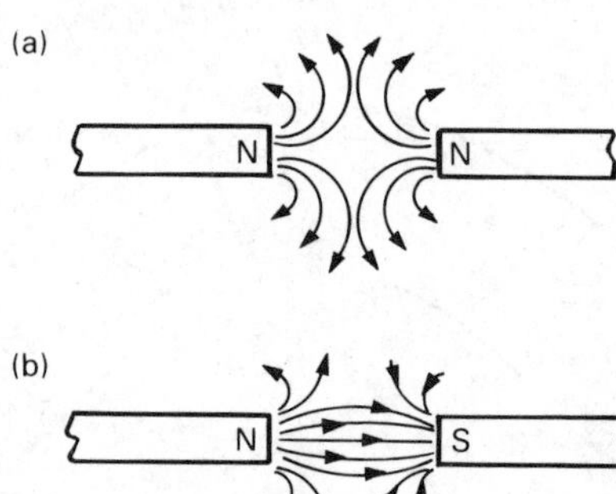

Figure 5.16 *Forces between magnetic poles: (a) like poles; (b) unlike poles*

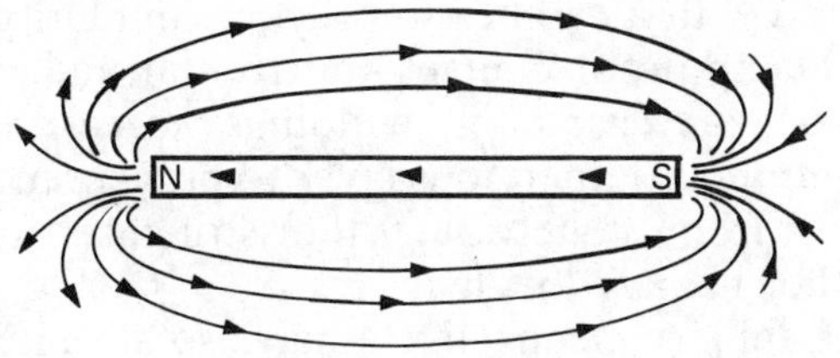

Figure 5.17 *Magnetic field around a bar magnet*

Capillarity and surface tension in building

One example where capillarity is put to good use is in the capillary type of joint used for copper tubing. This type of coupling usually has a channel on the inner face of each socket (see Figure 5.18) which contains a solder ring. After the cleaning of the joint surfaces and fluxing, the joint is slid together and heated, the solder ring melts and is drawn by capillarity into the space between pipe and socket. On cooling, a sealed watertight joint is created. An alternative to this form of joint is the 'end feed' pattern which requires the solder to be fed in from the end of the joint socket, as shown in Figure 5.19.

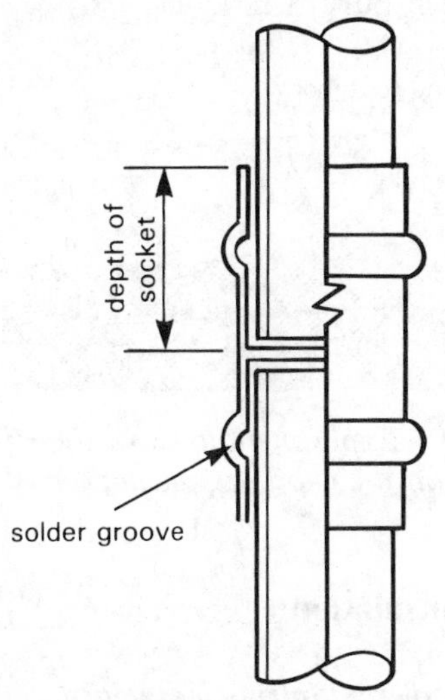

Figure 5.18 *Capillary fitting with integral solder ring*

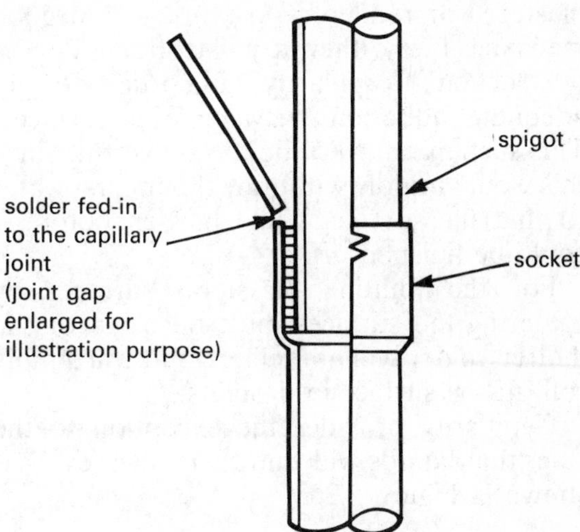

Figure 5.19 *End-feed capillary joint*

Another way that surface tension affects liquid is in the formation of drops. When a small drop of mercury is placed on a flat surface it forms a ball. This is the result first, of surface tension, and second, of the fact that it does not adhere to the surface. If a small quantity of water is dropped on a sheet of clean glass, adhesion causes it to spread out. If, however adhesion is prevented, say by the surface being oiled, round droplets are formed which will easily roll off. From this it can be seen that rainwater will be shed more easily as round droplets from surfaces that do not offer good adhesion. To achieve this, the surface may be treated by oiling (hardwoods, metals), waxing (masonry), painting, and the application of various sealers. The practice of oiling the surfaces of moulds to receive concrete is particularly important since it prevents the concrete from adhering while hardening.

In some circumstances, however, adhesion is an advantage; in painting, there must be sufficient adhesion for the wet paint to spread evenly, in a continuous film. If there is no adhesion, the paint will form into globules on the surface and the work will be unsatisfactory. If a wall which is to be plastered or rendered does not give a good mechanical key, then it must offer a degree of 'suction' (capillarity) in order to get adequate adhesion between the surfaces. This suction must not be too powerful, since this would rapidly withdraw the mixing water so that the surface would harden before it could be finished.

For the jointing of metal surfaces by soldering the surfaces must be clean or the molten solder will not adhere and will simply roll off the surface in droplets.

Capillarity provides the exception to the rule 'that liquids will find their own level' as shown in Figure 5.20.

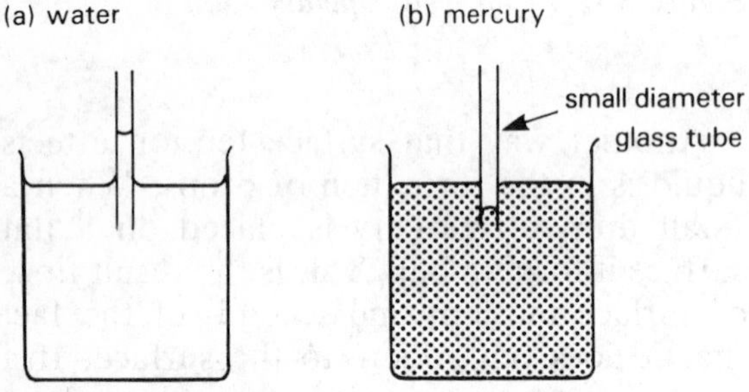

Figure 5.20 *Effect of capillary attraction*

When setting out the drip positions on a lead-covered roof or valley gutter, plumbers must impress on the carpenter the need for an anti-capillary groove in the vertical face of the drip. And when dressing the lead overcloak over the drip the plumber must be careful not to follow the contour of the undercloak, otherwise the whole point of the anti-capillary groove is lost.

Capillary grooves should also be formed if lap joints are being used on a ridge or hip roll, at the front edge of apron flashings, and on a lap joint on an easy pitched roof.

On slated roofs a 'tilting fillet' is used to permit a reasonable gap between the slates and chimney back-gutter to break any passage of water drawn up from the lower edge. (Plain lap tiles for roofing are manufactured with a camber to prevent capillarity).

Damp-proof courses (horizontal and vertical) are used in buildings to form a continuous impervious layer to prevent moisture from penetrating the structure from either high or low level.

Capillary means 'like a hair', so a capillary tube is one with a very small bore.

If a very small diameter glass tube, open at both ends, is placed vertically in a glass of water, the water will be seen to rise up the tube (see Figure 5.20(a)). This is due to 'capillarity' or capillary attraction.

The water rises because its molecules have a greater attraction to the glass than they have to each other. With mercury (see Figure 5.20(b)) the reverse is true and so the level in the capillary tube is below the surface of the liquid, not above it.

Any substances with pores of sufficient size are able to suck up water, for example, tissues, sponges, blotting paper. Moisture rises in the roots and stems of plants by the same means.

The height to which the water will rise depends on the diameter of the tube and becomes greater as the bore is diminished. This can be seen if two glass plates are held vertically and a small distance apart in the liquid (see Figure 5.21).

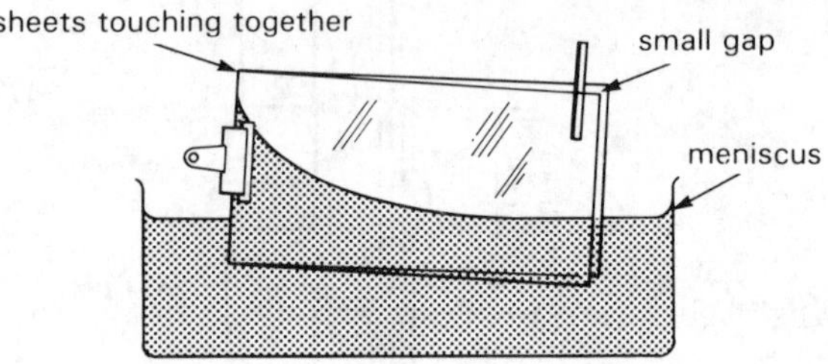

Figure 5.21 *Experiment to show the effect of size of the gap on the capillary attraction*

Thermal insulation

Properties of insulating materials

1 The material must have a *low thermal conductivity*. This means that the material

must resist the passage of heat through it. A poor conductor used as an insulator will reduce heat losses by conduction from the warm surfaces of pipes, cylinders, etc.

2 *Porosity* is another important property. Many modern insulating materials incorporate, and use, still air. Most of these materials are spongelike or fibrous in structure – containing millions of tiny air cells which trap air within the material and hold it still.

3 Insulating materials should be *incombustible*, or at least they should not catch fire easily. The reasons for this are obvious, especially for insulation in roof spaces and other inaccessible places where fire could cause havoc if encouraged to spread.

4 The material must be *resistant to fungal attack*. Fungi thrive in damp situations, and some insulating materials are made from organic materials which could nourish fungi and encourage them to grow. Where conditions may be damp, care must be taken to avoid using organic materials such as hairfelt. A better choice would be an inorganic material such as glass fibre, mineral wool, or foamed plastics material such as expanded polystyrene.

5 Although *weight* is not tremendously important in pipework insulation, it must be considered in cases where an additional load would be undesirable, such as if the material were being laid above ceilings.

6 The material should always be *resistant to vermin*, for the obvious reasons that vermin can give rise to insanitary conditions. Anti-vermin preparations can be applied to most materials likely to suffer from an infestation.

7 It is important that insulating materials should *not absorb moisture*, especially where they have to be used out of doors or in damp situations. If an open cellular material becomes waterlogged the air will be forced out of the cells and the insulating value of the material will be seriously affected. Many open-celled materials can be rendered moisture-proof by using a special water repellant covering or jacket.

8 A *good surface finish* is desirable in an insulating material. A surface finish which can be painted or decorated easily will help the insulation to blend with the general decorations and so be less conspicuous.

9 *Ease of application* is very important. The fixing of thermal insulations generally demands very careful attention, since coverage must be complete and uniform if it is to be fully effective. Materials that are fitted easily are more likely to be fixed with interest and care than those which are difficult to apply.

Table 5.2 *Thermal insulating materials in plumbers' work*

	Type	*Application*
A	*Loose fill* (In granular or fibrous form Vermiculite, Mica-Fill, etc.)	Infill to pre-formed cavities, for example between hot and cold store vessels and their prepared casings.
B	*Flexible* (In strip or blanket form, Cosywrap, Crown 75, Foamflex)	Wrap around covering for curved surfaces such as pipes and cylinders.
C	*Rigid* (In pre-formed, ready-to-fit sections, moulded polystyrene, etc.)	Made specially for pipework. Also in flat sections for hot water tanks and cold storage cisterns.
D	*Mouldable* (Workable until set hard, magnesium compound lagging)	For covering awkward or irregular shapes such as uncased boilers, large cylinders and valves.

Modes of heat transfer

Heat flows from a substance at a higher temperature to one at a lower temperature, and this effect is known as 'heat loss'. Heat loss continues until both substances are at the same temperature – unless some form of thermal insulation is used to isolate one substance from the other and thus prevent heat from being transferred.

Heat can travel as a result of one of the following 'modes of heat transfer': conduction, convection, and radiation, or by a combination of two or more of them.

Conduction occurs when heat flows through or along a material (see Figure 5.22) or from one material to another in contact with it. Conduction increases with the difference in temperature between the hot and cooler materials in contact. Some materials, such as metals, are good conductors of heat; others which resist the passage of heat by conduction are called poor conductors (thermal insulators). A good thermal insulating material must, among other things, be a poor conductor of heat.

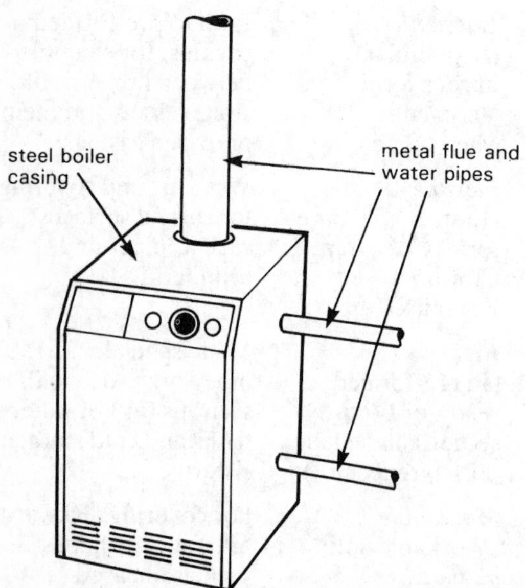

Figure 5.22 *Conductors of heat*

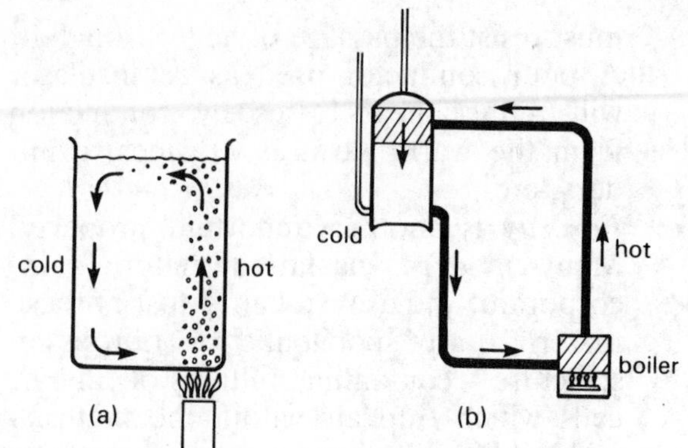

Figure 5.23 *Convection currents in liquid: (a) in an open vessel; (b) in a closed water heating circuit*

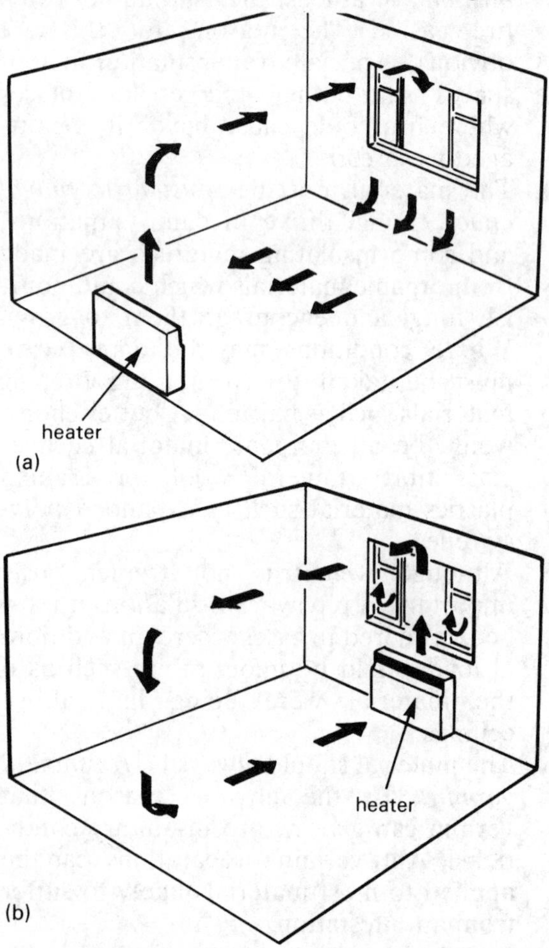

Figure 5.24 *Convection currents in a room: (a) heater at opposite end to window – air cooled by the window becomes a cold draught across the floor; (b) heater under window – cold air is warmed and carried up by convection currents*

Convection occurs in liquids and gases. When the liquid or gas is heated, the warmed particles expand and become lighter, the cooler heavier particles fall by gravity and push the ligher, warmed ones upward (see Figures 5.23 and 5.24). In this way convection currents will continue so long as heat is applied or so long as there is some difference in temperature in the liquid or gas.

Radiation heat is given off from a hot body to a colder one in the form of heat energy rays. Radiant heat rays do not appreciably warm the air through which they pass, but they do warm any cooler solid body with which they come into contact (see Figure 5.25). Radiant heat falling on a surface is partly absorbed and partly reflected. A dull black surface is a much better absorber of radiation than a polished surface. The latter is therefore a good reflector of heat.

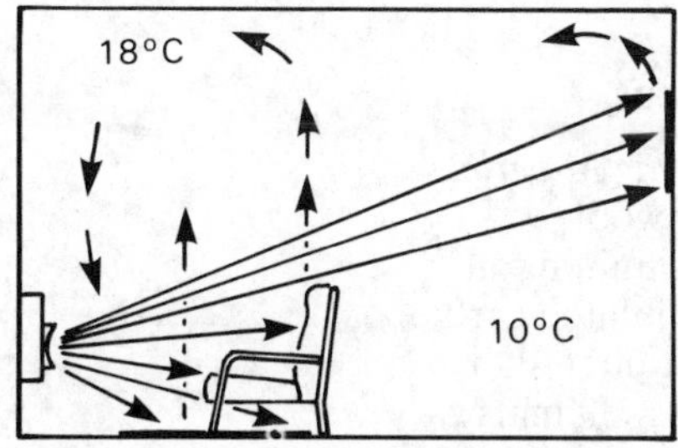

Figure 5.25 *Room heated by radiation from a gas fire*

Insulating values for building materials

It is important for the plumber to understand the thermal insulating values of building materials, and certain purpose-made insulating materials, if he or she is to be able to advise the customer on the best and most economic ways of using those materials to protect a plumbing system and keep the building warmer.

Transmittance coefficient (symbol U) is a term used to denote the quantity of heat which will flow from air on one side of the wall, roof, etc., to air on the other side, per unit area and for unit air temperature in unit time. The units employed are:

Area	square metres (m^2)
Temperature	degrees Celsius (°C)
Time	seconds (s)
Quantity of heat	watts (W)

The definitions are simply applied as shown in Figure 5.26, where 1 m^2 of 230 mm thick brickwork is shown – it can be seen that, for a 1 °C difference in temperature inside and outside, heat will pass from air on the warmer side, through the wall thickness and be lost to the cooler air on the other side at the rate of *2.7 watts/m^2/°C in 1 second*. This, then, would be referred to as the *thermal transmittance coefficient* or *'U-value'* for unplastered brickwork 230 mm thick.

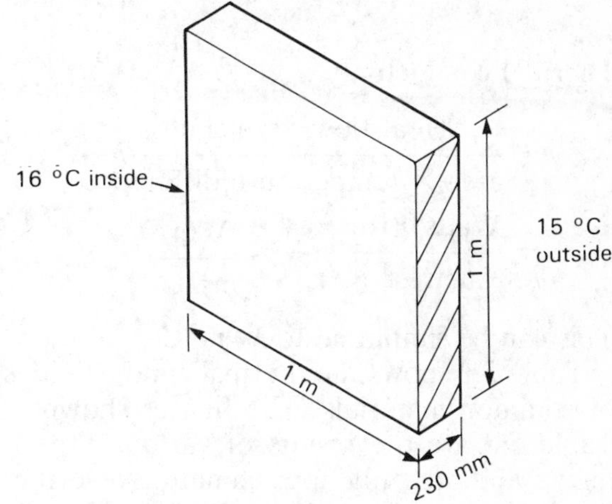

Figure 5.26

The thermal transmittance or U-value of a wall, roof or floor of a building is a measure of its ability to conduct heat out of the building: the greater the U-value, the greater the heat loss through the structure. The total heat loss through the building fabric is found by multiplying U-values and areas of the externally exposed parts of the building, and multiplying the result by the temperature difference between inside and outside.

The capacity of a material to conduct heat is called its 'thermal conductivity'. This is a measure of the amount of heat energy which can be conducted in 1 second through an area of 1 m^2 across a length of 1 m for 1 °C difference in temperature between the two ends, or 'faces' (see Figure 5.27).

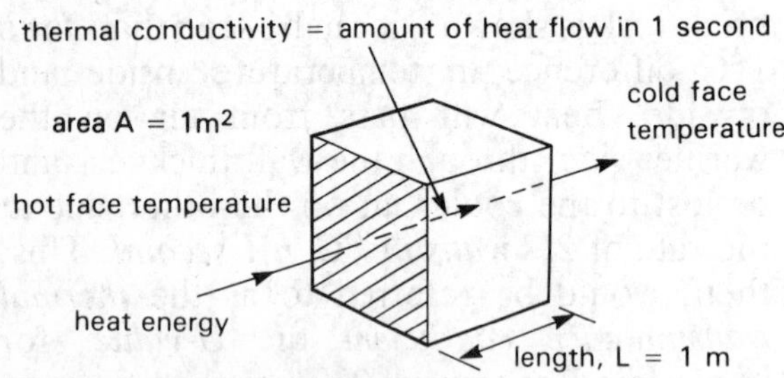

Figure 5.27 *Thermal conductivity*

Thermal conductivity

$$= \frac{\text{Heat flow} \times \text{thickness}}{\text{area} \times \text{temperature difference}}$$

$$= \frac{\text{Watts} \times \text{metres}}{\text{metres}^2 \times {}^\circ\text{C}} = \frac{\text{Wm}}{\text{m}^2\,{}^\circ\text{C}}$$

This can be simplified to W/m °C.

Table 5.3 shows the thermal conductivities of common materials. The figures shown in Table 5.3 are the results of various experiments and are only approximate. Nevertheless, they serve to show the considerable difference in conductivity between the insulators and the gases on the one hand and the solid materials and particularly the metals on the other. Copper is obviously the best and CO_2 is the worst of those shown.

Insulators generally have cellular, granular or matted thread construction. These forms of structure break up a solid path for heat flow and trap small pockets of still air which offer considerable resistance to conduction.

Table 5.3 *Thermal conductivities*

Material	*W/m °C*
Metals (at 18 °C)	
Copper	384.2
Brass	104.6
Aluminium	209.2
Steel	48.1
Cast Iron	45.6
Lead	34.7
Building Materials	
Brick	1.15
Concrete	1.44
Plaster	0.58
Glass	1.05
Deal boards	0.12
Fluids (at 0 °C)	
Methane	0.029
Hydrogen	0.16
Carbon dioxide	0.014
Steam	0.015
Air	0.022
Water	0.054
Oil	0.18
Mercury	8.37
Insulating Materials	
Slag wool	0.042
Aluminium foil	0.042
Granulated cork/ bitumen slab	0.15
Glass silk mats	0.040
Mineral wool slab	0.034
Fibre board	0.059
Vermiculite	0.067
Firebrick	0.61

Latent heat

There are three forms of state in which a substance can exist:

1 solid,
2 liquid,
3 gas.

Water can exist in three states, solid (ice), liquid (water), gas (steam), but because of the temperature normally prevailing we generally see the substance in only one of its states.

The change of state from solid to liquid or vice versa is termed the *lower change of state* and the change from liquid to gas is termed the *upper change of state*.

A solid material, brass for instance, can be changed to a liquid by the application of heat. Excessive heat will cause the zinc in the brass to be driven off in the form of a white vapour.

Gases can also be turned into liquid and solid states. Liquid oxygen is used as a propellent, while liquid butane gas is used in plumbers' blowlamps. Carbon dioxide is used to freeze the water in a pipe when carrying out an alteration or a repair.

Often when a change of state occurs, heat is being put into or given out by a substance without there being any change in the temperature of the substance.

Heat that brings a change in temperature is usually called sensible heat. Heat that does not bring about a change in temperature is usually called hidden or *latent heat*. This latent heat is used to 'unbind' from each other the particles that go to make up a substance. In a solid these are tightly packed giving the solid its characteristic shape. The extra energy (heat) moves the particles further apart and allows them to move about more freely, as in a liquid. Further energy (heat) causes them to become more 'unbound' and they can move even more freely – as in a gas. There are therefore two changes of state and two values of latent heat for any substance. These are:

1 Latent heat of fusion,
2 Latent heat of vaporisation.

Experiment to determine sensible and latent heat

If a piece of ice was placed in a pan and heated its temperature would gradually rise until it reached 0 °C. Then the ice would begin to melt. If the heating was continued until all the ice was melted, the pan would then contain water at 0 °C. There would be no increase in temperature while the ice was melting.

If heating was still continued, the temperature of the water would rise until it reached 100 °C. Then the water would begin to boil. The temperature would stay at 100 °C until all the water was turned into steam. If the heating process was continued with the steam, the temperature would begin to rise again.

Figure 5.28 shows, on a graph, what is happening.

From point A to point B the temperature is rising steadily. The heat being absorbed during that period is called 'sensible heat' because it can be 'sensed' by the thermometer.

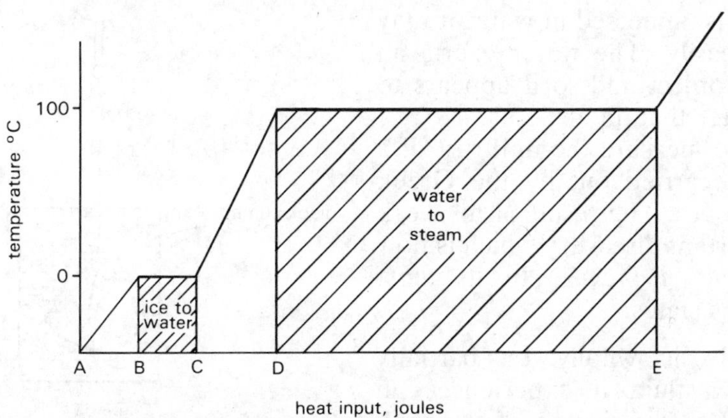

Figure 5.28 *Graph of sensible and latent heat of water (not to scale)*

From B to C it is apparent that heat is still being absorbed but it is not visible on the thermometer, so it is called 'latent heat' because it cannot be seen.

From C to D the thermometer shows the second increase in sensible heat and from D to E shows latent heat again.

The same sort of thing happens with most substances. When a solid melts into a liquid or a liquid cools to a solid the heat absorbed or given off is called the 'latent heat of fusion'.

When a liquid evaporates into a vapour or a vapour condenses into a liquid the heat involved is called the 'latent heat of vaporisation'.

Different substances require different amounts of heat to bring about these changes in state. For example:

Latent heat of fusion of ice = 334 kJ/kg

Latent heat of vaporisation of water = 2250 kJ/kg

Archimedes' Principle

When anything is placed in a liquid it is subjected to an upward force or *upthrust*.

A simple but striking experiment to illustrate the upthrust exerted by a liquid can be shown by tying a length of cotton to an object. Any attempt to lift the object by the cotton fails through breakage of the cotton, but if the object is immersed in water it may be lifted quite easily. The water exerts an upthrust on the object and so it appears to weigh less in water than in air.

Experiments to measure the upthrust of a liquid were first carried out by the Greek scientist Archimedes. The result of his work was a most important discovery which is now called *Archimedes' Principle*. In its most general form, this states:

> When a body is wholly or partially immersed in a fluid it experiences an upthrust equal to the weight of the fluid displaced (see Figure 5.29).

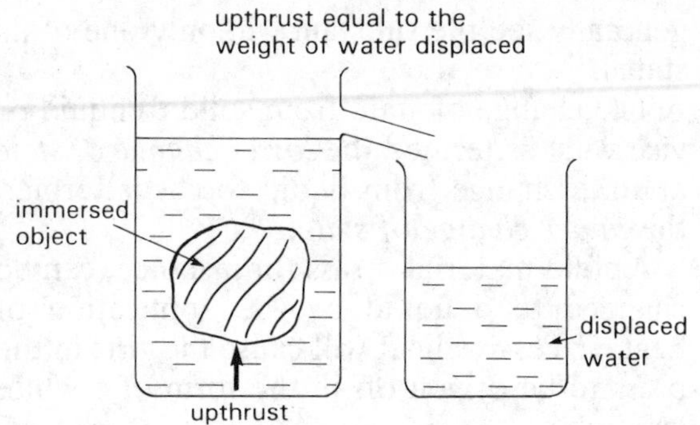

Figure 5.29 *Archimedes' Principle*

To verify Archimedes' Principle for a body in liquid

A eureka (or displacement) can is placed on the bench with a beaker under its spout (see Figure 5.30). Water is poured in until it runs from the spout. When the water has ceased dripping the beaker is removed and replaced by another beaker which has been previously dried and weighed.

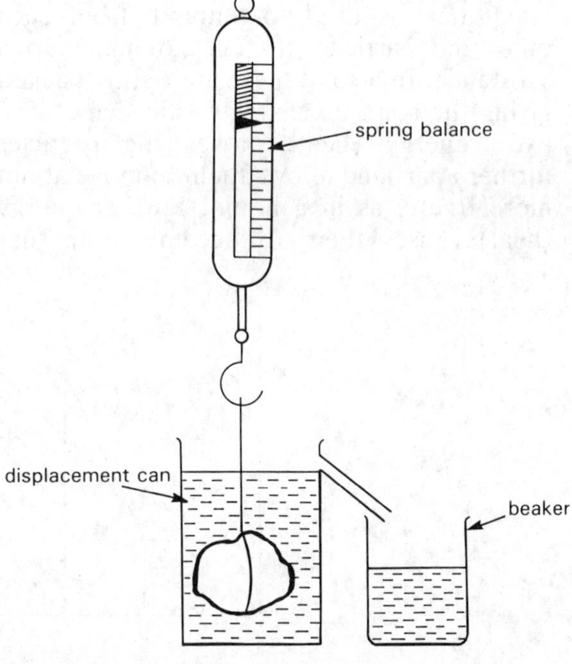

Figure 5.30 *Verification of Archimedes' Principle*

Any suitable solid body, e.g. a piece of metal, is suspended by thin thread from the hook of a spring-balance and the weight of the body in air is measured. The body, still attached to the balance, is then carefully lowered into the displacement can. When it is completely immersed its new weight is noted. The displaced water is caught in the weighed beaker. When no more water drips from the spout the beaker and water are weighed.

The results should be set down as follows:

Weight of body in air	=	g
Weight of body in water	=	g
Weight of empty beaker	=	g
Weight of beaker + displaced water	=	g
Apparent loss of weight of body	=	g
Weight of water displaced	=	g

The apparent loss of weight of the body, or the upthrust on it, should be equal to the weight of the water displaced, thus verifying Archimedes' Principle in the case of water. Similar results are obtained if any other liquid is used.

To measure the relative density of a solid by using Archimedes' Principle

Earlier we explained the meaning of the term relative density (or specific gravity), and its importance in the accurate measurement of density.

Relative density of a substance

$$= \frac{\text{mass of any volume of substance}}{\text{mass of an equal volume of water}}$$

or, since weight is proportional to mass

$$\text{RD} = \frac{\text{weight of any volume of substance}}{\text{weight of an equal volume of water}}$$

Archimedes' Principle gives us a simple and accurate method for finding the relative density of a solid. If we take a sample of the solid and weigh it first in air and then in water, the apparent loss in weight, obtained by subtraction, is equal to the weight of a volume of water equal to that of the sample. Therefore:

Relative density of a substance =

$$\frac{\text{weight of a sample of the substance}}{\text{apparent loss in weight of the sample in water}}$$

Density and relative density

One often hears the expression, 'as light as a feather' and 'as heavy as lead'. Equal volumes of different substances vary considerably in mass. Aircraft are made chiefly from aluminium alloys, which are as strong as steel but, volume for volume, weigh less than half as much. In plumbing we refer to the lightness or heaviness of different materials by the use of the word *density*.

The density of a substance is defined as its mass per unit volume.

One way of finding the density of a substance is to take a sample and measure its mass and volume. The density may then be calculated by dividing the mass by the volume. The symbol used for density is the Greek letter ρ (rho). Thus:

$$\text{Density} = \frac{\text{mass}}{\text{volume}} \text{ kg/m}^3$$

or in symbols $\rho = \dfrac{m}{v}$

The densities of all common substances, solids, liquids and gases, and all chemical elements have been determined and are to be found listed in books of physical and chemical constants.

Water has a density of about 1 g/cm^3 or 1000 kg/m^3 owing to the fact that the kilogramme was originally intended to have the same mass as 1000 cm^3 of water at 4 °C.

Mercury is a metal which is a liquid at ordinary temperatures and it has the very high density of 13.6 g/cm^3. It is a very useful substance in scientific laboratories and plays a part in many experiments.

Importance of density measurements

Architects and engineers refer to tables giving the densities of various building materials when engaged in the design of building works.

From the plans drawn up, they can calculate the volume of any part of the structure, which, multiplied by the density of the material, gives the mass and hence the weight. Such information is essential for calculating the strength required in foundations and supporting pillars.

Simple measurements of density

Liquids A convenient volume of the liquid is run off into a clean, dry, previously weighed beaker, using either a pipette or burette. The beaker and liquid are then weighed and the mass of the liquid found by subtraction.

Solids The volume of a substance of regular shapes e.g. a rectangular bar, cylinder or sphere, may be calculated from measurements made by vernier callipers or a micrometer screw gauge.

The volume of an irregular solid, e.g. a piece of lead, may be found by measuring its displacement volume in water. For solids soluble in water, e.g. certain crystals, some liquid such as white spirit would be used in the measuring cylinder. The mass of the solid is found by weighing.

In each case the density is calculated from:

$$\text{density} = \frac{\text{mass}}{\text{volume}}$$

The above descriptions have purposely been kept short as we shall be describing better methods later on.

Relative density (formerly called specific gravity)

In the last experiments described we had to make two measurements to find the density, namely, a mass and a volume. Now we can always measure mass more accurately than volume, and so, in the accurate determination of density, scientists have overcome the necessity to measure volume (hence eliminating one source of error) by using the idea of *relative density*.

The relative density of a substance is the ratio of the mass of any volume of it to the mass of an equal volume of water, or

$$\text{relative density} = \frac{\text{mass of any volume of the substance}}{\text{mass of an equal volume of water}}$$

In normal weighing operations the mass of a body is proportional to its weight, so it is also true to say

$$\text{relative density} = \frac{\text{weight of any volume of the substance}}{\text{weight of an equal volume of water}}$$

This will explain why relative density has also been called specific gravity: the word gravity implying weight. At the present time, by international agreement the term relative density is recommended rather than specific gravity.

Note that relative density has no units: it is simply a number or ratio. On the other hand, density is expressed in kg/m^3.

To measure the relative density of a liquid

The easiest way to ensure getting identically equal volumes of a liquid and water is to use a *density bottle* (Figure 5.31). This bottle has a ground glass stopper with a fine hole through it, so that, when it is filled and the stopper inserted, the excess liquid rises through the hole and runs down the outside.

Figure 5.31 *Density bottle*

So long as the bottle is used with the same liquid level at the top of the hole, it will always contain the same volume of whatever liquid is put in, provided the temperature remains constant.

The bottle is weighed empty and then when full of the given liquid. The liquid is then returned to the stock bottle. Having rinsed or cleaned the bottle thoroughly, it is filled with water and weighed again.

Two precautions are necessary. The outside of the bottle must be wiped dry before weighing. Secondly, the bottle should not be held in a warm hand or some of the liquid may be lost through expansion.

The results are set out as below. For neatness, decimal points, equals signs and so on should be arranged uniformly underneath each other.

Mass of empty bottle	=	g
Mass of bottle full of liquid	=	g
Mass of bottle full of water	=	g
Mass of liquid	=	g
Mass of water	=	g

$$\text{RD of liquid} = \frac{\text{Mass of liquid}}{\text{Mass of water}}$$

$$= \frac{\quad\text{g}}{\quad\text{g}}$$

$$= \underline{\qquad.\;}$$

Calculation of density from relative density

Suppose we have measured the relative density of a liquid as described above and found it to be 0.8. If we now assume that 1 cm^3 of water weighs 1 g, it follows that the mass of any volume of water is numerically the same as its volume in cm^3.

Consequently we can say straight away that the density of our liquid is 0.8 g/cm^3. However, a word of warning is necessary. To say that 1 cm^3 of water has a mass of 1 g is only a very close approximation. Owing to expansion, the density of water depends (like everything else) on its temperature.

In work of very high accuracy, scientists make due allowance for this when calculating density from relative density. Other corrections also have to be made concerned with the weighing process, but any further discussion would take us into the realm of more advanced studies.

Finally, it is worth noting that the figures expressing the density of a substance depend on the system of units used. For example, the density of lead is 11.4 g/cm^3 or 11400 kg/m^3, but *its density relative to water is a number or ratio, namely, 11.4 and this is the same whatever system of units is used.*

Advantage of the density bottle

We have already pointed out the advantage of the idea of relative density as a step towards the accurate measurement of density. There are no volume measurements to worry about. Weighings only are required and these can be carried out with a beam-balance to a fairly high degree of accuracy. Thus, a good value for the relative density, and hence the density of the liquid, can be obtained by this method.

Contrast this with the simple measurements of density described earlier: a measuring cylinder of the size ordinarily used can be read only to within about 0.5 cm^3; even a burette to within 0.1 cm^3 only. Consequently, unless a very large volume of substance is used the percentage error in the volume measurement will be large. It follows that density determinations of liquids which depend on the direct measurement of volume will be less accurate than those obtained by the use of a density bottle.

Relative density of a solid
An accurate method for finding the relative density of a solid, based on Archimedes' Principle, is described.

To measure the relative density of a powder
This method is useful for substances in powder or granular form, such as sand or lead shot. The procedure to be carried out in the case of sand is illustrated in Figure 5.32 which is self-expanatory.

Apart from the precautions already mentioned above, it is important to make sure that small air bubbles are not trapped in the sand. These can be removed by gently rotating or shaking the bottle. Violent shaking is to be avoided, or sand may become lodged in the stopper hole.

The relative density of sand is obtained as follows:

RD of sand

$$= \frac{\text{mass of any volume of sand in}}{\text{mass of an equal volume of water}}$$

$$= \frac{\text{mass of sand in (2)}}{\text{mass of water in (4)} - \text{mass of water in (3)}}$$

$$= \frac{(m_2 - m_1)}{(m_4 - m_1) - (m_3 - m_2)}$$

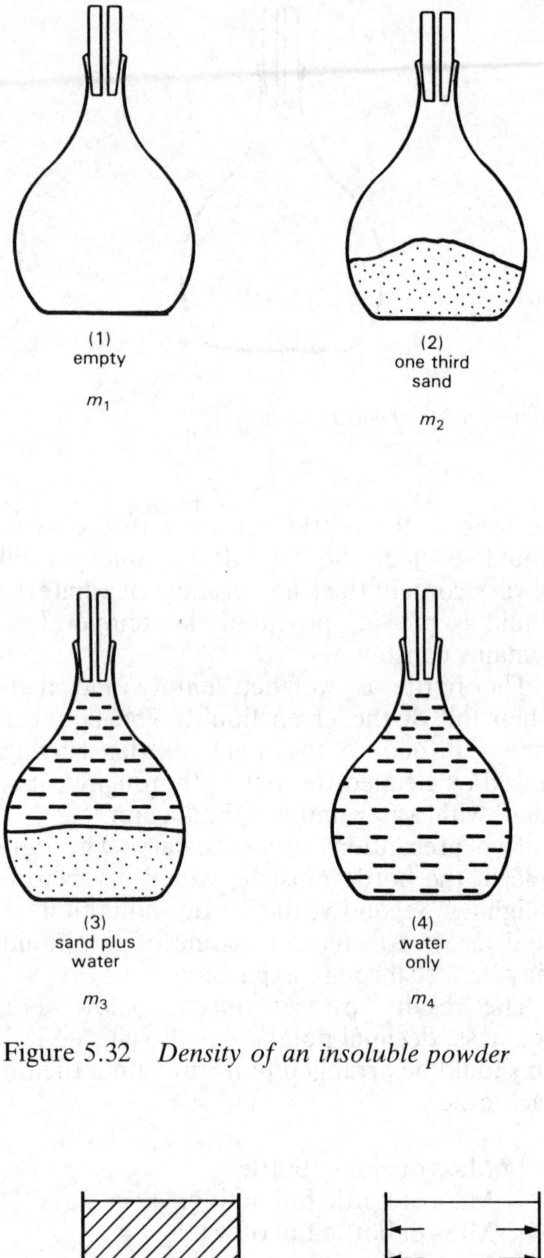

(1) empty m_1

(2) one third sand m_2

(3) sand plus water m_3

(4) water only m_4

Figure 5.32 *Density of an insoluble powder*

Pressure

Definition of pressure
'Pressure' is the same as stress, but whereas stress applies to solid objects, pressure is more concerned with fluids, that is, liquids or gases.

Fluids have the capacity to press themselves against the surface of the vessel that contains them (see Figure 5.33b) and so

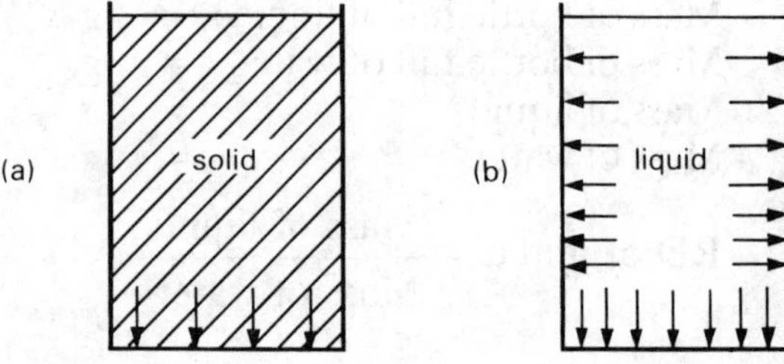

Figure 5.33 *Difference between pressures exerted by solids and liquids*

exert a force. The intensity of this force can be measured in relation to the area of the surface. Pressure is the relationship between force and area.

$$\text{Pressure} = \frac{\text{force}}{\text{area}} = \frac{\text{newtons}}{\text{square metres}}$$

$$\text{(in units)} \quad = \frac{\text{N}}{\text{m}^2} \text{ or N/m}^2$$

The unit N/m^2 has the special name 'Pascal' after the French mathematician. It is in more common use on the continent and has the symbol 'Pa'.

Pressure in fluids

Figure 5.33 illustrates the difference between solids and liquids. At (a) there is a solid block which just fits into a container. The only pressure is that exerted on the base.

At (b), the container is filled with liquid. This now exerts a pressure on the sides as well as the base. In fact, the pressure will increase as the depth increases.

If the weight of the block is 15.4 kg, then in the different positions the pressures exerted by the base would be different also.

It follows that, for the same weight of object, the smaller the base area, the higher the pressure that it will exert. For example, a boiler of shape (a) would exert more pressure per unit of area than a boiler of shape (b) or (c), as shown in Figure 5.34.

Pressure exerted by a solid

A solid object has the ability to exert a pressure on the floor on which it stands, but this is only a downward direction. Fluid pressure acts in all directions at the same time.

The pressure exerted by a solid object is the weight of the object divided by the area of its base. Take as an example the object shown in Figure 5.34.

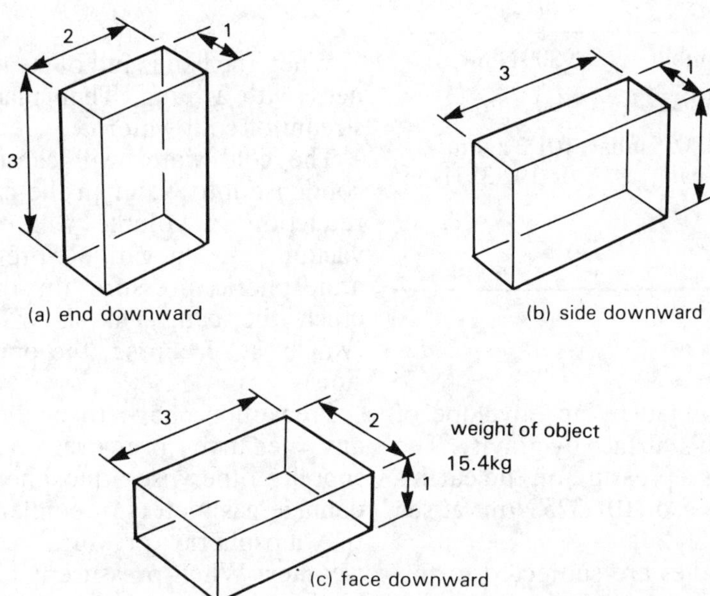

Figure 5.34 *Pressure exerted by a rectangular block: (a) end downward; (b) side downward; (c) face downward*

Units of pressure

Pressure can be calculated either by measuring the force exerted on a unit of area or by measuring the height of a column of liquid supported by the force. So its units are either those of force per unit area, e.g. Newtons per square metre, or they are metres or millimetres height or 'head' of liquid.

There are alternative units. In some areas of activity kilograms force per square centimetre may be used and in the gas industry, for example, pressure will be measured normally in 'bars' and 'millibars'. The symbols are 'bar' and 'mbar'.

Where liquids are used in pressure gauges, water is the most common for low pressures and mercury, which is 13.6 times as dense, for higher pressures. The gauges have scales graduated in millibars.

It may sometimes be necessary to convert a pressure reading from force/area units to height units or vice versa and Table 5.4 shows the comparison between them.

Table 5.4 *Comparison of units of pressure*

Height	*Bars*	*Force/area*
1 m head of water	98 mbar	9800 N/m^2
10.2 mm of water	1 mbar	100 N/m^2
1 atmosphere or 760 mm of mercury or 30 in of mercury	1013.25 mbar (1 bar)	101.3 kN/m^2 or 101,325 N/m^2

Atmospheric pressure

The earth is surrounded by an envelope of air, held to the earth's surface by gravity. The weight of air creates a pressure on the earth's surface of about 1 bar or 101,325 N/m^2 at sea level.

Although our bodies are subjected to this pressure they have an internal pressure which normally exactly balances atmospheric pressure so that we are unaware of it.

Atmospheric pressure must, however, be taken into consideration in a number of calculations involving gas pressure.

One effect of atmospheric pressure can be shown by a simple experiment (see Figure 5.35). Take a metal can which is open to the air so that the pressure inside is the same as that on the ouside. Put a small quantity of water in the can, place it over a gas flame and boil the water. The steam produced will push the air out of the can.

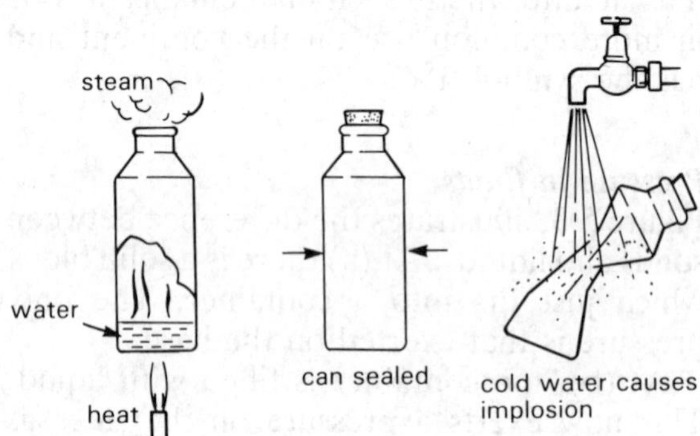

Figure 5.35 *Experiment to show the effect of atmospheric pressure*

When the can is full of steam only, seal the neck with a bung. Then place the can in a stream of cold water.

The cold water will cause the steam to condense into water in the can. The sudden reduction in volume will create a partial vacuum or 'negative' pressure and the atmospheric pressure on the outside will crush the can. This is called 'implosion' (which is, of course, the opposite of explosion).

Precautions have to be taken to prevent any negative pressures occuring in gas supplies otherwise atmospheric pressure can damage gas meters or similar components.

Atmospheric pressure varies with the weather. When pressure is high the day will be fine and dry. As pressure falls the weather becomes changeable, rainy and finally stormy.

Atmospheric pressure also varies with the height above sea level at which the reading is taken. On the top of a mountain the pressure is less than the pressure on the beach.

Barometers
Atmospheric pressure is measured by a 'barometer'. The simplest form of barometer is the mercury barometer devised by Torricelli (Figure 5.36).

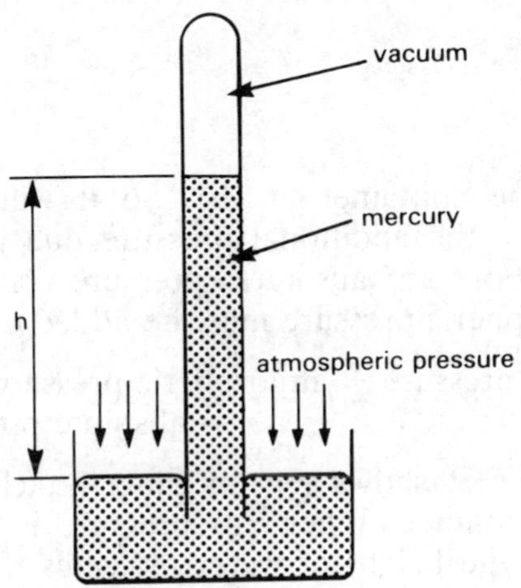

Figure 5.36 *Mercury barometer*

This consists of a glass tube, sealed at one end, which is first filled with mercury and then inverted into a trough also containing mercury.

The mercury will begin to pour out of the tube, leaving a vacuum at the top, until the height of the column of mercury exactly balances the atmospheric pressure which is being exerted on the surface of the mercury in the trough.

The reason for Torricelli using mercury becomes obvious when you calculate the length of tube which would be required if water was to be used.

$$h = \frac{101.3}{9.81} = 10.3 \text{ metres}$$

The mercury barometer offers a very accurate means of measuring atmospheric pressure and there are a number of ways in which provision can be made for the scale to be set to zero and compensation made for altitude. Most domestic barometers are, however, aneroid barometers.

Aneroid barometer
'Aneroid' means 'not liquid' and this type of barometer consists of a cylindrical metal box or bellows almost exhausted of air. The box has a flexible, corrugated top and bottom and is very sensitive to changes in atmospheric pressure (Figure 5.37). Such changes cause

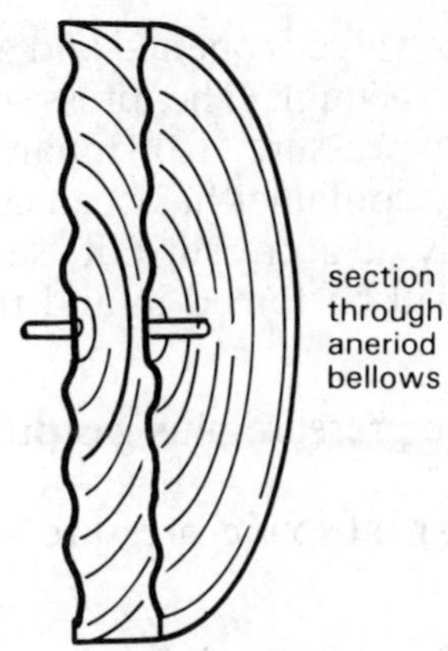

Figure 5.37 *Aneroid bellows*

inward or outward movements of the flexible top and bottom sections and this movement is made to rotate a pointer on a scale by means of a suitable mechanism attached to the bellows.

Absolute pressure
'Absolute' pressure is the pressure from zero to that shown on a gauge. Figure 5.38 shows a container under three different conditions. In all three cases the container is subjected to atmospheric pressure on all sides.

At (a) the air has been pumped out and the valve closed. The pressure inside is zero.

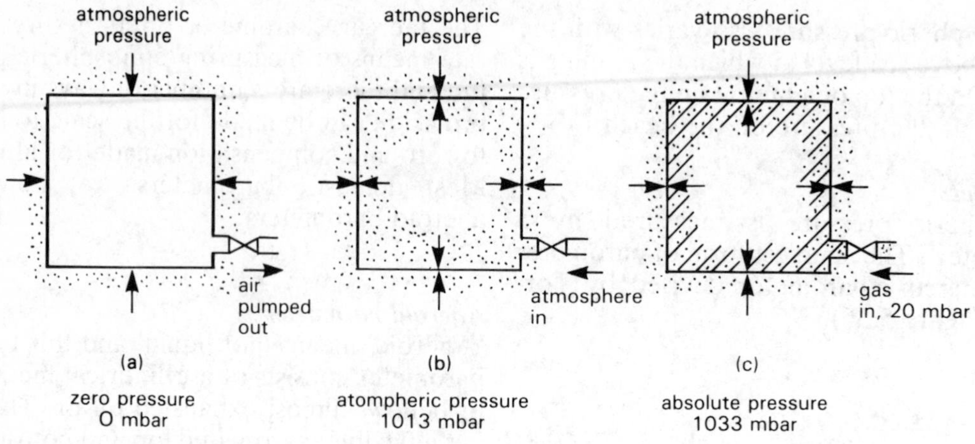

Figure 5.38 *Absolute pressure*

At (b) the valve is opened and air rushes in to fill the vacuum. The pressure inside is atmospheric pressure, 1013 mbar.

At (c) the container has been connected to a gas supply at a pressure of, say, 20 mbar. Some gas will be forced in and the pressure will now be

atmospheric pressure plus gas pressure
= 1013 + 20
= 1033 mbar, absolute pressure

So,

atmospheric pressure + gas pressure
= absolute pressure

Any pressure gauge will have atmospheric pressure inside it before the gas is turned on, like the container at (b). So it will only indicate the additional pressure due to the gas. For an absolute pressure reading, atmospheric pressure must be added on:

gauge pressure + atmospheric pressure
= absolute pressure

Figure 5.39 shows a water gauge attached to the container. When gas pressure is introduced the height of water indicates the gas pressure only.

Measurement of pressure

Pressure gauges, like barometers, can indicate height or force/area so they can be liquid or dry types. The liquid types commonly use water or mercury. There are two main

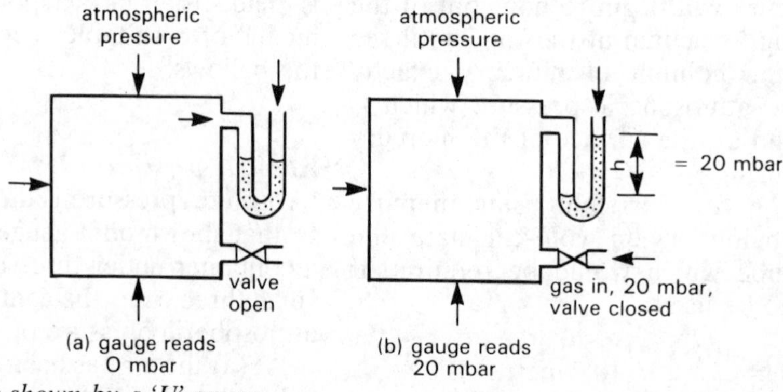

Figure 5.39 *Pressure shown by a 'U' gauge*

differences between these liquids, so far as their use in gauges is concerned.

1 Mercury has a specific gravity of 13.6, so it will measure pressures 13.6 times higher than the same height of water.
2 Water adheres to the side of the gauge tube and mercury does not. Figure 5.40 shows what effect this has on the 'meniscus' or liquid level which in both cases is crescent-shaped. When reading a gauge the true value comes from the bottom of a water meniscus and the top of a mercury meniscus (see Figure 5.41).

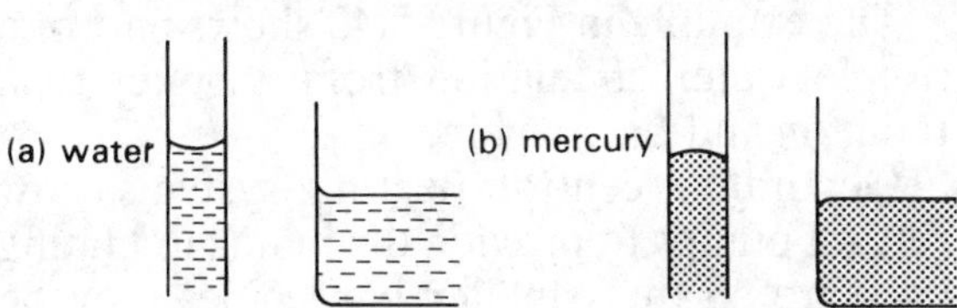

Figure 5.40 *Effect of adhesion and cohesion*

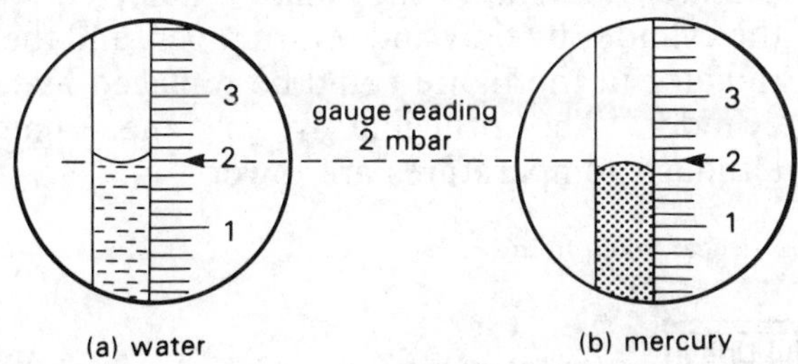

Figure 5.41 *Reading a gauge*

The reason for the difference in the meniscus lies in the difference in the force of attraction between the molecules. There are two forces on the molecules.

1 The attraction of one molecule of the substance to another of the same substance, 'cohesion',
2 The attraction of a molecule of the substance to a molecule of another substance, 'adhesion'.

In water, the adhesive quality is stronger than the cohesive quality. So water adheres readily to other substances. In mercury the reverse is the case. An example of the effect of adhesion and cohesion can be seen in capillary attraction.

Pressure in water heating systems

Hydrostatic pressure

'Hydrostatic pressure' is simply the pressure exerted in the system by the weight of the water.

Figure 5.42 shows a domestic water heating system consisting of a cistern A supplying cold water to a cylinder B, heated by a boiler C and with bath, basin and sink draw off taps at D, E and F.

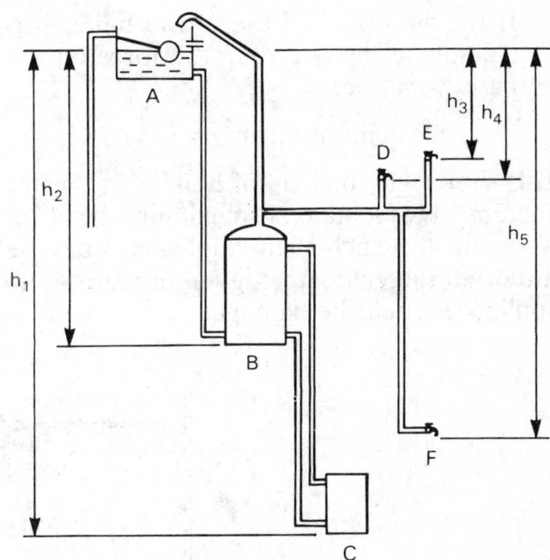

Figure 5.42 *Pressures in a domestic hot water system*

Typical pressures in an average house are shown in Table 5.5. From the table the following points can be seen:

1 The pressure at the sink tap, F, is more than twice the pressure at the basin tap, E. So you would expect to get a much faster flow of water at the sink (depending on the pipes and the size of taps).
2 If the area of the base of cylinder was 0.8

Table 5.5 *Pressures of different heads of water*

Head	*Head metres*	*Force/area kN/m²*	*Equivalent in millibars*
h_1	5	49(50)*	490
h_2	2.5	24.5(25)	245
h_3	1.5	14.7(15)	147
h_4	2	19.6(20)	196
h_5	4	39.2(40)	392

*Figures in brackets are approximate.

square metres, then the force on the base would be:

$25\,\text{kN/m}^2$ (pressure) $\times$ $0.8\,\text{m}^2$ (area)
= 20 kN or 20,000 Newtons

3 If the area of the base of the boiler was 0.5 square metres, then the force on the base would be:

$$50\,\text{kN/m}^2 \times 0.5\,\text{m}^2 = 25\,\text{kN}$$

This shows why makers of boilers and water heaters always state a maximum head of water up to which their appliances may be fitted. If subjected to higher pressures the appliances could be damaged.

Circulating pressure

Water 'circulates' or goes round and round in water heating or wet central heating systems. The pressure which causes this circulation is naturally called 'circulating pressure'.

The circulating pressure in a system depends on two factors:

1 The difference in temperature between the hot water flowing from the boiler and the cold water returning to it,
2 The height of the columns of hot and cold water above the level of the boiler.

The example in Figure 5.43 shows part of a simple water circulation from a boiler to a radiator and back again.

Generally, central heating systems now rely on pumps to provide the main circulating pressure so that smaller bore pipes may be used. But gravity can be used and is still the motive force in many water heating systems. (Figure 5.43 shows the boiler connected to the cylinder by flow and return pipes, and the radiator in the figure could be replaced by a cylinder. The principle is still the same although temperatures are lower.)

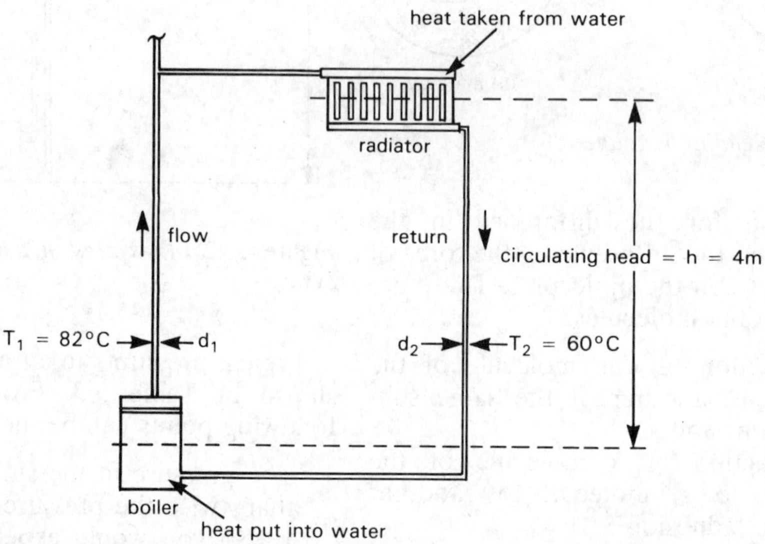

Figure 5.43 *Circulating pressure. On a water heating system the cylinder would be in the position occupied by the radiator. T1 would be 60°C and T2 would be about 10°C when heating began.*

All fluids expand when heated and water is no exception. When a fluid expands, the same weight of fluid occupies a bigger volume. So its density (weight ÷ volume) is reduced.

Put simply, a given volume of hot water weighs less than the same volume of cold water.

Tables are available giving the density of water at various temperatures and, if the flow and return temperatures and the circulation height are known, it is possible to calculate the circulating pressure in a system operating on gravity circulation.

Circulating pressure is proportional to the temperature difference between flow and return water and the circulating height.

Circulating pressure will increase:

1 If the temperature difference increases.
2 If the circulation height is increased.

The circulating pressure in Figure 5.43 is approximately 5 millibars. It can be obtained from the formula:

$$p = h\,\frac{(d_2 - d_1)}{d_2 + d_1} \times 196.1$$

where p = circulating pressure in millibars
h = circulation height in metres
d_1 = density of water in flow in kg/m^3
d_2 = density of water in return in kg/m^3 and 196.1 is a constant

From tables, d_1 at 82 °C = 970.40 kg/m^3
d_2 at 60 °C = 983.21 kg/m^3

$$\text{Therefore } p = 4 \times \frac{983.21 - 970.40}{983.21 + 970.40} \times 196.1$$

$$= 5.14\,\text{mbar (about 5 mbar)}$$

Other tables are available which give a reading of circulating pressure per metre height directly from a temperature difference.

Self-assessment questions

1 The poisonous gas liberated because of incomplete combustion of natural gas is:
(a) methane
(b) carbon dioxide
(c) hydrogen
(d) carbon monoxide

2 The main reason why thermal insulation is incorporated into a building is to prevent:
(a) vapour penetration
(b) sound transference
(c) condensation
(d) heat flow

3 The abbreviation 'PVC' refers to:
(a) protective varnish
(b) polythene
(c) polypropylene
(d) polyvinyl chloride

4 Steel cuttings left in a galvanised storage cistern are likely to cause:
(a) electrolytic action
(b) plumbo-solvency
(c) dezincification
(d) electrostatic action

5 Heat that does not bring a change in temperature is usually called:
(a) measured heat
(b) latent heat
(c) sensible heat
(d) heat loss

6 Which one of the following combinations is likely to produce the most electrolytic action?
(a) iron and zinc
(b) copper and lead
(c) lead and brass
(d) copper and zinc

7 Atmospheric pressure is usually measured on a:
(a) barometer
(b) thermometer
(c) spirit level
(d) hydrometer

8 The attraction of one molecule to another of the same substance is called:
(a) adhesion
(b) bonding
(c) condensation
(d) cohesion

9 A by-product of corrosion in a pressed steel radiator is:
(a) nitrogen
(b) carbon dioxide
(c) hydrogen
(d) oxygen

10 Warm, moist air coming into contact with a cold water pipe in a kitchen will form:
(a) surface tension
(b) condensation
(c) evaporation
(d) cold draughts

6 Electricity

After reading this chapter you should be able to:

1 Identify and name common electrical components.
2 Demonstrate knowledge of basic wiring and domestic electrical circuits.
3 Identify and name types of electrical cable.
4 Describe the function of a fuse.
5 Calculate the value of a fuse.
6 Identify and read an electricity meter.
7 Demonstrate knowledge of electricity distribution on building sites.
8 Identify electrical danger points.
9 Describe different installation arrangements for immersion heaters.
10 Demonstrate knowledge of termination methods for electrical cables.

Terminology

There is a growing need for mechanical engineering services operatives to have knowledge of electricity and electrical installation work. More and more plumbing systems and components depend on electricity to provide the energy for them to operate, and the 603 plumbing scheme includes several objectives related to electrical systems and components.

It is necessary for operatives to acquire as quickly as possible an understanding of the relevant terminology which is described in the glossary at the back of this book. It must be mentioned that the glossary does not give a complete coverage of electrical engineering and electrical installation terminology but contains sufficient information to enable plumbing students to make reference to specific topics of interest to them and to the electrical technology covered in the following pages.

Electric supply

Electricity is a form of energy which is produced by generating equipment at power stations. These generators may be powered by coal, oil, or gas turbines or a nuclear reactor. The power of flowing water may also be used to drive these generators (hydro electricity). The Central Electricity Generating Board is responsible for the generation and primary distribution of electricity, while area boards handle the regional and local

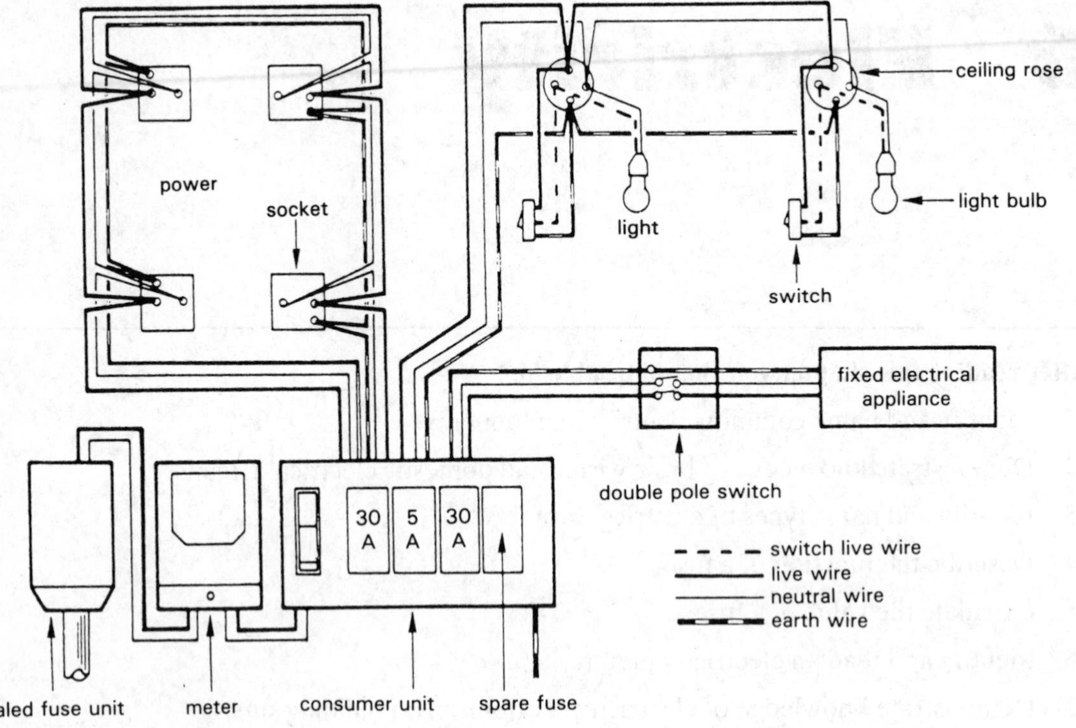

Figure 6.1 *Electric circuits*

distribution and supply of electricity to individual properties.

Electricity from the generators is supplied into the national grid (network of overhead cables on pylons) at very high voltages (electric pressure). This is transformed down in voltage and fed into the regional grids, which consist of both overhead and underground cables. Substations in the regional grid again reduce the voltage to suit the power requirements of the user.

Electricity from the grid is brought into a building to the meter position either by an overhead service cable from the Board's supply pole (common in rural areas) or an underground service cable connected to the Board's main, which is situated under the road. The service cable terminates at a sealed fuse unit which is connected via two short lengths of cable (live and neutral) to the meter fixed alongside. Electricity flows through the live wire and in order to complete the circuit returns along the neutral. In addition, an earthing connection is provided to an earth clamp, earth rod or an earth leakage circuit breaker. Some properties have a second meter which operates through a sealed time switch, to record the utilization of offpeak (cheaper) electricity.

The risks of electrocution and fire must be guarded against in all electrical installations. The precautions which are established practice are shown in principle in Figure 6.2.

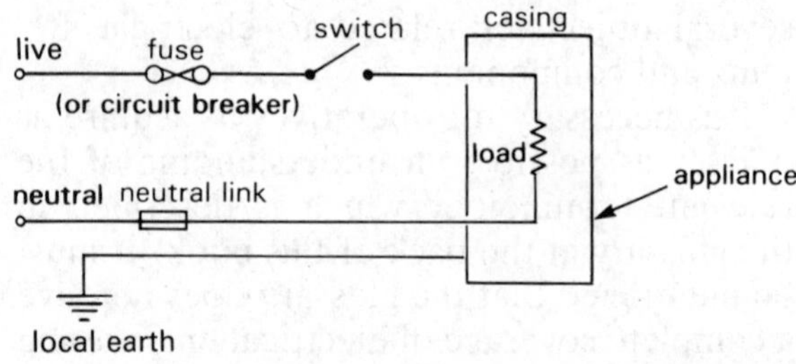

Figure 6.2 *Established precautions*

Types of Cable

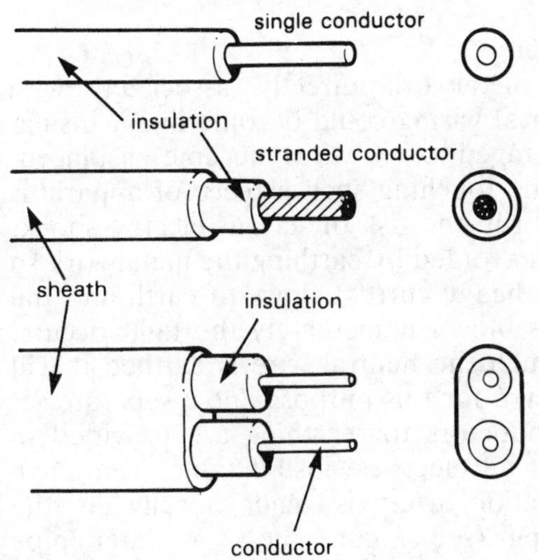

Single conductor, single core, insulated, non-sheathed.

Stranded conductor, single core insulated, sheathed.

Two-core flat, insulated and sheathed (also called flat twin).

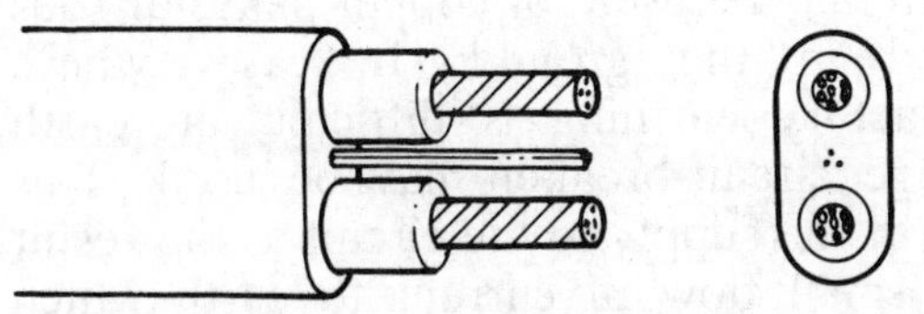

As above with an uninsulated earth continuity wire in same sheath.

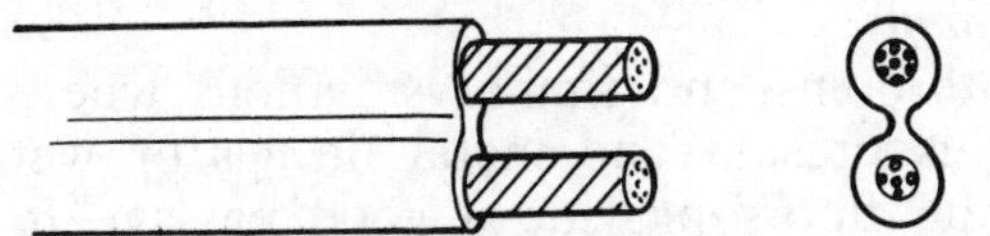

Parallel twin, cores easily separated without damage to insulation of either.

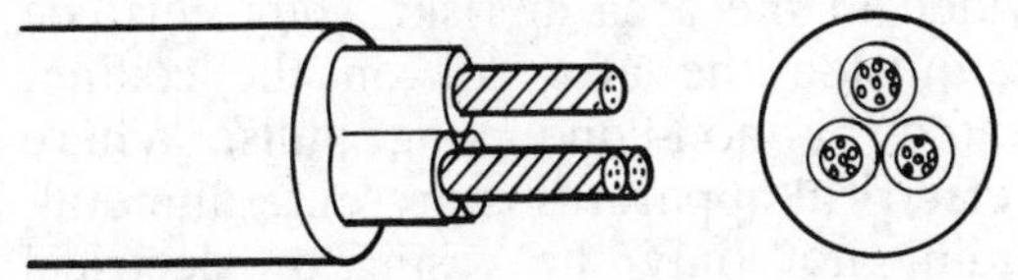

Three-core, insulated, sheathed, unarmoured.

Figure 6.3 *Electrical conductors*

Insulation

All conductors are covered with insulating material or supported on insulators within an earthed casing with a clear air gap round each conductor. Standards of insulation vary with voltage. If a number of wires carrying different voltages are enclosed in a trunking they must all be insulated to the standard of the highest voltage. To avoid this, systems at different voltages are usually run in separate trunkings (e.g. British Telecom wiring and electricity supply wiring). Figure 6.3 illustrates different types of conductor.

Fusing

Each section of wiring must be protected by having in the circuit a fuse wire which will melt if a current passes higher than that which is safe for the wiring (see Figure 6.4).

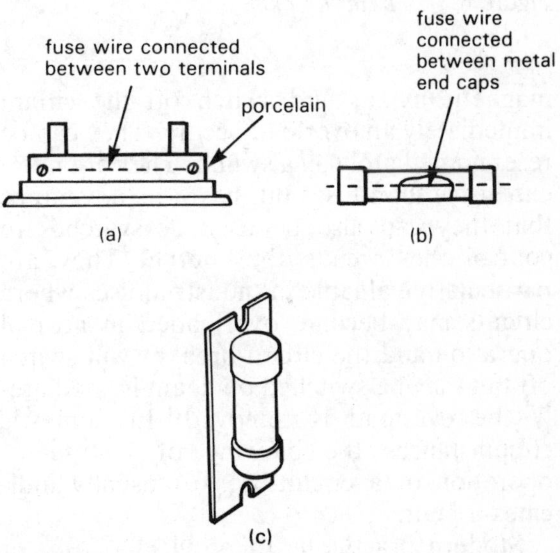

Figure 6.4 *Types of fuse: (a) rewirable, BS 3036; (b) cartridge, BS 1362; (c) high rupturing capacity, BS 88*

This prevents overheating of wiring with the possible risk of fire. Fuses may be of the traditional type where a fuse wire is stretched between terminals in a ceramic holder, or of the modern cartridge type where the wire is held in a small ceramic tube with metal ends.

The cartridge types of fuse are much easier and quicker to replace. It is also possible to use circuit-breakers instead of fuses (see Figure 6.5). These operate by thermal or magnetic means and switch off the circuit immediately an overload occurs. They can be reset immediately by a switch. They are more expensive than fuses but have the advantage that they can also be used as switches to control the circuits they serve. They are particularly valuable in industrial uses where circuits may become overloaded in normal operation and the circuit-breaker will switch off but may be switched on again immediately the overload is removed. In domestic circumstances the blowing of a fuse or operation of a circuit-breaker usually indicates a fault.

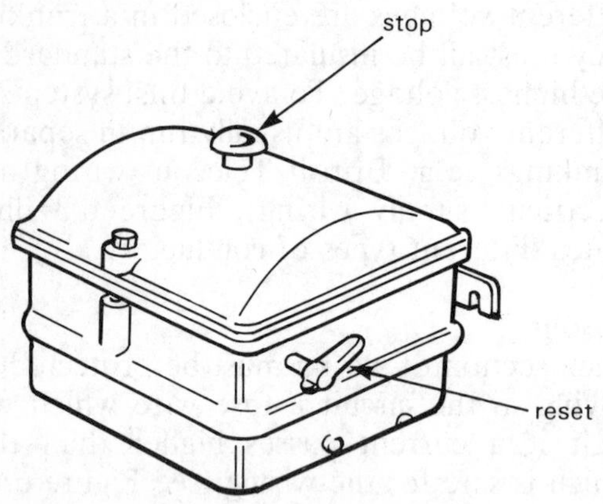

Figure 6.5 *Circuit-breaker*

Modern practice is to provide a fuse, or circuit-breaker, at the phase end of the circuit (called the line) and a simple link at the neutral end.

Switch polarity

The position of the switch has the same effect upon safety as that of the fuse. If the switch is fitted on the neutral side of the apparatus this will always be live, even when the switch is turned off. Switches are therefore always fitted on the phase side of the apparatus they control.

Earthing

Any metalwork directly associated with electrical wiring could become live if insulation frayed or if wires became displaced. Anyone touching such a piece of apparatus would run the risk of serious electric shock. This is avoided by earthing the metalwork so that a heavy current flows to earth and the fuse is blown immediately the fault occurs. Although the neutral wire is earthed it will not serve for this purpose and a separate set of conductors for earthing are provided in almost all electrical installations. The earth connection itself is made locally in the building (see Figure 6.6). A water pipe supplying the building is sometimes used, although some water companies do not approve, particularly where corrosive subsoil conditions exist, and care must be taken that the water pipes are of metal. Plastics pipes are increasingly used as service pipes and water mains. In the absence of a satisfactory pipe the electricity supply authorities' cable sheath may be used, or copper plates or rods buried in the ground. In cases where satisfactory earthing is difficult, an earth leakage circuit-breaker may be used. This cuts off the supply to the circuit as the result of a small flow of current to earth which would not be sufficient to blow the fuse and so overcomes the problem.

Bathrooms

In bathrooms and similar situations where water is present and metal fittings or wet concrete floors provide a good passage to earth, special safety precautions are called for. No socket outlets or switches may be provided in the area at risk. They must be sited outside the area, or on the ceiling, operated by non-conducting pulls. Where any electrical apparatus is present, all metal, including not only the casing of electrical apparatus but also pipework, baths, etc., must be bonded together electrically and earthed.

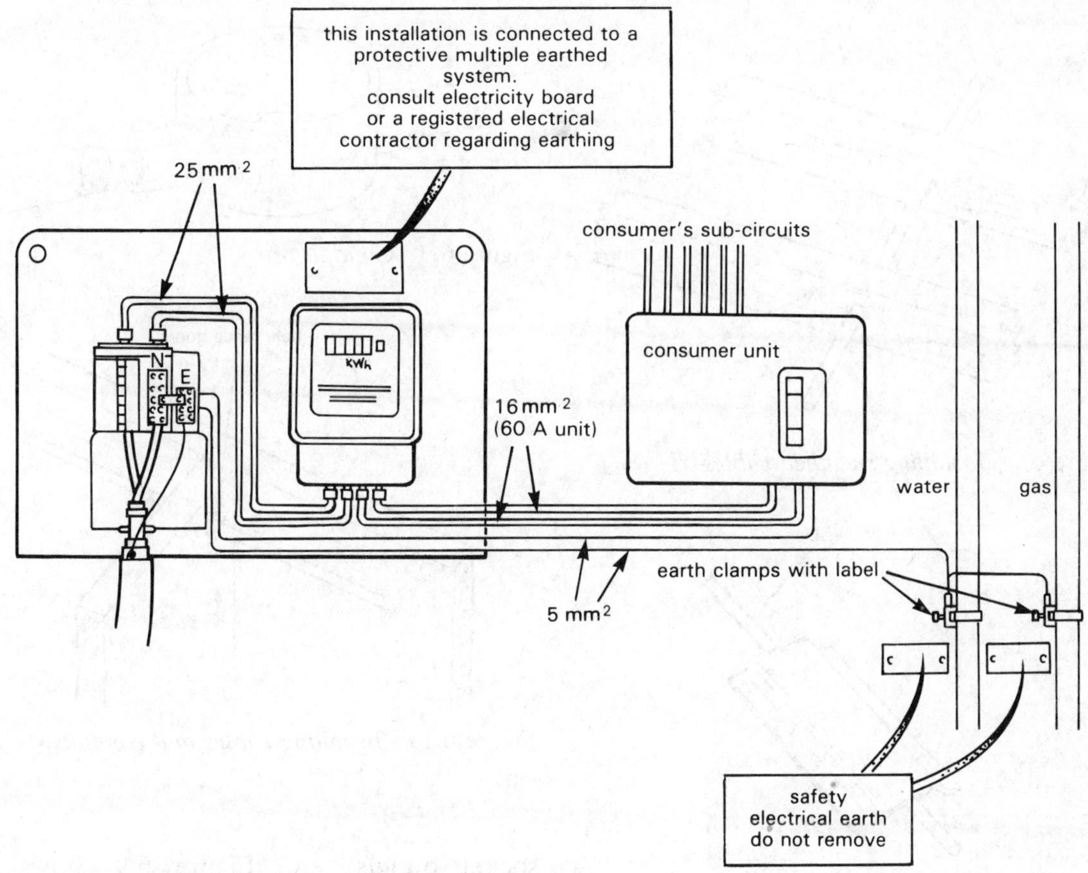

Figure 6.6 *Protective multiple earthing system, showing equipotential bonding of other services*

Basic wiring systems

Wiring in buildings is run either on the surface or concealed in the construction. Surface wiring is cheaper but its appearance limits its use to industrial conditions or alteration work where cost is a prime consideration. The following types of wiring system are available for use in building.

Sheathed

Two or more wires consisting of metal conductors, each having its own insulation, are enclosed in a protective sheath. In the past the insulation and sheathing was rubber and the system was referred to as TRS (tough rubber sheathed). Nowadays PVC is used as insulation and sheathing, giving a slightly smoother and neater cable. This type of wiring is well suited to surface use and can be concealed in timber floors and joinery (see Figure 6.7). It must be protected by conduit or metal channels where it is to be covered by plaster or screed (see Figure 6.8).

Conduit

This is a system used for electrical distribution in some buildings. A system of tubing is laid to the points where electricity is required and insulated cables are subsequently drawn through. Conduit systems either may be used on the surface (see Figure 6.10) or concealed in the construction. One of the principal advantages of the conduit system is the ease with which modifications and rewiring may be carried out by drawing in new cables. Conduit is available in a variety of materials.

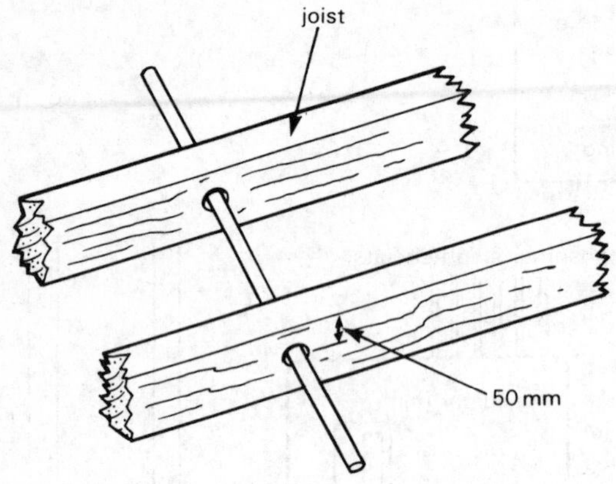

Figure 6.7 *Installing sheathed cables*

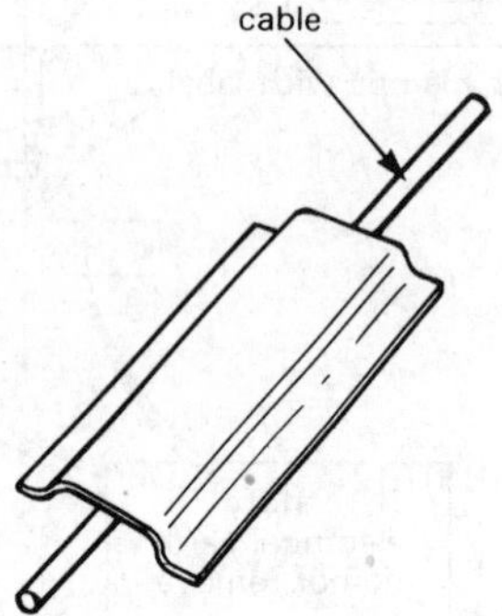

Figure 6.8 *Protective chanelling*

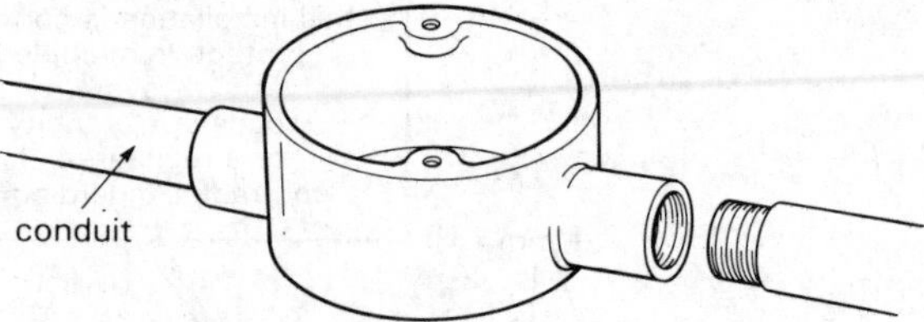

Figure 6.9 *Conduit box*

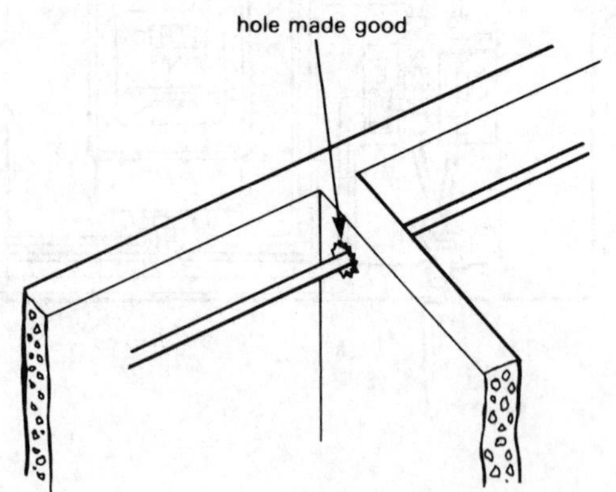

Figure 6.10 *Installing cables and conduits*

Steel is widely used but aluminium may be employed and plastic systems are available. Plastic conduit is usually flexible and easy to fix and airtight, which protects against condensation, but an additional cable is required for earthing, which in metal conduit is achieved by the conduit itself. The best metal conduit is of drawn tube with screwed joints; this gives good earth continuity and makes a waterproof system. Lighter tubes made from strip metal with welded longitudinal joints, or even open seams, are available with grip rather than screwed joints. These types give less satisfactory earth continuity and less protection from damp.

Ranges of bends and fittings are available for most conduit systems, which include special boxes to take switches, ceiling roses, socket outlets, etc. Figure 6.9 shows a conduit box.

Conduit is normally laid on the surface of the slab in concrete floors and covered by the floor screed. In walls the thickness of plaster is not usually enough to contain the conduit and vertical chases in the bricks or blocks have to be made so that there will be sufficient plaster depth over the conduit. In timber construction, floor joists and wall nogging pieces are notched to take the conduit.

Conduit gives a good mechanical protection to wiring and can itself be arranged to span from point to point, whereas the other basic wiring systems require frequent support.

MICC cables (mineral insulated copper covered)

These cables consist of one or more copper conductors, embedded in powdered magnesium oxide and sheathed with copper (see

Figure 6.11). The magnesium oxide must be protected from damp and cable ends are sealed with 'pots' containing an insulating sealer. MICC cables can give very high quality installations; they are heat resisting and consequently can be used with underfloor heating. They can be buried in concrete and since they are small in diameter compared with other types of cable, they can often be run in joints of brickwork, or in plaster, without charring the wall. Mineral insulated cables are also available in aluminium.

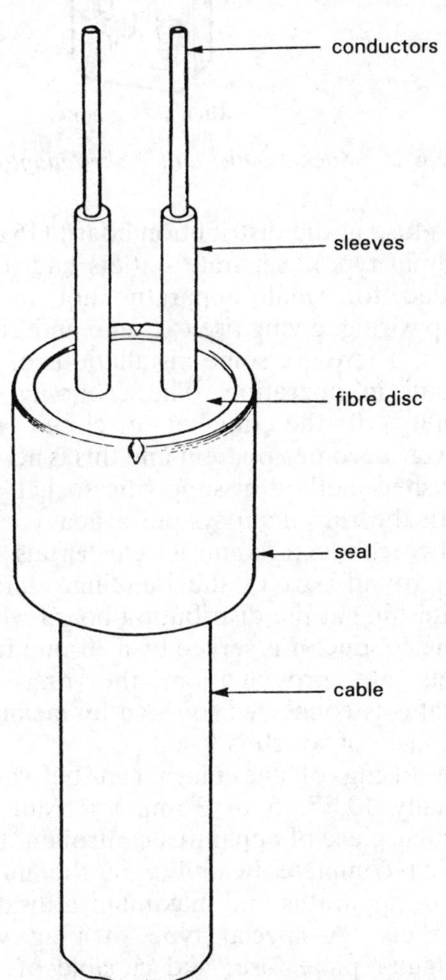

Figure 6.11 *Mineral insulated copper covered cable*

Fire risk

When cables, conduits, ducts and trunkings pass through walls and ceilings, the hole through which they pass must be made good with fire-resisting material to the full thickness.

Also where they pass through walls and ceilings in trunkings, rising mains etc., fire barriers must be provided internally.

Termination of the cable

The cable is cut to length allowing for the brass gland to be put on. When the conductors have been thoroughly cleaned a brass pot type seal is screwed on the cable. This is a self-tapping seal and any slivers of metal should be removed from the inside of the pot.

An insulating plastic compound is then pressed into the seal from one side so that there are no air pockets left in the seal.

A fibre cap with insulating sleeves is threaded over the conductors and then crimped with a crimping tool. This compresses the insulating compound in the seal.

The whole sealed end of the cable is then contained in a brass gland so that it may be terminated into a distribution board or consumer unit.

The bare sheath of the cable must not be allowed to come into contact with other metal when used in damp situations to avoid electrolytic action.

Distribution circuits

In most buildings electricity supply is divided finally into three types of circuit:

1 Lighting,
2 Socket outlets,
3 Fixed apparatus.

Lighting

Lighting is normally carried out in circuits with 5-amp wiring and fuses although 15-amp circuits for lighting are permitted and used mainly in industrial situations and large buildings. Each circuit consists of a cable with two conductors, live and neutral, which link a number of ceiling roses or wall lighting points. The lamp and switch are connected in the way shown in Figure 6.12 thereby allowing independent control of the lamps although they are part of a single circuit.

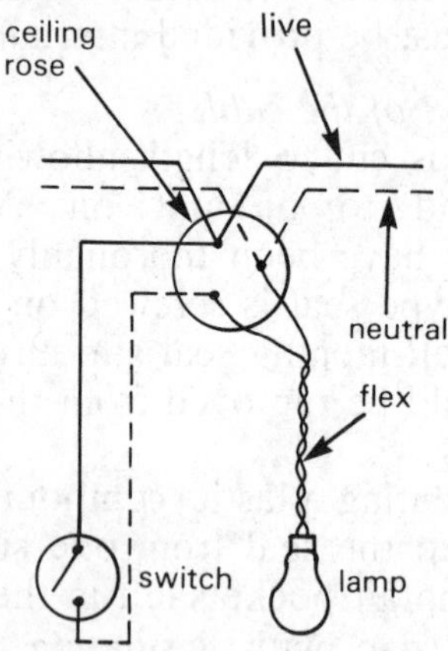

Figure 6.12 *One-way switch control of lighting*

Although Figure 6.12 shows only one lamp operated by the switch, it is clearly possible to have several lamps controlled by the one switch. The converse is also the case and in many instances it is desirable to have one or more lamps controlled from more than one switch. Figure 6.13 shows a method of achieving this from two switch positions.

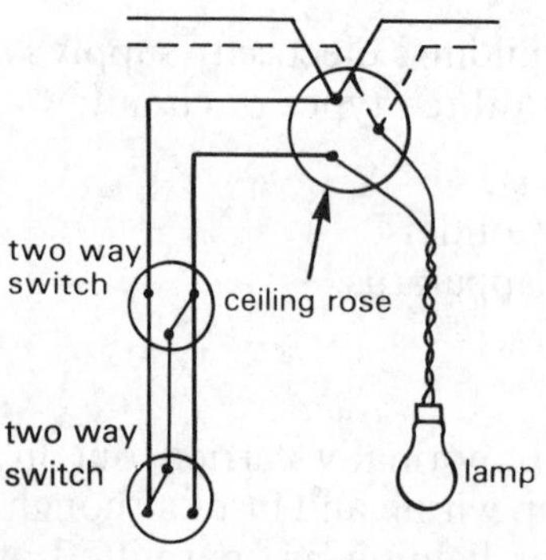

Figure 6.13 *Two-way switch control of lighting*

Socket outlets

Socket outlets into which portable electrical apparatus can be plugged are a vital part of almost all installations in buildings (see Figure 6.14). At one time each socket outlet was served by an individual circuit leading

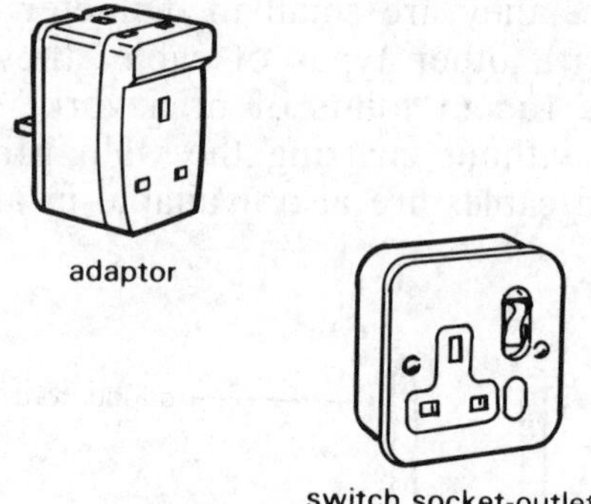

Figure 6.14 *Socket outlet and socket adaptor*

from a fuse at the distribution board (15 amp round-pin type). Separate outlets had to be provided for small apparatus not having 15 amp wiring, giving rise to 5 amp and 2 amp sockets. There are some installations of this sort still in operation. The economy and flexibility of the ring main circuit has, however, become apparent and this is now an established method of supplying socket outlets. In the ring main system a heavy cable with two conductors and an earth runs in a circuit round part of the building, starting and finishing at the distribution board where the line conductor is served by a 30 amp fuse. Sockets are provided on the ring and apparatus is connected to them by means of plugs, each of which is fused.

The fusing of the plugs can be varied (normally 13, 7, 5 or 3 amps) to suit the particular piece of apparatus controlled, thus enabling complete flexibility in the movement of apparatus and maximum utilisation of sockets. A special type of plug with rectangular pins, fuse, and capable of carrying a 13-amp current, has been developed for use with ring mains. When a fault develops in a piece of apparatus the fuse in the plug blows, leaving the main circuit unaffected. The ring circuit is for most applications very much more economical of wiring than individual connections to the distribution board. The installation of ring circuits is governed by a number of rules mainly concerned with electrical efficiency. The main point of general planning interest is that the maximum number of 13-amp socket outlets allowed on one ring main is 10.

Normal domestic operation does not overload ring mains but the economical provision of an adequate number of socket outlets is of very great convenience to inhabitants of buildings.

Spurs may be provided from the ring, thereby economising wiring, provided there are not more than two sockets per spur and that no more than 50 per cent of the sockets on the ring are served by spurs. Figure 6.15 shows the principles of wiring to a ring main.

Fixed apparatus

Pieces of fixed equipment such as cookers, water heaters, boilers etc., usually have their own individual fuses and circuits (see Figure 6.16). There should be an isolating switch adjacent to the piece of apparatus in each case so that work on it may be carried out in safety.

Figure 6.17 shows a number of lighting, ring main and fixed circuits served from a consumer's unit.

Bathroom

Because of the presence of water there is increased risk of electrocution in bathrooms

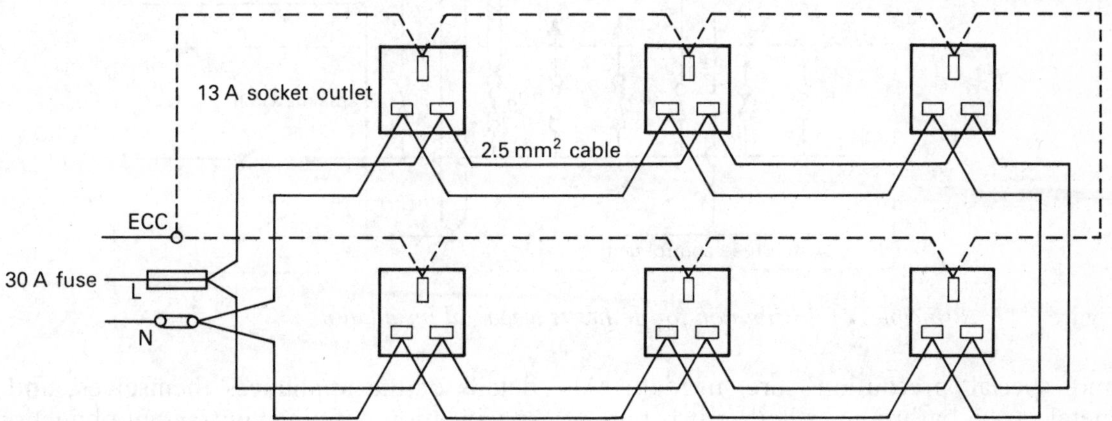

Figure 6.15 *Ring circuit*

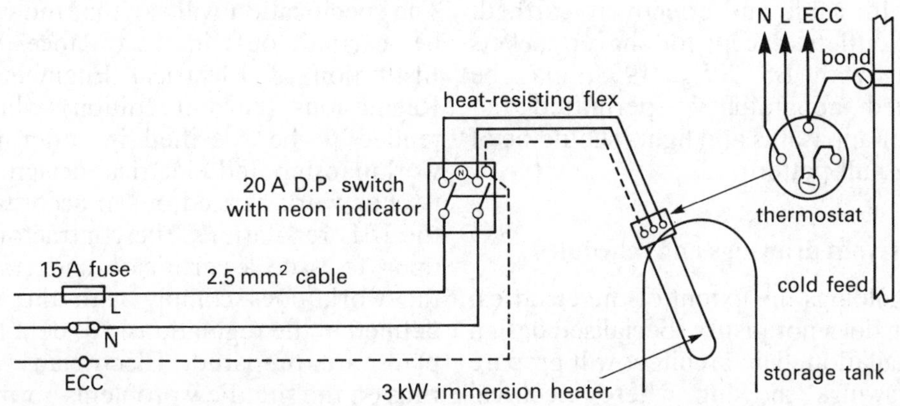

Figure 6.16 *Immersion heater circuit*

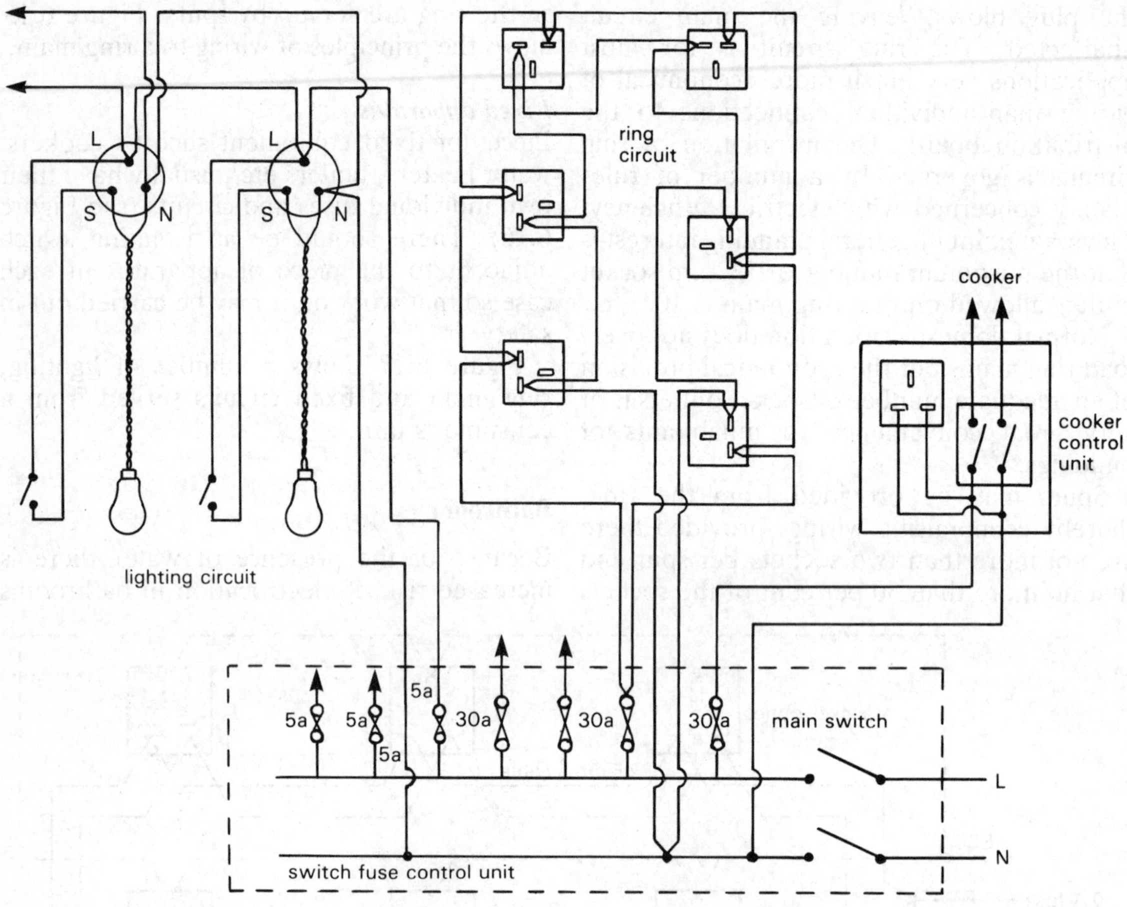

Figure 6.17 *Principles of distribution for domestic electrical installation*

and special precautions are needed. All metal in the bathroom, whether it is part of electrical apparatus or not, must be bonded together with cable and effectively earthed. No socket outlets, except for shaver sockets conforming with BS 3052: 1958, may be used. Fixed apparatus is permitted but switches for apparatus and lights should be of the ceiling pull pattern.

Electrical layout drawings and schedules

In small buildings the extent of the electrical installation does not justify specialised design of the installation. The architect will prepare layout drawings showing where electrical appliances are to be fitted, a schedule giving details of the appliances themselves, and a specification defining contractual obligations and quality of materials and workmanship. The specification will say that the work must be carried out in accordance with the Institution of Electrical Engineers Wiring Regulations (current edition) which is intended to be specified in order to define workmanship and electrical design and safety. For work carried out in accordance with the IEE regulations, the contractor is called upon to issue a certificate undertaking that the work does comply with the standards defined in the regulations. No detailed wiring plans are prepared. Electricians work out runs on the site. Few problems normally arise but if architects wish to have all wiring concealed (e.g. in screeds and plaster) they must ensure that practicable wiring routes exist to all the pieces of electrical apparatus.

Cable
A cable is a length of insulated conductor (solid or stranded) or of two or more such conductors, which are laid together (see Figure 6.18).

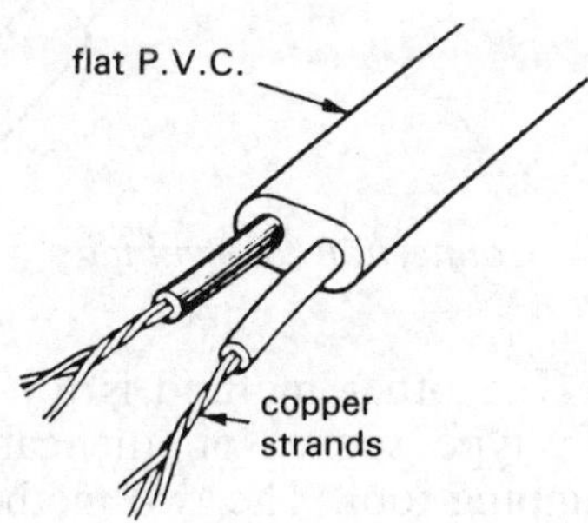

Figure 6.18 *Two core cable*

Cord
A cord is a flexible cable of one or more cores where the cross-section area of each conductor does not exceed 4 mm^2. Each core consists of a group of wires of small diameter. Cords are used on all portable electrical appliances. Figure 6.19 illustrates a cord with two cores.

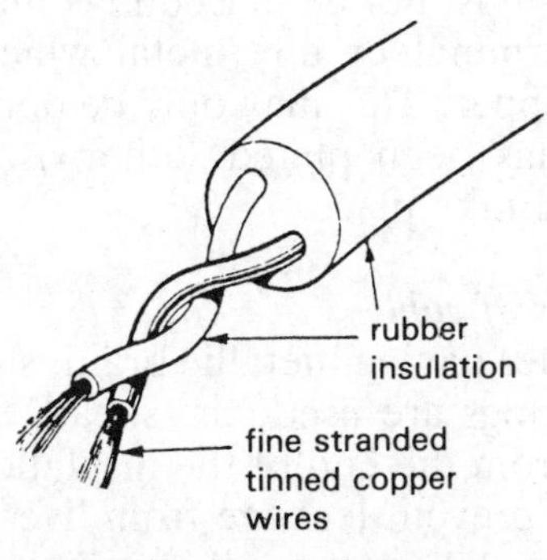

Figure 6.19 *Two core flexible cord*

Bus-bars
A bus-bar system is usually made up of a rectangular section of conducting material to which many connections may be made. These systems are usually employed in situations where a large number of interconnections are to be made in a confined space.

Conducting materials
Copper is preferred for general use since apart from silver, it has the lowest resistivity. Copper is used extensively for conductors in all the above applications.

Aluminium is used primarily as a conductor in cables. It has a higher resistivity than copper, which means that for a given current-carrying capacity, the cross-sectional

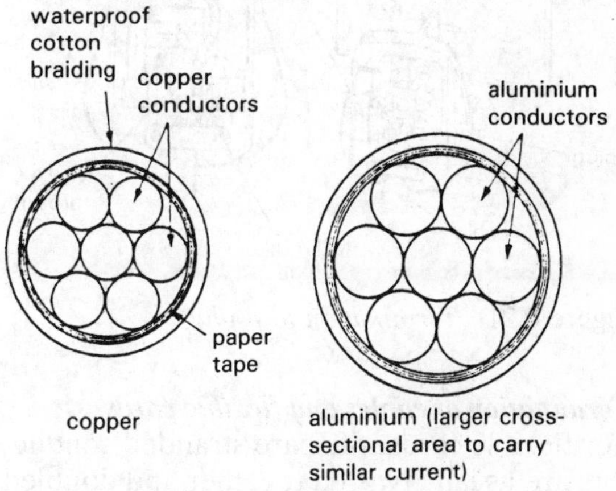

Figure 6.20 *Illustrating that aluminium cables need to be larger than copper to have the same current-carrying capacity*

area must be greater (Figure 6.20). It possesses these advantages over copper:

1 It is cheaper,
2 It is much lighter.

The latter advantage is of importance when installing large cables as the cost to install is considerably reduced. Aluminium is also used for the protective sheath of mineral-insulated cables.

There are, of course, applications where other conducting materials are used, for instance, brass is used for terminations and connecting blocks because it is harder than copper.

Terminations and connections

The type of termination used depends upon the type and size of cable and the kind of connection to be made. Figure 6.21 illustrates termination to a plug.

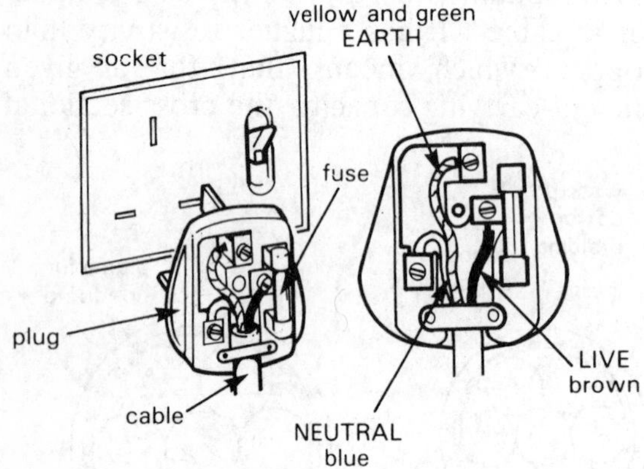

Figure 6.21 *Termination to a plug*

Termination of cables and flexible cords

For flexible cords, the bare stranded conductors are usually twisted together and doubled back, if room permits, then screwed firmly in a pillar terminal, as shown in Figure 6.22.

For cables having single-stranded conductors which are to be terminated and connected under screwheads or nuts, it is convenient to shape the bare end into an eye to fit over the thread as shown in Figure 6.23.

Cables with more than one stranded conductor may be terminated in a similar manner to that shown for the flexible cord.

Sockets (soldered and crimped)

For conductors of larger cross-sectional area, a socket may be used. There are two methods of fixing a socket to a cable end. One method is by soldering a tinned copper socket on the

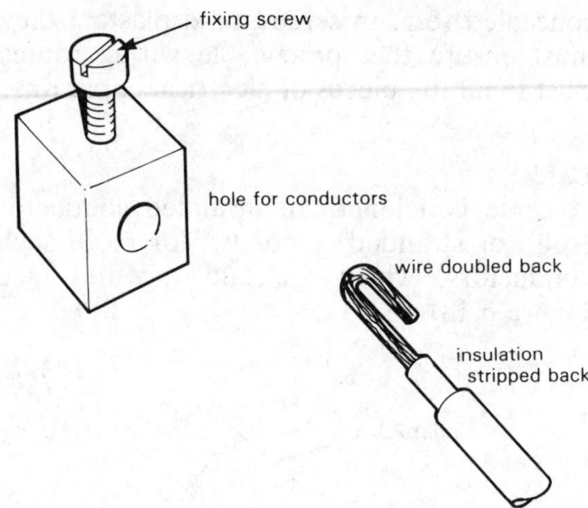

Figure 6.22 *Connection to terminal*

cable end. The other method is to fasten a compression-type socket on the cable end using a crimping tool. The two methods are shown in Figure 6.24.

The crimped socket is used to a large extent for terminating smaller cables. In smaller sizes they are even being used for flexible cords. Hand crimping tools are usually used for smaller size sockets which are capable of crimping a range of different sizes.

Termination of an aluminium conductor

Where an aluminium conductor is to be connected to a terminal it must be ensured that the conductor is not under excessive mechanical pressure. Further, an aluminium conductor must not be placed in contact with a brass terminal or any metal which is an alloy of copper. This may only be done if the terminal has been plated, otherwise corrosion would take place.

Other types of cables

When cables having metallic braid, sheath or tape coverings are used, the sheath must be cut back from the end of the insulation. This is done to prevent leakage from live parts to the covering. However, this is unnecessary for mineral-insulated cables. The ends of

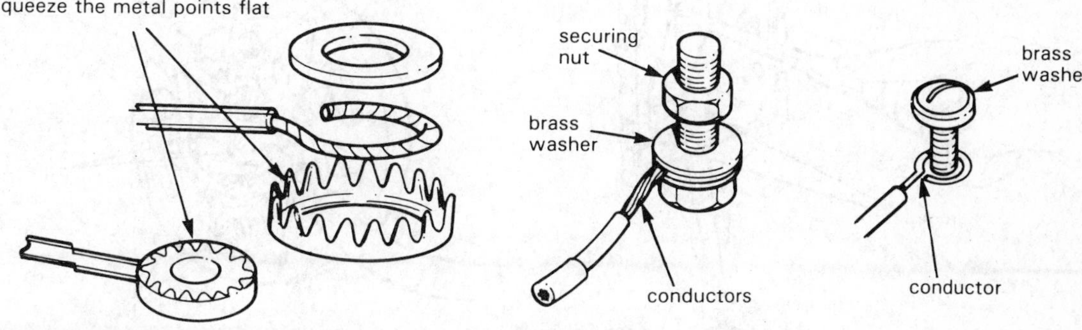

Figure 6.23 *Termination under screwheads or nuts*

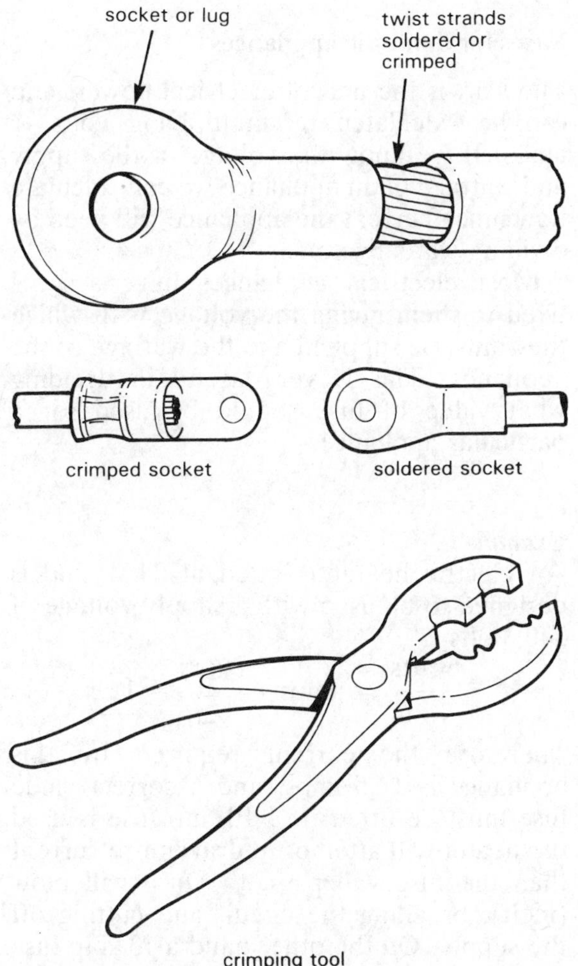

Figure 6.24 *Soldered and crimped sockets*

mineral-insulated copper-sheathed cables must be protected from moisture by sealing, ensuring that the mineral insulation is perfectly dry before sealing.

Wiring to socket outlets, spur boxes and junction boxes

Wiring a socket outlet (see Figure 6.25)

1 Cut the cable loop to make two tails 75 mm long and thread the cables through grommets in the mounting box.
2 Screw the mounting box to the wall and strip back the cable insulation.
3 Twist together and connect:
   (a) Brown wires to terminal L.
   (b) Blue wires to terminal N.
4 Fit green/yellow insulated sleeving over both earth wires.
5 Twist earth wires together and connect to terminal E.
6 Tighten all securing screws.

Note Ensure that the wires are pushed into terminals up to the insulation.

Spur boxes (see Figure 6.26)
These are fitted in a similar manner to socket outlets but have two sets of terminals. The lead to the appliance must be clamped in the cord grip.

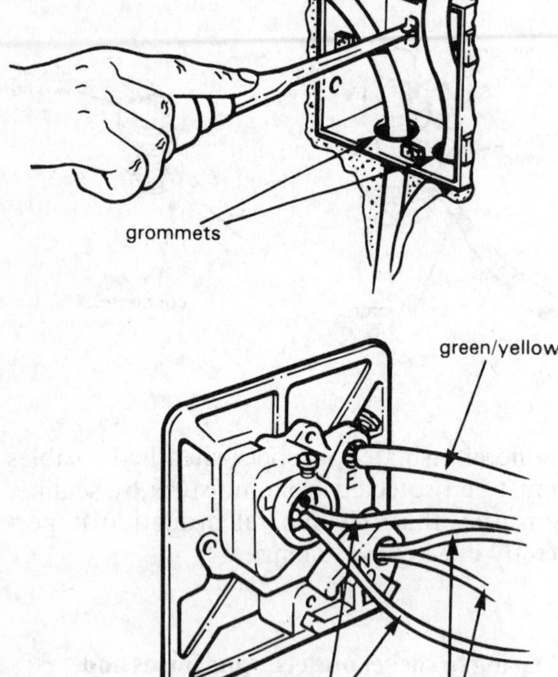

Figure 6.25 *Wiring a socket outlet*

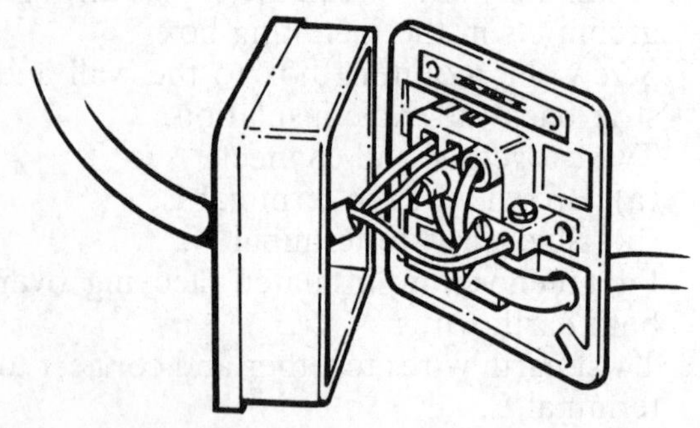

Figure 6.26 *Spur box*

Junction boxes (see Figure 6.27)
Junction boxes have four knockout sections for cable entry and exit. The central terminal is always used for the earth connection.

It is good practice to run spurs from the back of a socket but junction boxes can be used for this purpose.

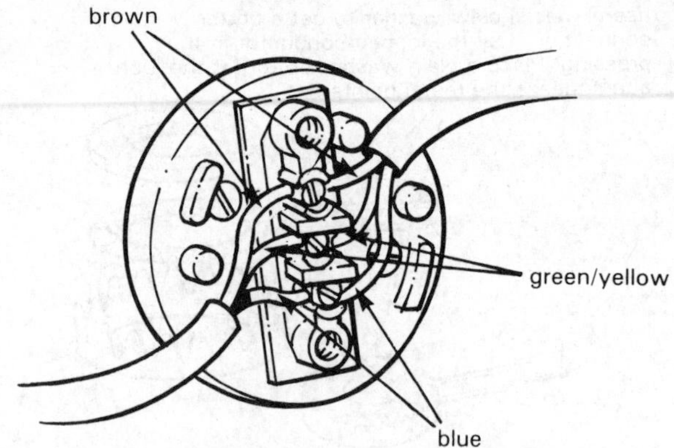

Figure 6.27 *30A Junction box*

Fuses for domestic appliances

The watt is the unit of electrical power, and can be calculated by multiplying volts × amps. If we know the voltage of the supply and wattage of an appliance we can calculate how much current the appliance will need by dividing watts by volts.

Most electrical appliances have a label fixed to them giving the voltage with which they must be supplied and the wattage of the appliance. This is very useful in deciding what value of fuse should be used for a particular appliance.

Example
An electric heater is rated at 3 kW and is designed to be used with a supply voltage of 240 Volts

$$= \frac{\text{watts}}{\text{volts}} = \text{amps} = \frac{3000}{240} = \underline{\underline{12.5}}$$

Therefore the current required by this appliance is 12.5 amps, and a correct value fuse must be fitted. If a 10 amp fuse is used the heater will attempt to draw more current than the fuse will pass, the fuse will blow (melt), breaking the circuit and cutting off the supply. On the other hand a 13 amp fuse will permit the heater to operate normally and still provide a safety factor if needed.

How to read an electricity meter

An electricity meter records the number of units (kilowatt hours) of electricity used. It is quite easy to read, and daily or weekly consumption records are simple to keep.

Remember that an electricity meter is a precision instrument. You can be confident of its accuracy as the equipment has to be certified before installation by independent meter examiners appointed by the Secretary of State for Energy.

There are basically two kinds of meter: the digital meter and the dial meter.

Digital meter

This is the modern type of meter. The number of units of electricity used is shown by a simple row of figures. With special day/night tariff meters, two digital indicators are used: one for lower priced night rate and the other for the day rate.

Always remember that the reading on the digital meter is the total number of units used since the meter was installed. To keep a check on the consumption, subtract the previous reading from the new reading.

Dial meter

This type of meter appears to be more difficult to read but after a little practice you will find it quite straightforward.

When reading the dial meter, always remember that adjacent dials revolve in opposite directions as shown in Figure 6.28. First, ignore the dial marked $\frac{1}{10}$ (it is only there for testing purposes). Start by reading the dial showing single units and write down the figure. Then read the dial showing tens of units, then the one showing hundreds, then thousands, then tens of thousands, working from right to left and writing them down in that order.

Points to note

1 Always write down the number the pointer has passed (this is not necessarily the nearest number to the pointer). So if the pointer is anywhere between, say, 3 and 4, write down 3.

2 If the pointer is directly over a figure, say 7, look at the pointer on the dial immediately to the right. If the pointer is between 9 and 0, write down 6. If it is between 0 and 1, however, write down 7.

The sample meter reading in Figure 6.28 will help you to understand the procedure involved. This meter reads 94.694 units.

As with the digital meter, when you have worked out the reading, subtract the previous reading shown on your bill to find the number of units of electricity used.

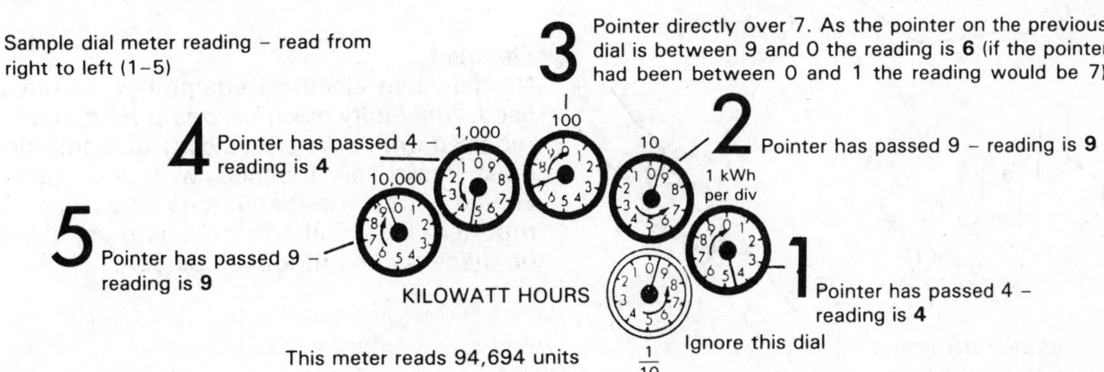

Figure 6.28 *Electricity meter dial*

Electricity distribution on building sites

Safety requirements and supply voltage

Construction work makes considerable use of electrical supplies. The risk of serious accident by electrical causes is greatest in damp conditions, such as are common on building sites; in addition, equipment and cabling are subject to frequent repositioning, and this increases the danger (see Figure 6.29).

For site lighting other than floodlighting, and for portable and hand-held tools and lamps, the personal injury risk is reduced if a lower operating voltage is employed. A 110 V supply has been selected as the most satisfactory compromise between safety requirements and working efficiency. It is still necessary, even at this voltage, to take precautions, which should include a high standard of routine maintenance and periodic testing of all apparatus used.

With a properly designed system, voltage drop is not serious and the slightly heavier cables needed do not prove to be a disadvantage. It is now possible to obtain the whole range of portable power tools for 110 V working. All power tools should be double insulated (see Figure 6.30).

Contract planning

Temporary electrical supplies for construction sites should be planned with as much care as is normally given to permanent installations. Electricity is required for motive power, heating and lighting for construction work and for space and water heating,

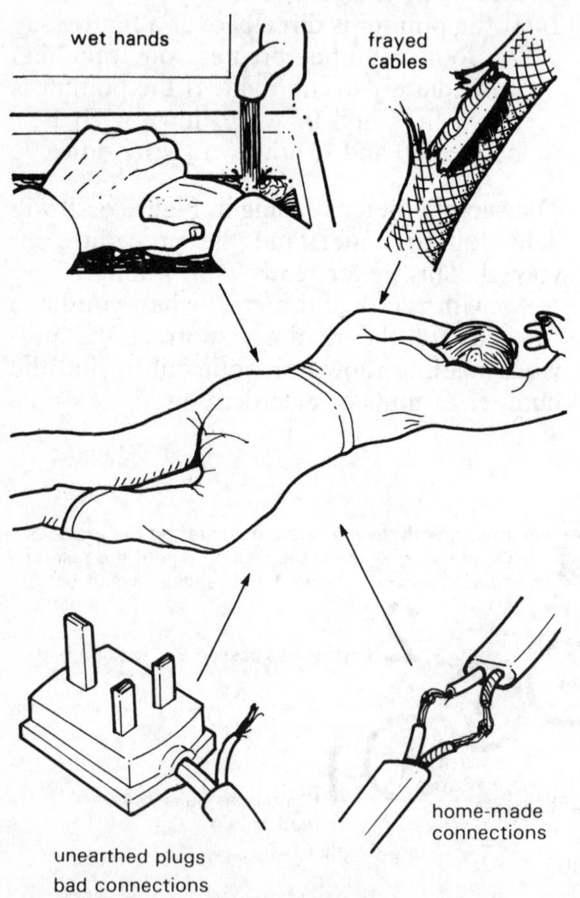

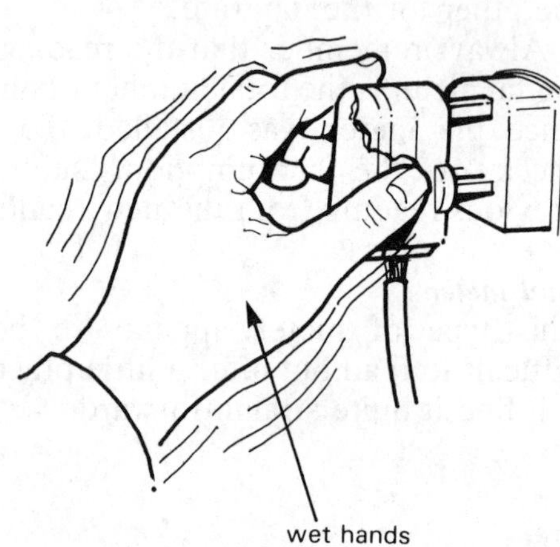

Electricity
No defective electrical equipment should be used. *Any faulty electrical equipment must be reported immediately.* Electrical equipment must never be touched with wet hands. Frayed wires are dangerous and must be reported. Electrical connections must always be made by an authorised person.

Figure 6.29 *Electrical danger points*

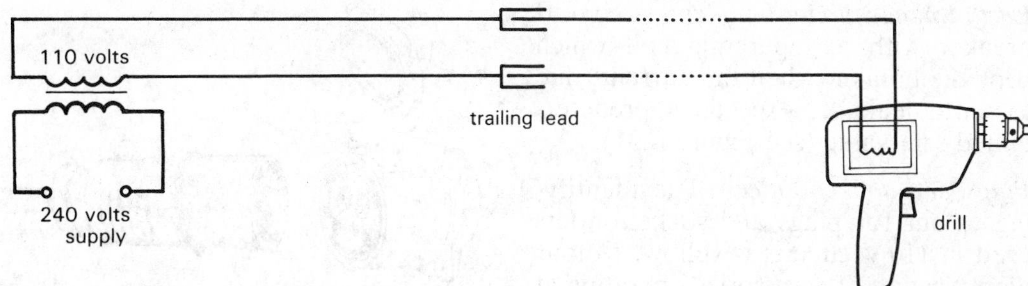

Figure 6.30 *Supply to a double-insulated drill*

cooking and lighting for staff and office accommodation. Demands on electricity are extensive; smaller portable tools such as saws, hammers, drills, sanders and grinding wheels are also widely used. The use of electricity for aggregate heating and drying-out processes to facilitate winter working and speedier construction is increasing; site lighting is an appreciable part of the total load in winter months.

Distribution equipment

Electricity supplies should be distributed to the building site through distribution units meeting the requirements of BS 4363 : 1968. The units comprise the following:

1 *Supply incoming unit (SIU)* A unit to house the supply undertaking's incoming cable, service fuses, neutral link, current transformers and metering equipment and with provision for one outgoing circuit of 300 A, 200 A or 100 A maximum current, controlled by switch and fuse or circuit-breaker.

2 *Main distribution unit (MDU)* Equipment for control and distribution at voltages of up to 415 V three-phase and 240 V single-phase and fitted with 300 A, 200 A or 100 A isolator, lockable in the off position. The available outgoing supplies are controlled by moulded case circuit-breakers (MCCB).

3 *Supply incoming and distribution unit (SIDU)* The two foregoing units may be combined in a single unit.

4 *Transformer unit (TU)* A unit incorporating a transformer and with provision to distribute electricity at reduced voltage (usually 110 V); available for 240 V single-phase input (TU/1), or 415 V three phase input (TU/3), or both (TU/1/3).

5 *Outlet unit (OU)* Facilities for the control, protection (by miniature circuit breakers) and connection of final subcircuits, fed from a 32 A supply and operating at 110 V. The single-phase outlet unit (OU/1) has up to eight 16 A double-pole socket outlets; a 32 A socket-outlet may be added, but this is not controlled by a circuit-breaker.

6 *Extension outlet unit (EOU)* An extension outlet unit is similar to an outlet unit but is fed from a 16 A supply and does not incorporate protection by circuit-breakers.

7 *Earth monitor unit (EMU)* Flexible cables used to supply power at mains voltage to movable plant incorporate a separate pilot conductor in addition to the main earth continuity conductor. A very low voltage current passes between the portable equipment and a fixed monitoring device (EMU) through the pilot conductor and the earth continuity conductor. Failure of the latter will interrupt the current flow; the fixed monitoring device will detect this and automatically isolate the circuit.

8 *Earth leakage circuit-breaker* A circuit-breaker with an operating coil which trips the breaker when the current, due to earth leakage, exceeds a predetermined safe value (see Figure 6.31).

9 *Plugs and socket-outlets* The identifying colour for plugs and socket-outlets used in 110 V circuits is yellow. Colour identification of accessories operating at other voltages should be as shown in Table 5. Accessories should meet the requirements of BS 4343: 1968 Industrial plugs, socket-outlets (see Figure 6.32) and couples for AC and DC supplies which is based on International Standard CEE 17.

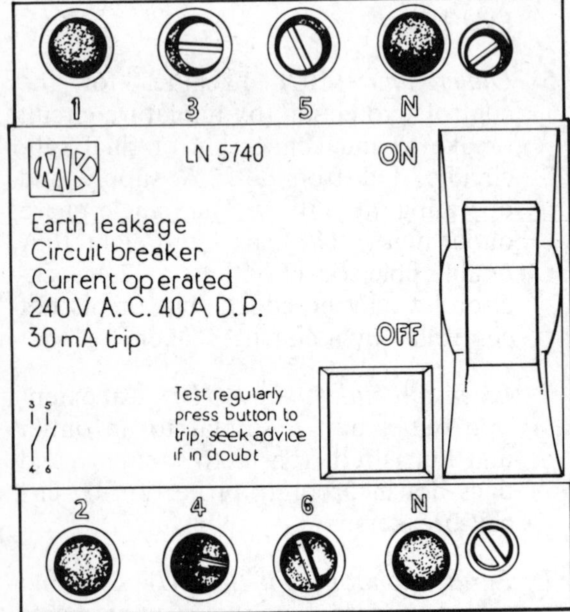

Figure 6.31 *Earth leakage circuit-breaker*

Table 6.1 *Identification colours*

Operating voltage	*Colour*
25	violet
50	white
220–240	blue
380–415	red
500–650	black

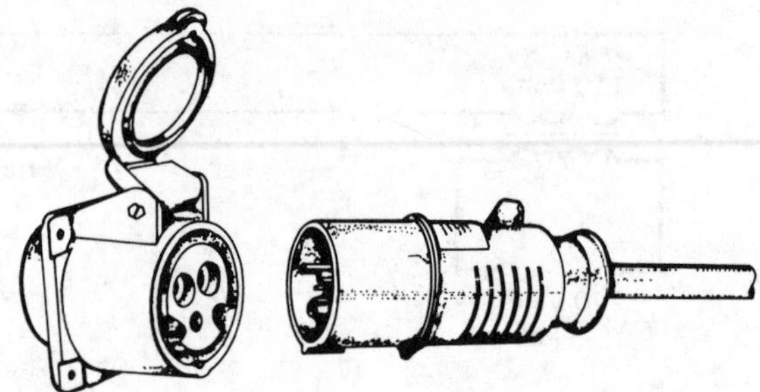

Figure 6.32 *Socket outlet and plug to BS 4343*

10 *Cabling* For semi-permanent parts of the installation, e.g. site offices and ancillary buildings, the IEE Regulations for the electrical equipment of buildings state all the necessary requirements.

Cables should have a metal sheath and/or armour, which must be continuously earthed in addition to the earth core of the cable. The earthed sheath or armour must not be used as the sole conductor. Except where the risk of mechanical damage is slight, the cables should have an oversheath of PVC or oil-resisting and flame-retardant compound. If the voltage applied to a cable will not normally exceed 65 V, it may be of a type insulated and sheathed with a general purpose or heat-resisting elastomer.

Cables should be fixed on site so as to be clear of constructional operations and not to be a hazard to operatives. They should be kept clear of passageways, walkways, ladders, stairs and the like and should be at least 150 mm clear of piped services such as steam, gas and water. Where they pass under roadways and access routes for transport, they should be laid in ducts at a minimum depth of 0.6 m with a marker at each end of the crossing.

The use of overhead cables is not recommended but if used, they should be fixed at a minimum height of 5.8 m, or 5.2 m in areas where motor transport and mobile plant are prohibited. They should be marked conspicuously either by yellow and black binding tape or with freely moving fabric or plastic strips.

Maintenance

Site work is, of necessity, in a constant state of change and because of this the associated electrical installation is subject to risk of damage or misuse. Testing, strict maintenance and frequent checking of control apparatus and the wiring distribution system, by a competent person, are essential to promote safety and efficient operation.

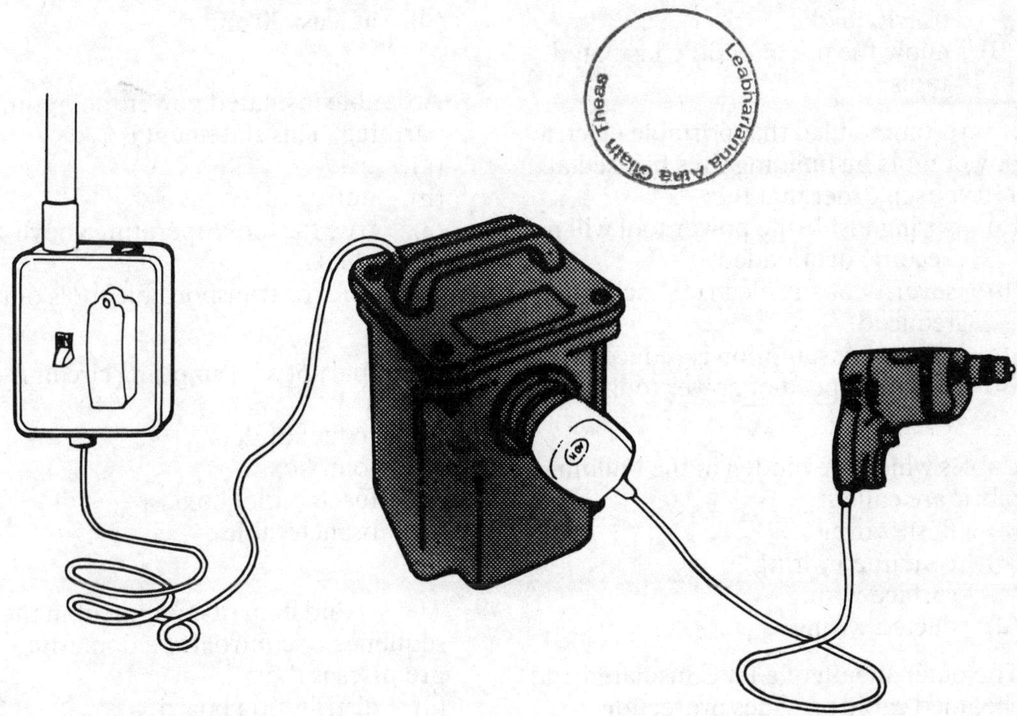

Figure 6.33 *Method of reducing voltage on site*

Self-assessment questions

1 The fuse used in a 13 amp plug is intended to:
(a) maintain a regular voltage
(b) avoid the use of an earth line
(c) fail as soon as the system is overloaded
(d) allow the use of double insulated tools

2 It is recommended that portable electric power tools on building sites be used at 110 volts in order that the:
(a) wiring inside the power tool will not become overloaded
(b) severity of any electric shock is reduced
(c) power consumption is reduced
(d) working speed of power tools is not too fast

3 Cables which are hidden in the building fabric are called:
(a) flush wiring
(b) contained wiring
(c) surface wiring
(d) ducted wiring

4 The outer sheath of a PVC insulated and sheathed cable provides protection against:
(a) sunlight
(b) vermin
(c) excess heat
(d) mechanical damage

5 The 'live' conductor marking on a 3 core flex is:
(a) blue
(b) green
(c) brown
(d) yellow

6 Insulation should be removed from a cable:
(a) at least 10 mm
(b) no further than 10 mm
(c) no further than is necessary
(d) at least 20 mm

7 A double insulated power tool requires earthing. This statement is:
(a) true
(b) untrue
(c) true for tools operating above 110 volts
(d) true for transportable tools only

8 What part of a 13 amp ring circuit is called a 'spur'?
(a) socket outlet
(b) joint box
(c) fused outlet box
(d) branch cable

9 The second item of equipment in the sequence of control for a domestic premises is the:
(a) distribution board
(b) service fuse
(c) energy meter
(d) circuit breaker

10 Which method listed below is NOT recognised as an effective means of earthing?
(a) cable sheath
(b) water pipe
(c) protective multiple earthing
(d) earth – leakage circuit – breaker

7 Glossary

Absolute Zero The coldest possible temperature (−273°C)
Accessory Any device connected and used with other appliances/apparatus
Access Pipe A pipe provided with a removable cover for inspection and maintenance
Acid A corrosive substance
Adhere Stick (hold) together
Adhesion The force of attraction between molecules of different substances
Air Gap The physical break between the water inlet or feed pipe to an appliance and the water in that appliance or cistern
Air Lock The restriction or the stopping of the flow of a liquid by trapped air
Alloy A mixture of metals, or metal and non-metal substances
Alnico A carbon-free alloy used for permanent magnets
Ambient Temperature The temperature of the surrounding element, i.e. air or water
Ammeter An instrument used for measuring direct electric current
Anode The electrode to which negative ions move, the positive electrode
Anti-splash Device A device fitted to the nozzle of a tap or valve to ensure that the discharge will be non-splashing
Anti-splash Shoe A rainwater fitting fixed at the lower end of a rainwater pipe and so shaped as to reduce splashing when rainwater is discharging
Atmospheric Pressure Pressure due to the atmosphere surrounding the earth

Back Fill This term is used to indicate the return of excavated soil (spoil)
Back Flow Flow in a direction opposite to the intended direction of flow
Back Siphonage This is caused by negative pressure within the water system which could result in foul water being drawn back into the service pipe or even the main
Bar A unit of pressure, equivalent to the weight of a column of mercury of unit area and 760 mm in height. 1 bar = 1000 millibars
Barff Process An anticorrosion treatment of iron or steel by the action of steam on the red-hot metal; the layer of black oxide formed gives protection
Barometer A device for measuring atmospheric pressure
Bath A sanitary appliance in which the human body may be immersed in water
Bell-type Joint This is used on copper tube; one end of the pipe is opened to form a bell shaped socket which is then filled with bronze rod
Bends (A) *Set* A single bend
(B) *Offset* A double bend
(C) *Passover/Crank* A pipe formed to pass over a pipe fixed at right-angles to it
Bidet A sanitary appliance used for washing the excretory organs
Blender Also known as a mixer valve, is used on a hot water heating system to mix the hot flow and cooler return water to give a blended temperature flow
Bobbin Specially designed wooden bobbins are used to ensure uniformity of a pipe
Boiler An enclosed vessel in which water is heated by the direct application of heat
Bonding (A) This term used in soldering

or welding refers to the adhesion of the solder to the parent metal

(B) Where the electrical system is earthed through the cold water supply pipe, the term bonding refers to the earth connection

Borax A white crystalline solid used as a flux

Bore The internal diameter of a pipe or fitting

Boss An attached fitting on a vessel or pipe which facilitates the connection of a pipe or pipe fitting

Bush A pipe fitting used for reducing the size of a threaded or spigot connection

Cable One or more conductors provided with insulation. The insulated conductor(s) may be provided with an overall covering to give mechanical protection

Calorifier This is a hot storage vessel containing a coil or inner cylinder and is used in an indirect hot water system

Calorific Value The number of heat units which can be obtained from a measured quantity of fuel

Candela A unit of luminous intensity based on electric lamps

Capacity (A) *Actual* is the volume of water that a vessel contains when filled up to its water line

(B) *Nominal* is the theoretical capacity of a vessel calculated using overall dimensions

Cathodic Protection A sacrificial metal (anode), usually a block of magnesium, is fixed in the system

Caulking Compound A cold jointing substance used on cast-iron drains, i.e. lead wool or asbestos cement

Cavitation This is the recessing of the metal by erosion (eating away). In welding the term is used to signify holes or pitting

Cement A powder used in conjunction with sand to form a mortar used in jointing. It is produced by a breaking down process of a special rock

Chamfer When the square end of the pipe is cut (worked) to a specific angle, i.e. 45°

Circuit An assembly of pipes and fittings, forming part of a hot water system through which water circulates

Circuit-breaker A mechanical device designed to open or close an electrical circuit

Circulation (A) *Primary* is the circulation between the boiler and the hot storage vessel

(B) *Secondary* is the circulation of the hot water to supply the appliances, and is taken from the vent pipe, returning into the top third of the cylinder

(C) *Gravity* could also be called natural circulation because it is brought about naturally by the difference in weight between two columns of water (hot water is lighter than cold water and is therefore displaced, i.e. rises)

(D) *Forced* is the movement of water brought about by the introduction of an impellor (pump)

Cleaning Eye An access opening in a pipe to facilitate the clearing of obstructions

Cohesion The force of attraction between molecules of the same substance

Collar A pipe fitting in the form of a sleeve for joining the spigot ends of two pipes in the same alignment, also called a socket

Combined System See Systems

Combustion The burning of a substance or fuel brought about by rapid oxidation

Compression This term refers to the compressive stress applied to metal when jointing, or the throat of the pipe when bending

Conductor The conducting part of a cable or functioning metalwork which carries current

Connector A means of connecting together pipes and cables

Consumer Unit A combined fuseboard and main switch controlling and protecting a consumer's final sub-circuits

Corrosion This is the term given to the destruction of a metal, i.e. the rusting of metal by oxidation. Water containing free

free oxygen being conveyed in unprotected steel pipes will cause corrosion

Cover (A) Air-tight cover
(B) Material above pipe

Cradle A support shaped to fit the underside of a pipe, cylinder or appliance

Cross Vent A short relief vent between a main discharge pipe and a main ventilating pipe

Crown The highest point of the inside surface of the pipe (also known as the soffit)

Curtilage The area attached to a dwelling

Cylinders (A) *Ordinary* Cylindrical closed containers in which hot water is stored under pressure from the feed cistern

(B) *Indirect* Sometimes called calorifiers, they are used where domestic hot water and heating are fed from the same boiler or where the water is of a temporary hard nature. They come as cylinders within a cylinder or with annular rings designed to keep the primary and secondary waters separate (two cisterns are required)

(C) *Patent indirect* These do not require two cisterns, the filling, venting and expansion being taken care of in the unique patent cylinder

Dead Leg This is the length of pipe from the source of supply, i.e. storage or secondary circulation, to the appliance

Density The density of a substance is defined as its mass per unit volume. Relative density of a substance is the ratio of the mass of any volume of it to the mass of an equal volume of water

De-oxidised This is the removal of oxygen from the material, i.e. the use of flux in soldering or welding

Dew Point Dew point is defined as the temperature at which the water vapour present in the air is just sufficient to saturate it (thus forming condensation when this air is in contact with colder surfaces)

De-zincification The term used when the zinc used in the manufacture of brass (copper-zinc) is partically destroyed (becomes a porous mass)

Discharge Pipe A pipe which conveys the discharge from sanitary appliances

Duckfoot Bend A bend having a foot formed integrally in its base, also known as a rest bend

Duct A closed passage way formed in the structure or underground to receive pipes and cables

Earth Refers to the facility of a system being connected to a general mass of earth, i.e. by a wire and rod or through a metal pipe laid in the ground

Elasticity The term referring to the elongation of a substance and its ability to return to its normal position when the load has been removed. It is a relationship of stress to strain within the elastic range of the material

Electric Immersion Heater This is manufactured as either a single or dual element heater to be fitted in the hot storage vessel; it can be used as the sole means of heating or as a booster to the existing method

Electrolysis This is a detrimental action between dissimilar metal. Increased action is brought about if there is an electrical contact, i.e. earth. One metal acts as a cathode, the other as an anode

Emitter The name given to a radiator or appliance emitting heat for space heating

Equilibrium Refers to equal pressures

Expansion Joint A joint fitted in pipework to allow for the linear expansion of the pipe material when the water temperature is raised

Feed Cistern Supplies cold water to the sanitary appliances and also cold water to the hot water system

Ferrous This is an iron-based material

Flexible Joint A joint designed to allow small angular deflection without loss of water tightness

Float A buoyant device floating on a water surface and actuating a mechanism or valve by its response to rise or fall of the water surface

Flushing Cistern A cistern provided with a device for rapidly discharging the contained water and used in connection with a sanitary appliance for the purpose of cleansing the appliance and carrying away its contents into a drain

Flux A substance used in soldering and/or welding to prevent oxidation

Foul Water Water contaminated by domestic and/or foul or trade effluent

Fuse A form of safety device in a system. Part of the fuse is designed to melt if a fault or overloading occurs

Galvanising Applying a protective surface coating of zinc on steel to prevent oxidation (rusting)

Gutter A channel for collecting surface rainwater

Hard Water (A) Temporary hard water is water that can be softened by boiling. The hardness is due to carbonates of lime and magnesium

(B) Permanent hard water is water that cannot be softened by boiling. The hardness is due to sulphates of lime and magnesium

Head of Water This is the height to which the water will rise in a pipe under atmospheric pressure, i.e. in a domestic water system it is the water level in the cistern

Heating See Systems

Heel of Bend This is the back of the bend which is under tension during the bending operation

Hopper Head A flat or angle-backed rainwater fitting used to collect discharge of rainwater

Hose Union A fitting consisting of a coupling nut for screwing to the external threaded outlet of a hose union tap and a serrated tail for insertion into a hose pipe

Humidity The measure of the water vapour present in a gas; it is measured by a hygrometer or wet and dry bulb thermometer

Hydraulic Gradient The 'loss of head' in liquid flowing in a pipe or channel, expressed per unit length of the pipe or channel

Hydraulic Pressure Fluid pressure

Hydrometer An instrument used to find densities of liquids

Immersion Heater An electrically heated rod type heater inserted in the hot water storage vessel, the heater is usually thermostatically controlled

Insulation The opposite of conduction. Insulators are bad conductors of electricity or heat (e.g. rubber, PVC, glass, wood, cork)

Integral Means 'part of', i.e. an overflow in a wash basin, or an 'O' ring in a drain pipe joint

Interstitial Occurring within the tissues of the fabric or structure of a building

Invar Steel A metal with a very low rate of expansion, used in thermostats

Invert The lowest point of the inside of the pipe or channel

Isolating Valve Any valve so positioned that a part of the system can be isolated

Joint Box A box forming part of an electrical installation in which the cables are joined

Jig An accessory usually purposely made to assist in the carrying out of an activity

Junction A pipe incorporating one or more (junctions) branches

Key (A) A roughening or indentation made on a surface to provide better adherence of another surface, filler or jointing medium

(B) A tool or device provided to enable a person to operate a valve, tap, cock or device or to lift an access cover

Kite Mark This is the British Standards Mark; it is placed on all items manufactured up to an approved standard (ensures good quality)

Knuckle Bend This is a short radius bend

Lagging Material used for thermal or acoustic insulation

Lead Tack A lead casting or piece of sheet lead soldered or welded onto lead pipe or lead sheet and used to secure the lead to its supporting surface

Lime (A) Carbonates of lime cause temporary hard water. When water containing these carbonates of lime is heated to a high temperature (approximately 70 °C or above) the lime is deposited as a scale or fur in the boiler or flow pipe

(B) Sulphates of lime cause permanent hard water. These sulphates of lime are not removed when the water is heated, the water therefore remains hard after boiling

Longscrew A piece of low carbon steel tube threaded externally at each end, one end having the thread sufficiently long to accommodate a backnut and the full length of a socket. It is used to join two pieces of steel tube, neither of which can be rotated. (It is also called a connector)

Loop Vent (A) The continuation of a discharge pipe which is taken up as a vent to connect back into the stack vent or other ventilating pipe

(B) The branch ventilating pipe, which, after being carried up above the flood level of the appliance it serves, is connected back into the vertical section of the discharge pipe as near to the branch discharge pipe as is practicable.

Magnet Mass of iron or other metal which attracts or repels other similar masses and which exerts a force on a current–carrying conductor placed near it

Main The pipe which carries the public water supply

Mandrel A wooden tool (elongated bobbin) which is driven through a lead pipe to ensure uniformity of bore

Mechanical Advantage The mechanical advantage of a machine is defined as the ratio of the load to the effort used

Meniscus The curved surface of a liquid when it touches a solid object rising above the liquid level

Mixing Valve A valve used to mix hot and cold water to give a temperature controlled outflow; it may be thermostatically controlled

Multimeter An instrument used for measuring electrical current, voltage, etc.

Nozzle The open ended portion of a tap, draw-off cock or swivel arm from which water is discharged

Overflow An overflow is sometimes confused with a warning pipe. Some appliances have an overflow as an integral part of the fitment, the overflow water discharging direct into the waste outlet. All cisterns must have either an overflow pipe or a warning pipe or both. A warning pipe is an overflow pipe which discharges in an obvious position to warn the householder of some malfunction of the ball-valve. An overflow pipe can discharge into a drainage system provided the cistern is still fitted with a warning pipe

Oxidation This is an action brought about by the element oxygen, it is generally of a destructive nature, i.e. iron oxide (rusting)

Oxygen A gas ever present in the atmosphere; it is a supporter of combustion

Partially Separate System see Systems

Penetration This refers to the position of the jointing material in relation to the parent metal in a joint

Performance Test A test for the stability of the trap seals in above or below ground drainage discharge systems

Pictorial Views (A) *Isometric projection* In this projection all horizontal lines are drawn at an angle of 30° to the horizontal, all vertical lines remain vertical

(B) *Axonometric projection* In this projection all vertical lines are drawn vertical while all horizontal lines are drawn at an angle of 45°

(C) *Planometric projection* In this the horizontal lines are drawn horizontal while all vertical lines are drawn at an angle of 60° to the horizontal

Pillar Tap This type of tap is fixed to the top surface of the appliance with the outlet above the highest possible water level in the fitment

Plastic Memory Plastic pipes when heated and bent, and most plastic mouldings, have a residual strain in them. When this strain is released, i.e. by heating, the pipe will try to revert to its original position. This is plastic memory

Plug (A) A fitting used to seal off a pipe or section of pipeline, usually fitting into the bore of the pipe

(B) An electrical device intended for connection to a flexible cord or flexible cable

Pressure Defined as the force acting normally per unit area (here the word normally means vertically)

Pressure (Bar) The reading of the air pressure is recorded as so many bars (one bar is equivalent to one atmosphere)

Pump A mechanical device for causing a liquid to flow

Purging The cleaning out of the system

PVC (polyvinyl chloride) This is a tough lightweight plastic material produced from oil

Rainwater Pipe A pipe for conveying rainwater from a roof or other parts of a building

Relief Valve An additional ventilating pipe connected to a discharge pipe at any point where excessive pressure fluctuation is likely to occur

Rest Bend see Duck Foot Bend

Return Pipe A pipe in a hot water system which conveys water back to the boiler, or a pipe in a secondary hot water circuit through which the water flows back to the hot storage vessel

Sanitary Appliance A fixed soil appliance or waste appliance

Scale see Lime

Self-aligning Pipe A pipe which by means of the shape of the socket and spigot is naturally held in a straight line

Self-cleansing Velocity The velocity of the flow of the liquid in the pipe or channel necessary to prevent the deposition of solids

Self-siphonage The extraction of the water from a trap by siphonage set up by the momentum of the discharge from the sanitary appliance to which the trap is attached

Sleeve A length of pipe built into the fabric of a building to allow for the passing through of a smaller pipe, thus allowing for movement and giving protection to the pipe and building fabric

Socket The enlarged part of a pipe which receives the spigot

Socket Outlet A device connected to the electrical installation to enable a flexible cord or cable to be connected by means of a plug

Soffit The top of a pipe

Soil The discharge from water closets, urinals, slop hoppers and similar appliances, i.e. any water containing human or animal excrement

Soil Pipe A pipe which conveys to a drain the waste matter from a W.C., urinal, slop hopper, etc.

Solvent Cement (weld) A liquid used in the jointing of plastic; it has the property of eating into the plastics material so producing a homogeneous bond

Soldering (A) *Soft* The joining of metals using a lead-tin solder

(B) *Hard* The joining of metals using a copper-zinc solder (bronze)

Sparge Pipe A perforated pipe used for flushing a urinal slab or stall

Specific Heat The specific heat of a substance is the heat required to raise a unit mass of that substance through 1°C

Spigot The plain end of a pipe, inserted into a socket to make a spigot and socket joint

Spreader A fitting which connects to the flush pipe and which spreads the flushing water around the surface of a bowl or slab urinal

Stratification The term given to the layers of hot water in the hot water vessel

Surface Tension The elastic skin effect present in liquids; forces between the molecules cause a state of tension in the surface of the liquid

Surface Water Water that is collected from the ground, paved areas and roofs, etc.

Systems Hot Water (A) *Direct* In this system the water that is heated in the boiler is drawn off at the appliance. Water in this system is continually being changed (causing excessive furring and/or corrosion)

(B) *Indirect* In this system the water in the boiler is not changed, the same water being heated again and again and used to indirectly heat the water that is drawn off at the appliances

(C) *Space Heating* A heating system using heat emitters (radiators) to heat the air in the building

(D) *Single Feed* An indirect hot water system using only one cold water feed cistern and a patented hot water storage vessel. They are also known as self-venting systems

(E) *Small-bore* A domestic space heating system involving the use of small diameter pipes (usually 15 mm) with pumped circulation

(F) *Mini-bore* As for the small-bore system only in this instance the pipes are even smaller, i.e. 6 mm diameter

(G) *Unvented* The unvented hot water system is a comparatively new system to the United Kingdom, it is also known as the 'pressurised system', the cold water supply to the hot water system and all the appliances being fed direct from the authorities' water main

Switch A means of connecting or disconnecting an electrical supply

Template This is a purpose-made aid used when forming bends, i.e. a piece of wire or sheet metal cut and formed to the required radius bend

Tension This is the stretching of the material

Thermostat A device used for automatically controlling the flow of energy to an appliance

Throat of a bend The surface of the pipe on the inside of the bend

Tinning The coating of a metal with solder

Transformer A device used in electrical work to achieve a change of voltage

Trap A separate fitting or part of an appliance or pipe arranged so that a small quantity of water is retained to act as a seal, and so prevent the passage of foul air

'U' Gauge A glass or plastic 'U' tube half filled with water; one end of the 'U' gauge is connected by a flexible tube to a system or pipe under test. It is also used in the reading of the working pressure of gas appliances. It is also known as a manometer

uPVC This abbreviation means 'unplasticised polyvinyl chloride' and is a plastic material used in the manufacture of pipes and fittings

Valves (A) *Check valve* This valve will allow the flow of liquid in one direction only. It is also known as a non-return valve or anti-gravity flow valve

(B) *Economy valve* As the name implies its function is to save fuel. It is used to divert the flow of water in the hot water system to heat only the top half of the storage

(C) *Reflux valve* An automatic valve for preventing reverse flow, being open when flow is normal and closing by gravity when flow ceases

Ventilation This is the movement of air in a system or work place by forced or natural means

Ventilating Pipe A pipe connected to the drainage system carried up above the ground terminating above the highest window. Its function is to facilitate the removal of air from the system and maintain equilibrium of pressure so protecting the water seals

Viscosity This is the measure of the ease of flow of a liquid

Voltage The term used to designate an electrical installation, i.e. normal domestic consumers' supply is 240 volts

Water (types) (A) *Soft water* is a water that lathers readily. This is because it contains only a small amount of lime

(B) *Hard water* is a water with which it is difficult to obtain a lather. This is because it contains a large amount of lime

(C) *Temporary hard water* is a water that contains carbonates of lime together with carbon dioxide gas which dissolves the lime. This type of hardness can be softened by boiling

(D) *Permanently hard water* is a water that contains sulphates of lime which are dissolved in the water without the assistance of gases. This type of hardness cannot be softened by boiling

Water Hammer This is a hammering sound in a water pipe caused by violent surges and the sudden arresting of the flow of the water

Water Line A line marked inside a cistern to indicate the highest level of water at which the supply valve should be adjusted to shut off the supply

Waste Pipe A pipe which conveys the discharge from a sanitary appliance used for ablutionary, culinary or drinking purposes

Waste Water This is the discharge from appliances that do not contain human or animal excrement, i.e. wash basins, baths, sinks and similar

Wetted Perimeter This is the line of contact between the pipe or channel and the liquid flowing through it (used in calculating flow through pipes)

Wire Balloon A bulbous-shaped wire guard fitted at the top of a ventilating pipe to prevent birds nesting there

Answers to self-assessment questions

Chapter 1 Domestic hot water

1 (a)
2 (a)
3 (b)
4 (d)
5 (d)
6 (c)
7 (b)
8 (d)
9 (b)
10 (d)

Chapter 1 Heating

1 (b)
2 (c)
3 (d)
4 (a)
5 (c)
6 (c)
7 (b)
8 (c)
9 (c)
10 (b)

Chapter 2 Sanitation and sanitary pipework systems

1 (b)
2 (d)
3 (c)
4 (c)
5 (a)
6 (d)
7 (a)
8 (c)
9 (d)
10 (a)

Chapter 3 Working processes

1 (a)
2 (a)
3 (a)
4 (d)
5 (d)
6 (d)
7 (b)
8 (b)
9 (b)
10 (a)

Chapter 5 Science

1 (d)
2 (d)
3 (d)
4 (a)
5 (b)
6 (d)
7 (a)
8 (d)
9 (c)
10 (b)

Chapter 6 Electricity

1 (c)
2 (b)
3 (a)
4 (d)
5 (c)
6 (c)
7 (b)
8 (d)
9 (c)
10 (b)

Index